Lecture Notes in Computer Science 6621

Commenced Publication in 1973
Founding and Former Series Editors:
Gerhard Goos, Juris Hartmanis,

T0074381

Sara Silva   James A. Foster
Miguel Nicolau   Penousal Machado
Mario Giacobini (Eds.)

Genetic Programming

14th European Conference, EuroGP 2011
Torino, Italy, April 27-29, 2011
Proceedings

 Springer

Volume Editors

Sara Silva
INESC-ID Lisboa, Rua Alves Redol 9, 1000-029 Lisboa, Portugal
E-mail: sara@kdbio.inesc-id.pt

James A. Foster
University of Idaho, Department of Biological Sciences
Moscow, ID 83844-3051, USA
E-mail: foster@uidaho.edu

Miguel Nicolau
University College Dublin, UCD CASL, Belfield, Dublin 4, Ireland
E-mail: miguel.nicolau@ucd.ie

Penousal Machado
University of Coimbra, Faculty of Sciences and Technology
Department of Informatics Engineering
Pólo II - Pinhal de Marrocos, 3030-290 Coimbra, Portugal
E-mail: machado@dei.uc.pt

Mario Giacobini
University of Torino, Department of Animal Production Epidemiology and Ecology
Via Leonardo da Vinci 44, 10095 Grugliasco (TO), Italy
E-mail: mario.giacobini@unito.it

Cover illustration:
"Globosphere" by Miguel Nicolau and Dan Costelloe (2010),
University of Dublin, Ireland

ISSN 0302-9743   e-ISSN 1611-3349
ISBN 978-3-642-20406-7   e-ISBN 978-3-642-20407-4
DOI 10.1007/978-3-642-20407-4
Springer Heidelberg Dordrecht London New York

Library of Congress Control Number: 2011925001

CR Subject Classification (1998): D.1, F.1, F.2, I.5, I.2, J.3

LNCS Sublibrary: SL 1 – Theoretical Computer Science and General Issues

Typesetting: Camera-ready by author, data conversion by Scientific Publishing Services, Chennai, India

Printed on acid-free paper

Springer is part of Springer Science+Business Media (www.springer.com)

Preface

The 14th European Conference on Genetic Programming (EuroGP) convened during April 27–29, 2011 in Torino, Italy, in the year Italy celebrates the 150th anniversary of its unification. EuroGP is the only conference worldwide exclusively devoted to genetic programming and the evolutionary generation of computer programs.

For almost two decades, genetic programming (GP) has been recognized as a dynamic and important form of evolutionary computation. With over 7,000 articles in the online GP bibliography maintained by Bill Langdon, it is clearly a mature field. EuroGP has contributed to the success of the field substantially, by being a unique forum for expressing new ideas, meeting, and starting up collaborations.

This year we received 59 submissions from 25 different countries across four continents. The papers underwent a rigorous double-blind peer-review process, each being reviewed by at least three members of the International Program Committee from 22 countries on four different continents. The selection process resulted in this volume, with 20 papers accepted for oral presentation (33.90% acceptance rate) and nine for poster presentation (49.15% global acceptance rate for talks and posters combined). The wide range of topics in this volume reflects the current state of research in the field, including theory, representations, novel operators and techniques, self organization, and applications.

EuroGP 2011 was part of the Evo* 2011 event, which included three additional conferences: EvoCOP 2011, EvoBIO 2011, and EvoApplications 2011. Many people helped to make this event a success.

Firstly, we would like to thank the great community of researchers and practitioners who contributed to the conference by both submitting their work and reviewing others' as part of the Program Committee. Their hard work resulted in a high-quality conference.

The papers were submitted, reviewed, and selected using the MyReview conference management software. We are sincerely grateful to Marc Schoenauer of INRIA, France, for his great assistance in providing, hosting, and managing the software.

Many thanks to the local organizer Mario Giacobini of the University of Torino, and to the following local sponsors: Human Genetics Foundation of Torino (HuGeF) and the following institutions of the University of Torino: School for Biotechnologies, Molecular Biotechnology Center, Museum of Human Anatomy "Luigi Rolando," and Museum of Criminal Anthropology "Cesare Lombroso."

Also a big thanks to Penousal Machado of the University of Coimbra, assisted by Pedro Miguel Cruz and João Bicker, for creating and maintaining the official Evo* 2011 website.

We would also like to express our sincerest gratitude to our invited speakers, who gave the inspiring keynote talks: Craig Reynolds, senior researcher at Sony Computer Entertainment, USA, and Jean-Pierre Changeux, professor emeritus at the Collège de France and at the Institut Pasteur, France.

Last but certainly not least, we especially want to express a heartfelt thanks to Jennifer Willies and the Centre for Emergent Computing at Edinburgh Napier University. Ever since its inaugural meeting in 1998 this event has relied on her dedicated work and continued involvement and we do not exaggerate when we state that without her, Evo* could not have achieved its current status.

April 2011

Sara Silva
James A. Foster
Miguel Nicolau
Penousal Machado
Mario Giacobini

Organization

Administrative details were handled by Jennifer Willies, Edinburgh Napier University, Centre for Emergent Computing, Scotland, UK.

Organizing Committee

Program Co-chairs	Sara Silva, INESC-ID Lisboa, Portugal
	James A. Foster, University of Idaho, USA
Publication Chair	Miguel Nicolau, University College Dublin, Ireland
Publicity Chair	Penousal Machado, University of Coimbra, Portugal
Local Chair	Mario Giacobini, University of Torino, Italy

Program Committee

Lee Altenberg	University of Hawaii at Manoa, USA
Lourdes Araujo	UNED, Spain
R. Muhammad Atif Azad	University of Limerick, Ireland
Wolfgang Banzhaf	Memorial University of Newfoundland, Canada
Xavier Blasco	Universidad Politecnica de Valencia, Spain
Anthony Brabazon	University College Dublin, Ireland
Nicolas Bredeche	Université Paris-Sud XI / INRIA / CNRS, France
Stefano Cagnoni	University of Parma, Italy
Pierre Collet	LSIIT-FDBT, France
Ernesto Costa	University of Coimbra, Portugal
Luis Da Costa	Université Paris-Sud XI, France
Michael Defoin Platel	Rothamsted Research, UK
Antonio Della Cioppa	University of Salerno, Italy
Ian Dempsey	University College Dublin / Virtu Financial, Ireland
Stephen Dignum	University of Essex, UK
Federico Divina	Pablo de Olavide University, Spain
Marc Ebner	Universität Tübingen, Germany
Anikó Ekárt	Aston University, UK
Anna Esparcia-Alcázar	S2 Grupo, Spain
Daryl Essam	University of New South Wales @ ADFA, Australia
Francisco Fernandez de Vega	Universidad de Extremadura, Spain

Table of Contents

Oral Presentations

Posters

A Sniffer Technique for an Efficient Deduction of Model Dynamical Equations Using Genetic Programming

Dilip P. Ahalpara and Abhijit Sen

Institute for Plasma Research, Near Indira Bridge, Bhat, Gandhinagar-382428, India
dilip@ipr.res.in

Abstract. A novel heuristic technique that enhances the search facility of the standard genetic programming (GP) algorithm is presented. The method provides a dynamic *sniffing* facility to optimize the local search in the vicinity of the current best chromosomes that emerge during GP iterations. Such a hybrid approach, that combines the GP method with the sniffer technique, is found to be very effective in the solution of *inverse* problems where one is trying to construct model dynamical equations from either finite time series data or knowledge of an analytic solution function. As illustrative examples, some special function ordinary differential equations (ODEs) and integrable nonlinear partial differential equations (PDEs) are shown to be efficiently and exactly recovered from known solution data. The method can also be used effectively for solution of model equations (the *direct* problem) and as a tool for generating multiple dynamical systems that share the same solution space.

1 Introduction

Most mathematical modeling of complex phenomena in the natural sciences and engineering applications lead to ordinary and/or partial differential equations (ODEs and PDEs). The solutions of such equations provide a temporal and/or spatial description of the evolution of the physical variables of the system. A wide variety of reliable and well established techniques, both analytical and numerical, are presently available for the solution of ODEs and PDEs and hence provide a direct means of solving the model equations e.g. Press *et al* (2007), Tsoulos and Lagaris (2006). Numerical simulation of model ODE/PDEs is therefore now a well developed field and extensively employed in a variety of applications. Often however it is not possible to derive model equations from first principles and one would like to infer such equations from experimental time series data. Such an *inverse* problem has received much attention in recent times particularly in areas like weather modeling, turbulence studies, financial market analysis etc., where the inference of a model would be very useful for prediction or control purposes. The inverse problem is mathematically quite challenging and the available solution methods are fewer and less well developed compared to the techniques available for the direct problem. In the absence of well established direct methods, soft computing methods that do not demand any domain specific knowledge

S. Silva et al. (Eds.): EuroGP 2011, LNCS 6621, pp. 1–12, 2011.

have an important role to play. Of the available methods, the Genetic Programming (GP) method, that is based on the pioneering work of Koza (1992) in the area of evolutionary algorithms, has had relatively good success and shows much promise. The GP approach primarily involves finding a solution in symbolic form that fits a given set of defined 'data' and follows a 'fitness driven evolution' path that minimizes the error in the fitting data. It is also classified as a 'symbolic regression' technique and has been successfully employed in a wide variety of applications including finding solutions of ODE/PDEs and solving the *inverse* problem. One of the shortcomings of the GP technique however is that its 'stochastic' method of evolution creates a large search space and slows down its approach to the correct solution. In addition, it often fails to arrive at exact solutions (with a fitness value of 1) which can be a serious limitation in some applications. This has led to modifications in the standard GP method by addition of supplementary tools to improve its efficiency and accuracy. Sakamoto and Iba (2001), Sugimoto *et al* (2001) and Iba (2008) have used a least mean square method for inferring ODEs for a chemical reaction process. Their method however does not deduce the equations exactly but does achieve a fair degree of accuracy. Cao *et al* (2000) have employed a method that they term as 'Hybrid Evolutionary Modeling Algorithm (HEMA)' for the deduction of a system of 1st order ODEs by embedding a genetic algorithm (GA) into genetic programming (GP), where GP is employed to optimize the structure of a model, while a GA is employed to optimize its parameters. Gray *et al* (1997) have used a simulated annealing process along with Nelder-simplex optimization method for inferring a system of 1st order ODEs.

In this paper, we report our development of a new heuristic approach that uses concepts similar to a memetic algorithm (Moscato (1999)) to enhance the search capability of the standard genetic programming (GP) algorithm, and also makes it more efficient and effective for the solution of the *inverse* problem. The basic innovation is the introduction of a suite of 'local search' functions - which we call *sniffing* routines - that are turned on at predetermined intervals to refine the currently available best chromosomes that have emerged during the GP iterations. A Mathematica code based on this algorithm and incorporating a variety of sniffer routines has been developed. We have then tested this hybrid approach on several problems, including deduction of model ODEs from a finite time series data and deduction of a nonlinear PDE from an analytic solution. In all these cases we observe that the presence of the sniffer routines bring about a significant improvement in the search procedure that helps in cutting down the convergence time and to also improve the accuracy of the deduced model parameters. We will provide some quantitative estimates of these improvements in our discussion of the test problems on which we have employed the hybrid approach. One of the interesting and surprising finding of our present work has been that in the search for a model nonlinear PDE from a given analytic solution we have often found multiple model equations (all with exact fitness value of unity) that have the same solution. This raises interesting possibilities for using our method as a useful tool in mathematical explorations of independent equation systems that share the same solution space. Some of these newly found model equations are presented along with results of GP experiments.

The paper is organized as follows. In the next section 2 we briefly describe the sniffer routine enhanced GP algorithm. This is followed by its application to some *inverse* problems in Section 3. The specific examples discussed include deduction of the set of model chemical reactions considered by Iba (2008) as a benchmark study, followed by deduction of the Laguerre ODE from a finite digitized data series, and finally the deduction of the nonlinear Korteveg-de Vries equation from its known analytic solution, along with some newly found model equations that share the same solution space. Section 4 provides a concluding summary and some observations.

2   Sniffer Enhanced GP Search Technique

As mentioned in the Introduction, the basic idea in our heuristic enhancement of the GP technique is the incorporation of a collection of 'sniffer routines' that can improve and refine the GP search process by carrying out a local search around a chromosome that emerges during a given GP iteration. The local search domain is defined by a *sniffer radius* of magnitude $\epsilon_{sniffer}$ which forms a maximum bound within which stochastic variations on various components of a chromosome are carried out. The sniffer method, comprising of a number of sniffer operators, is best described with reference to an example. Consider an inverse problem for the inference of an ODE of nth order, which in general can be expressed as,

$$\frac{d^n y}{dx^n} = f\left(x, \; y, \; \frac{dy}{dx}, \; \frac{d^2 y}{dx^2}, \; \cdots \; , \; \frac{d^{n-1} y}{dx^{n-1}}\right) \qquad (1)$$

where $y(x)$ is a function involving independent variable x, and it is assumed that the system is describable by derivatives up to n^{th} order. The aim then is to infer, by the GP method, the exact mathematical form of the function f occurring on the right hand side of (1). For simplicity let us take $n=3$ and use symbols a, b, c, d to denote $x, y, \frac{dy}{dx}$, and $\frac{d^2 y}{dx^2}$ respectively. Suppose further, as an illustration, that the following symbolic expression arises as a best chromosome during a particular GP iteration,

$$1.1d + 2c(c - 3b) + a \qquad (2)$$

At this stage we subject the chromosome to the following sniffing operations with the stochastic variations restricted to a window $[1-\epsilon_{sniffer}, \; 1+\epsilon_{sniffer}]$.

- *Sniff1: Stochastic perturbation of variables.* The positions of each variable a, b, c and d in the chromosome are found out (using the *Position* function of Mathematica), and these variables are selected and then varied stochastically. For variable b, for example, a random number r (say 0.9822), is used to effect the following variations,

$$b \rightarrow b * r, \quad b \rightarrow b/r, \quad b \rightarrow b + (r - 1), \quad b \rightarrow b - (r - 1) \qquad (3)$$

By carrying out each of the above mathematical operations of (3), four new chromosomes are generated as follows,

$1.1d + 2c(c - 3 * (0.9822b)) + a$, $1.1d + 2c(c - 3 * (b/0.9822)) + a$, $1.1d + 2c(c - (3 * (b + 0.9822 - 1))) + a$, $1.1d + 2c(c - (3 * (b - 0.9822 + 1))) + a$

- *Sniff2: Stochastic replacement of each variable.* The positions of each variables a, b, c and d in the chromosome are found out, and the occurrence of these variables is selected and then changed to a different variable stochastically. Thus, for example, for variable a, such a change over of the variable would generate the following new chromosomes,

$1.1d + 2c(c - 3b) + b$, $1.1d + 2c(c - 3b) + c$, $1.1d + 2c(c - 3b) + d$

- *Sniff3: Stochastic perturbation of numerical values.* The positions of each number occurring in the chromosome are found out, and the numbers are varied stochastically. For number 3, for example, a random number r within the window, say 1.02, is used to vary it using mathematical operations similar to those in (3). The new chromosomes thus obtained would be,

$1.1d + 2c(c - 3 * 1.02b) + a$, $1.1d + 2c(c - (3/1.02)b) + a$, $1.1d + 2c(c - (3 + 1.02 - 1)b) + a$, $1.1d + 2c(c - (3 - 1.02 + 1)b) + a$

- *Sniff4: Perturbing/ removing a sub-tree within S-expression.* The S-expression for the chromosome is first converted to a tree structure. A subtree is then selected stochastically, and is applied with two operations: 1) dropping it to see whether it is a redundant sub-structure, and 2) varying it stochastically by a random number r within the window, and applying mathematical operations similar to those in (3). Thus for example, if the subtree selected is 3b, the resulting new chromosomes with $r=0.9824$ say, would be,

$1.1d + 2c(c) + a$, $1.1d + 2c(c - 3b * 0.9824) + a$, $1.1d + 2c(c - 3b/0.9824) + a$, $1.1d + 2c(c - (3b + (0.9824 - 1))) + a$, $1.1d + 2c(c - (3b - (0.9824 - 1))) + a$

- *Sniff5: Perturbing the chromosome as a whole.* A random number r is selected, and the chromosome is perturbed by applying each of the mathematical operations similar to those in (3). With $r=0.9931$ say, this would generate following new chromosomes,

$(1.1d + 2c(c - 3b) + a) * 0.9931$, $(1.1d + 2c(c - 3b) + a)/0.9931$, $(1.1d + 2c(c - 3b) + a) + (0.9931 - 1)$, $(1.1d + 2c(c - 3b) + a) - (0.9931 - 1)$

After applying each of the above sniffer operations, the fitness of the resulting new chromosomes is evaluated. Moreover, to give enough variety in the sniffing operation, each of the above operation (Sniff1 to Sniff5) is carried out for a predefined number of times (say 100), and the chromosome corresponding to the best fitness value obtained from these new chromosomes is selected. If this trial fitness happens to be better than the fitness of the current best chromosome, it is used to replace the current best chromosome, otherwise it is neglected. In fact, the refinements of chromosomes by sniffer operators is programmed such as to store and use refined chromosome from each sniffer pass, so that multiple sniffer

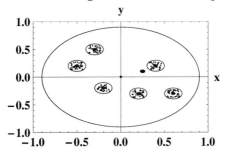

Fig. 1. Schematic diagram showing the usefulness of sniffer technique, described in text

passes may give rise to multiple refinements in the current best GP chromosome. It should be noted that the Sniff4 operator partly helps to get rid of a possible redundant sub-tree in the S-expression thereby curtailing the bloating effect to some extent.

It must be emphasized that the sniffer method is not aimed at (and in fact it cannot) replace the global search achieved by the basic GP method. The sniffer method is introduced primarily to improve on the slow convergence of the GP iterative procedure and to increase its accuracy. Collectively they constitute a more efficient method. On the one hand the GP iterations provide a variety of candidate solutions, and on top of this basic search procedure, the sniffer operators carry out a stochastic (and not an exhaustive) local search in the vicinity of the best solution achieved so far by the GP procedure. This is further explained by a schematic diagram in Fig. 1. The outer circle represents the search domain of a hypothetical problem having two independent variables x, y. The true solution is represented by a disk shown at (0.25, 0.1). The centers within smaller circles represent the best solutions found by GP at various iterations. The small circles represent the locality of search domains surrounding the best chromosome with a small sniffer radius $\epsilon_{sniffer}$. The dots within these circles represent the search carried out by various sniffer operators. The heuristic method thus sniffs around a given best chromosome, thereby making a focussed search around a given chromosome. During GP iterations, even when it does not hit upon the true solution exactly, the sniffer method increases the chances of finding the true solution, especially when the GP solution falls close to the true solution within the distance of $\epsilon_{sniffer}$. It should also be noted that when the best solution is not within the range of the true solution by $\epsilon_{sniffer}$, the sniffer operators are still useful to facilitate a gradual shift towards the true solution by possibly making an improvement in the vicinity of the current best solution.

As regards the choice of the sniffer radius $\epsilon_{sniffer}$, it should be mentioned that making it bigger would help the best GP solution to fall close (within the $\epsilon_{sniffer}$ radius) to the true solution. However, a local search within the resulting bigger search domain would decrease the probability of finding the true solution

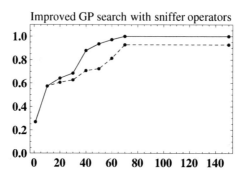

Fig. 2. Fitness versus generations curve obtained by a standard GP calculation (dashed line) having best fitness=0.9284, as against the exact solution achieved when sniffer operators are activated (thick line) having fitness=1 for deduction of the KdV equation

stochastically. Thus a judicious compromise is required to set a reasonably small value for the sniffer radius $\epsilon_{sniffer}$. In the present study we have found that $\epsilon_{sniffer}$=0.1 is a useful choice for the GP experiments considered.

We next discuss the quantitative gain in efficiency brought about by the introduction of the sniffer operators. For a typical run for the deduction of the KdV equation (discussed later in Sec. 3.3), Fig. 2 shows the growth of fitness values at successive 10th iterations. The figure serves as an illustration for the enhancement of GP search when sniffer operators are activated. It is noted that the standard GP calculation, in this case, is stuck at the best fitness=0.9284, indicating that the exact solution for the differential equation has not been obtained. On the other hand, when the sniffer operators are activated, the GP calculation finds the best fitness=1.0, and the exact solution for the differential equation is found. This clearly highlights the usefulness of the sniffer technique as well as demonstrates its efficiency. More details showing the growth of fitness and the related chromosome structure for this example are given in Sec. 3.3. It may also be noted that sniffer methods are compute intensive, and hence it is not advisable to invoke them at each successive GP iteration. In view of this, in our GP experiments sniffer methods are invoked at the interval of several GP iterations (typically 10), so as to allow sufficient relaxation for GP to explore a variety of chromosome structures.

Thus the local search carried out by the sniffer technique may be viewed as an *adaptive mutation operator*, that is activated stochastically and intermittently after allowing GP to relax for a given number of iterations.

The parameters of the standard GP framework used in various GP experiments are shown in Table 1. The chromosome template shown in Table 1 is used for populating the initial pool of chromosomes, which then grows to complex structures with a fitness-driven standard GP search. Further implementation details of the standard GP algorithm used here are given in Ahalpara *et al* (2008 and 2009).

Table 1. GP parameters used in various GP experiments

Parameter	Values
Terminal set	Numbers and x, y, dy/dx, d^2y/dx^2, ... $d^{n-1}y/dx^{n-1}$
	(where 0.5 is the cutoff probability to select either a number
	or any other member from the terminal set, using random
	numbers in the range $[0, 1]$).
Function set	$+, -, *$ and $\backslash$
Chromosome template	$((A \oplus B) \oplus (C \oplus D))$
	(where $\oplus$ denotes an operator from function set,
	and A,B,C,D denote an element from terminal set)
Elitism	Best 5% chromosomes are copied to next generation
Selection	Tournament selection
Crossover	One point crossover
Probabilities	Crossover:0.9, Mutation:0.1
Fitness cases	200 to 500
Fitness	Mean Squared Error (scaled to 1.0 for an exact fit)
	between given n^{th} derivatives that are fitted by the
	right hand side of equation (1)
Maximum bounds	Population size:200, Generations:300, Fitness runs:30
Sniffer activation	After 10 GP iterations, with $\epsilon_{sniffer}=0.1$

3 Deduction of Model Differential Equations

We now discuss the results of the application of our method to a few test *inverse* problems involving the deduction of model differential equations.

3.1 Deduction of Chemical Reaction Model Equations

Our first application was to a set of three coupled first order equations representing the concentration dynamics of three interacting chemicals, given as,

$$X_t = -k_1 X, \quad Y_t = k_2 X - k_3 Y, \quad Z_t = k_4 Y \tag{4}$$

where X, Y, Z are the chemical concentrations and the constant coefficients are given as $k_1=k_2=1.4000$ and $k_3=k_4=4.2000$. This set of equations has been previously studied by Iba (2008) using the least squares modification of the GP method. We therefore chose this as a benchmark problem to test out our method and to also compare our results with those of Iba. In order to deduce these ODE equations, we first obtained the time series data by solving the system of equations using the NDSolve function of Mathematica. The data for $X(t)$, $Y(t)$ and $Z(t)$, as well as their derivatives $X'(t)$, $Y'(t)$ and $Z'(t)$ thus calculated for the time domain $t \in [0, 2]$ are shown in Fig. 3. GP experiments were then carried out using these time series data, and inferring the S-expressions for the functional forms for $X'(t)$, $Y'(t)$ and $Z'(t)$, each on individual basis, to arrive at the underlying system of differential equations. Our method converged very quickly to arrive at exact results with a *fitness*=1 such that the sums of square errors

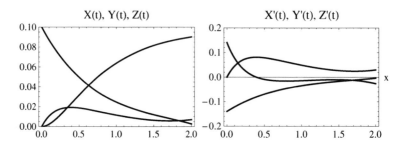

Fig. 3. Time series data for the 3 components (X(t), Y(t), Z(t)) of a chemical reaction, and their numerical derivatives (X'(t), Y'(t), Z'(t)) used in the GP experiments for inferring the underlying system of ODE equations

(SSE) for (X, Y, Z) were (0, 0, 0). Our deduced coefficients were accurate up to 8 significant digits. In comparison, the results obtained by Iba (2008) were $k_1 = 1.4000$, $k_2 = 1.4004$, $k_3 = 4.2006$, $k_4 = 4.19980$.

3.2   Deduction of a Special Function ODE

Encouraged by the success of our hybrid method in deducing the simple set of chemical equations we next undertook a more challenging task, namely the deduction of an ODE representing a special function - in this case the Laguerre equation. The GP experiment was carried out in as unbiased and blindfold manner as possible. One of us generated a digitized data series by solving the Laguerre equation numerically and gave it to the other without any other additional information, in the form of a table $[x_i, \ f(x_i)]$ for x_i values in the range $x_i \in [0.0, \ 10.0, \ 0.0001]$. In order to solve this problem, the numerical derivatives up to $3rd$ degree were calculated using a standard 5-*point* central difference formulae. Fig. 4 shows the digitized unknown function (a) and its numerically

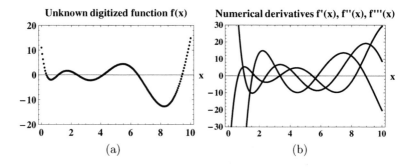

Fig. 4. (a) An unknown function $f(x)$ digitized at successive $x=x_i$ values in the range $x_i \in [0.0, \ 10.0]$, (b) Numerical derivatives up to 3rd degree

calculated derivatives (b). Since the order of the equation was not known a priori, GP experiments were initially made with derivative up to 1st degree, i.e. equation (1) with $n=1$. Thus the calculation amounted to searching for an S-expression for function f in equation $\frac{dy}{dx}=f(x,y)$. This calculation converged at the most to a poor fitness value of 0.4045. Since for obtaining an exact solution of ODE, a fitness value of 1.0 is required, this GP experiment was considered a failure, and it was extended to also consider derivative up to 2nd degree. Thus the S-expression for function f was now searched corresponding to equation (1) with $n=2$, namely $\frac{d^2y}{dx^2}=f\left(x,y,\frac{dy}{dx}\right)$. The resulting GP experiment aided by sniffer technique was then successful in obtaining the desired $fitness=1.0$, and gave the exact ODE equation,

$$xY_{xx} + (2-x)Y_x + 10Y = 0 \tag{5}$$

The GP experiments were further extended to test whether the generalized Laguerre equations as well as higher order ODEs can be deduced. Specifically, the experiments were carried out in following two ways,

- Two more digitized data sets, again given in a blindfold manner, were analyzed in a similar manner and the ODE equations corresponding to the generalized Laguerre equation $xY_{xx}+(\nu+1-x)Y_x+10Y=0$ for $\nu=2$ and $\nu=3$ respectively were deduced.
- Derivatives up to 3rd degree were included and S-expression was tried out corresponding to equation (1) with $n=3$, namely $\frac{d^3y}{dx^3}=f\left(x,y,\frac{dy}{dx},\frac{d^2y}{dx^2}\right)$. This GP experiment again successfully converged to give $fitness=1$, and gave the following ODE equation,

$$xY_{xxx} + (3-x)Y_{xx} + 9Y_x = 0$$

This is not an independently new ODE equation, and as can be easily verified, this ODE can be generated from differentiation of equation (5).

3.3 Deduction of KdV Equation

Our final example concerns the application of the sniffer technique to deduce a nonlinear PDE in this case the well known Korteveg-deVries (KdV) equation, given as,

$$Y_t(x,t) + 6Y(x,t)Y_x(x,t) + Y_{xxx}(x,t) = 0 \tag{6}$$

The KdV equation was originally derived to describe the nonlinear propagation of shallow water waves (Korteweg and Vries (1895)) but has subsequently found applications in many areas of physics (Zabousky and Kruskal (1965) and Miles (1981)). It is also one of those remarkable nonlinear PDEs that are exactly integrable and admit soliton solutions. An exact analytic single soliton solution of this equation is given by,

$$Y(x,t) = \frac{v}{2}Sech^2[(\sqrt{v}/2)(x-vt)] \tag{7}$$

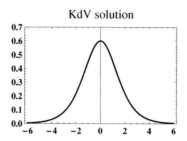

Fig. 5. Soliton solution of the KdV equation

which represents a pulse like solution propagating with a velocity v. For $v=1.2$, the KdV solution $Y(x,t)$ at $t=0$ is shown in Fig. 5.

Our objective is to use the analytic soliton solution (7) and use the sniffer aided GP technique to deduce the KdV equation. For this we note that the soliton solution belongs to a special class of solutions of the KdV equation, known as traveling wave solutions that are functions of the combined variable $\xi=x-vt$. In terms of this new variable ξ,

$$\frac{\partial Y}{\partial t} = -v\frac{\partial Y}{\partial \xi}; \quad \frac{\partial Y}{\partial x} = \frac{\partial Y}{\partial \xi} \tag{8}$$

Such a transformation converts the nonlinear PDE into a nonlinear ODE and can be then cast in the form of (1). In order to obtain the S-expression on right hand side of (1), data for $Y(\xi)$, Y_ξ and $Y_{\xi\xi\xi}$ are calculated with velocity $v=1.2$ for 200 equi-spaced ξ-values in the range $\xi \in [-6, 6]$. It may be noted that the derivatives are calculable with very good precision from its analytical expression (7). Denoting $\{\xi, Y, Y_\xi, Y_{\xi\xi}, Y_{\xi\xi\xi}\}$ by $\{a, b, c, d, e\}$, it is noted that GP calculation essentially aims at deducing a S-expression for function f in $e=f(a, b, c, d)$ (see (1)). It is found that a standard GP search achieves a maximum fitness of 0.9284 (Fig. 2) up to 300 generations corresponding to the best chromosome structure $-c + \frac{2cd}{b}$. In fact, when we repeat the standard GP runs a 100 times for 200 generations, the mean fitness value achieved is 0.98698 without a single run giving a fitness=1.0, whereas the Sniffer approach consistently achieves a fitness value of 1.0. With the activation of sniffer operators, the fitness improvements in a typical run during GP iterations go as follows,

$$\{0.27, -1.4c - 2bcd\}, \ \{0.57, -c + bcd\}, \ \{0.64, -c - cd\}, \{0.69, c - 4bc + 0.4d\},$$
$$\{0.88, c - 6bc + 0.4d\}, \ \{0.94, c - 6bc + 0.25d\}, \{0.97, c - 6bc\},$$
$$\mathbf{\{1.0, 1.2c - 6bc\}} \tag{9}$$

where each brace shows $\{fitness, chromosome\}$ values (numbers shown with 2 significant digits only), at some selective GP iterations. As can be seen, the best chromosome corresponds to $fitness=1.0$ (shown in bold in (9)), meaning that an exact ODE solution has been obtained corresponding to $Y_{\xi\xi\xi}=1.2Y_\xi - 6YY_\xi$. Since the data is generated using $v=1.2$, it is seen that the soliton solution satisfies Y_t=-1.2Y_ξ, and hence it is seen that the exact PDE equation (6) for KdV, namely $\mathbf{Y_{xxx}} = -\mathbf{Y_t} - 6\mathbf{YY_x}$, is successfully deduced by the GP experiment. It may be noted that the GP experiment has also independently been successful in generating the solution Y_t=-1.2Y_ξ used in above transformation of the equation.

For the KdV equation we have also extended the GP calculations to try and find out more differential equations involving higher derivatives. Such GP experiments have also proved successful, and the new equations that share the same analytic solution along with (6), for a specific value for velocity v=1.2, are shown in (10).

$$Y_t + Y_x(15Y^2 + 5Y_{xx}) + Y_{xxx} = 0$$
$$Y_t(24Y - 75Y^2) + 1.2Y_{xxx} - Y_{xxxxx} = 0$$
$$1.8Y_t + Y_{xxx}(3 - 15Y) - Y_{xxxxx} = 0 \qquad (10)$$

These differential equations are not found in the literature, and it is also straight forward to check that each of the above equation (10) satisfies the analytic solution (7). We have not found a simple transformation by which any of these equations are derivable from the original KdV equation (6), and in that sense they appear to be non-trivial independent equations that share the same solution space with the KdV equation. The discovery of these new equations by the sniffer technique aided GP method opens up exciting possibilities for future mathematical explorations as well as physical analysis of these systems.

4   Conclusion and Outlook

It has been shown that the sniffer technique enhanced Genetic Programming approach is quite useful in inferring ODE/ PDE equations. The intermittent local search in the vicinity of the best solution achieved by GP at a given instance of its iterative procedure greatly facilitates the overall convergence and accuracy of the search algorithm. We have successfully applied the method to deduce model ODE/ PDE equations from given digitized finite data series as well as an analytic solution function. A novel finding is the generation of multiple model equations for a given analytic solution which can have exciting mathematical and physical implications. The method is not limited to applications for the *inverse* problem. We have also successfully used the method to efficiently solve ODE and PDEs, the results of which will be reported in a future publication.

The authors are thankful to the referees for providing very interesting and useful comments and making suggestions for the improvement of the paper.

References

Ahalpara, D.P., Parikh, J.C.: Genetic Programming Based Approach for Modeling Time Series Data of Real Systems. Int. J. Mod. Phys. C 19, 63 (2008)

Ahalpara, D.P., Arora, S., Santhanam, M.S.: Genetic programming based approach for synchronization with parameter mismatches in EEG. In: Vanneschi, L., Gustafson, S., Moraglio, A., De Falco, I., Ebner, M. (eds.) EuroGP 2009. LNCS, vol. 5481, pp. 13–24. Springer, Heidelberg (2009)

Cao, H., Kang, L., Chen, Y., Yu, J.: Evolutionary Modeling of Systems of Ordinary Differential Equations with Genetic Programming. Genetic Programming and Evolvable Machines 1, 309–337 (2000)

Gray, G.J., Murray-Smith David, J., Li, Y., Sharman Ken, C.: Nonlinear Model Structure Identification Using Genetic Programming. In: Conference Proceedings Genetic Algorithms in Engineering Systems: Innovations and Applications GALESIA 1997 at Glasgow, pp. 308–313 (1997), ISBN: 0-85296-693-8

Iba, H.: Inference of Differential Equation Models by Genetic Programming. Information Sciences, an International Journal archive 178(23), 4453–4468 (2008)

Korteweg, D.J., de Vries, F.: On the Change of Form of Long Waves Advancing in a Rectangular Canal, and on a New Type of Soliton Waves. Philosophical Magazine 39, 422–443 (1895)

Koza, J.R.: Genetic Programming: On the Programming of Computers by Means of Natural selection and Genetics. MIT Press, Cambridge (1992)

Miles, J.W.: The Korteweg-De Vries equation: A Historical Essay. Journal of Fluid Mechanics 106, 131–147 (1981)

Moscato, P.: New ideas in optimization. In: Corne, D., et al. (eds.) pp. 219–234. McGraw-Hill Ltd., UK (1999)

Press, W.H., Teukolsky, S., Vetterling, W., Flannery, B.: Numerical Recipes in C++: The Art of Scientific Computing, 3rd edn. Cambridge University Press, Cambridge (2007), ISBN: 0-521-75033-4

Sakamoto, E., Iba, H.: Evolutionary Inference of a Biological Network as Differential Equations by Genetic Programming. Genome Informatics 12, 276–277 (2001)

Sugimoto, N., Sakamoto, E., Iba, H.: Inference of Differential Equations by Using Genetic Programming. Genome Informatics 12, 276–277 (2001)

Tsoulos, I.G., Lagaris, I.E.: Solving differential equations with genetic programming. Genetic Programming and Evolvable Machines 7(1), 33–54 (2006)

Zabusky, N.J., Kruskal, M.D.: Interaction of Solitons in a Collisionless Plasma and the Recurrence of Initial States. Physical Review Letters 15, 240–243 (1965)

Robustness, Evolvability, and Accessibility in Linear Genetic Programming

Ting Hu[1,*], Joshua L. Payne[1,*], Wolfgang Banzhaf[2], and Jason H. Moore[1]

[1] Computational Genetics Laboratory, Dartmouth Medical School,
Lebanon, NH 03756, USA
{Ting.Hu,Joshua.L.Payne,Jason.H.Moore}@Dartmouth.edu
[2] Department of Computer Science, Memorial University of Newfoundland,
St. John's, NL, Canada
banzhaf@mun.ca

Abstract. Whether neutrality has positive or negative effects on evolutionary search is a contentious topic, with reported experimental results supporting both sides of the debate. Most existing studies use performance statistics, *e.g.*, success rate or search efficiency, to investigate if neutrality, either embedded or artificially added, can benefit an evolutionary algorithm. Here, we argue that understanding the influence of neutrality on evolutionary optimization requires an understanding of the interplay between robustness and evolvability at the genotypic and phenotypic scales. As a concrete example, we consider a simple linear genetic programming system that is amenable to exhaustive enumeration, and allows for the full characterization of these properties. We adopt statistical measurements from RNA systems to quantify robustness and evolvability at both genotypic and phenotypic levels. Using an ensemble of random walks, we demonstrate that the benefit of neutrality crucially depends upon its phenotypic distribution.

1 Introduction

Redundant mappings between genotype and phenotype are common in Genetic Programming (GP). Such encodings often produce neutrality [1], where many mutational variants of a genotype produce identical phenotypes. Conflicting opinions have been proposed with regard to the effects of neutrality on evolutionary search. While some studies have found no benefit [2,3,4], others have claimed that neutrality provides a buffer against deleterious genetic perturbation [5,6] and reduces the risk of premature convergence through an expansion of the search space [7,8]. As argued in [9], the lack of consensus regarding the benefits of neutrality largely stem from the overly complex problems, representations, and search algorithms used in these analyses, which make it difficult to tease apart the effects of neutrality from other confounding factors. Further, neutrality is often artificially added to the problem representation and little attention is paid to how this alters the fitness landscape [9].

---

* These authors contributed equally to this work.

S. Silva et al. (Eds.): EuroGP 2011, LNCS 6621, pp. 13–24, 2011.

In natural systems, neutrality is often discussed in terms of robustness, which can be defined as an organism's ability to maintain its phenotype in the face of genetic perturbation. Of particular interest is the relationship between robustness and evolvability [10,11], which can be defined as an organism's ability to generate novel phenotypes. At first glance, robustness and evolvability appear contradictory. While the former requires genetic alterations leave the phenotype intact, the latter requires these alterations be used for the exploration of new phenotypes. Despite this apparent paradox, empirical analyses of several natural systems have revealed that robustness can actually facilitate evolvability [12,13,14]. In the cytochrome P450 BM3 protein, for example, increased robustness increases the probability that mutants can exploit new substrates [12].

To elucidate the relationship between robustness and evolvability in natural systems, several theoretical models have been put forth (*e.g.*, [15,16,17,18]). Central to many of these analyses is the concept of a genotype network (a.k.a. a neutral network), where vertices represent genotypes and edges connect genotypes that share the same phenotype and can be interconverted via single mutational events [10]. In this framework, robust phenotypes correspond to large genotype networks, where most mutations are neutral and therefore leave the phenotype unchanged. Robust phenotypes are evolvable because a population can diffuse neutrally throughout the genotype network and build up genetic diversity [16], which facilitates access to novel phenotypes through non-neutral mutations into adjacent genotype networks [10].

One of the primary advantages of discussing neutrality in terms of robustness is that an exact measure of neutrality can be specifically assigned to each genotype and phenotype. This in turn allows for an assessment of the distribution of neutrality at both the genotypic and phenotypic scales. Further, it allows for a quantitative analysis of the relationship between robustness and evolvability. For GP, this offers a marked improvement over previous neutrality analyses, which typically use only performance statistics, such as success rate or search efficiency, to estimate how neutrality affects evolutionary search.

The exact distributions of neutrality, at the level of the genotype and phenotype, have never been explicitly characterized in a GP system. As such, the influence of robustness on evolvability, and in turn on search performance, is not well understood. Further, the ease with which a phenotype can be accessed in an evolutionary search, and how this relates to robustness and evolvability at the genotypic and phenotypic scales, has not been addressed. To elucidate these relationships, we extend the work of [19] by exhaustively characterizing the search space of a simple Linear Genetic Programming (LGP) representation used to solve a Boolean search problem. This system offers several advantages. First, the representation is compact; the set of all genotypes is finite and computationally enumerable. Second, neutrality is intrinsic to the system; a total of 2^{28} genotypes map to 16 phenotypes. Third, there is a clear delineation between genotype, phenotype, and fitness, allowing for a full description of their interplay.

By capitalizing on recent developments in the characterization of robustness and evolvability in RNA [11,20], we provide a comprehensive quantitative analysis of the genotype and phenotype spaces in this LGP system. Using a large ensemble of random walks, we then conduct a preliminary exploration of the relationships between robustness, evolvability, and mutation-based search. We discuss the implications of our results and present directions for future work.

2   Methods

2.1   Linear Genetic Programming

In the LGP representation, an individual (or program) consists of a set of L instructions, which are structurally similar to those found in register machine languages. Each instruction is made up of an operator, a set of operands, and a return value. In the programs considered in this study, each instruction consists of an operator drawn from the set {AND, OR, NAND, NOR}, two Boolean operands, and one Boolean return value. The inputs, operands, and return values are stored in registers with varying read/write permissions. Specifically, R_0 and R_1 are calculation registers that can be read and written, whereas R_2 and R_3 are input registers that are read-only. Thus, a calculation register can serve in an instruction as an operand or a return, but an input register can only be used as an operand. An example program with $L = 4$ is given below.

$$R_1 = R_2 \ \text{ OR } \ R_3$$
$$R_0 = R_1 \ \text{ AND } \ R_2$$
$$R_1 = R_0 \ \text{ NAND } R_1$$
$$R_0 = R_3 \ \text{ NOR } \ R_1$$

These instructions are executed sequentially from top to bottom. Prior to program execution, the values of R_0 and R_1 are initialized to FALSE. After program execution, the final value in R_0 is returned as output.

2.2   Genotype and Phenotype Networks

To facilitate the enumeration of the entire genotype and phenotype spaces, we consider a two-input, one-output Boolean problem instance with $L = 4$ instructions. This sequence of instructions is referred to as the *genotype*. Letting C and I denote the numbers of calculation and input registers, respectively, and O the cardinality of the operator set, there are a total of $(C \times (C + I)^2 \times O)^L$ genotypes in the LGP representation. We refer to this set of programs as the *genotype space*. In the system considered here $(L = 4, C = 2, I = 2, O = 4)$, the genotype space comprises 2^{28} unique programs.

These genotypes map to a considerably smaller set of phenotypes, which are defined by the functional relationship between the input and output registers. Specifically, the *phenotype* is defined by the set of outputs observed across each of the four possible combinations of Boolean inputs. Since the outputs are also

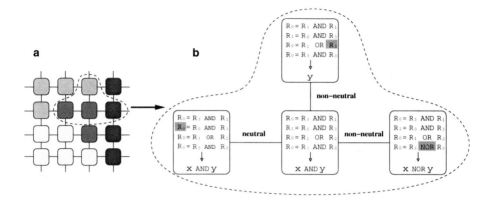

Fig. 1. (a) Schematic diagram of a subset of genotype space in linear genetic programming. Vertices correspond to genotypes, their color to phenotypes, and edges connect genotypes that can be interconverted via point mutations. (b) These point mutations can be neutral or non-neutral, depending on whether the phenotype is preserved. For visual clarity, we only depict a small subset of the potential point mutations to each genotype.

Boolean, there are $2^4 = 16$ unique phenotypes, the set of which we refer to as the *phenotype space*.

The mapping from genotype to phenotype is highly redundant. As previously mentioned, a convenient formalism for understanding this redundancy is a *genotype network*, in which genotypes are represented as vertices and edges connect genotypes that can be interconverted via neutral point mutations. In this study, we define a point mutation as a single change to an element of the instruction set of a program. This point mutation is neutral if it does not lead to a change in phenotype (Fig. 1).

The number of point mutations v_{ij} between the genotype networks of phenotypes i and j captures the number of unique mutational events that interconvert these two phenotypes. By considering the adjacency of all genotype networks in the genotype space, we can construct a phenotype network. Vertices correspond to phenotypes and are weighted according to the size of their underlying genotype network, and edges correspond to the adjacency of genotype networks and are weighted according to the number of non-neutral point mutations between genotype networks (Fig. 2). By characterizing the genotype and phenotype spaces in this way, we can describe the distribution of neutrality in this LGP system and gain insight into how this distribution influences robustness and evolvability.

2.3   Robustness, Evolvability, and Accessibility

Several measures of robustness and evolvability exist in the literature, at both the genotypic and phenotypic scales. Following [11], we define *genotypic robustness* as the fraction of the total number of possible point mutations to a given genotype that are neutral. *Genotypic evolvability* is defined as the fraction of the total

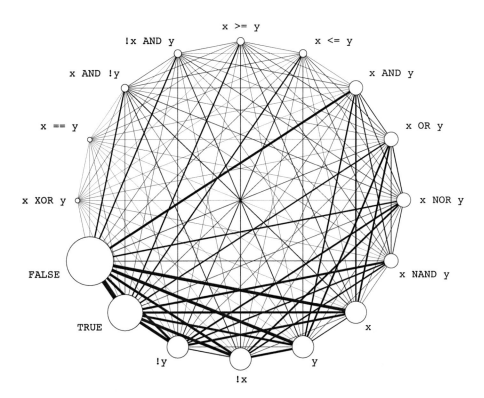

Fig. 2. Phenotype network for linear genetic programming with two inputs, one output, and four instructions. Each vertex comprises a genotype network, as depicted in Fig. 1, and thus vertex size corresponds to phenotypic robustness. Edge width denotes the number of non-neutral point mutations between two phenotypes (v_{ij}). Phenotypes are labeled according to their functional relationship between input and output, where x and y denote the inputs stored in registers R_2 and R_3, respectively.

number of possible phenotypes that are accessible through non-neutral point mutations to a single genotype. *Phenotypic robustness* is defined as the size of the phenotype's underlying genotype network.

For *phenotypic evolvability*, we consider two metrics. The first E_1 is simply the proportion of the total number of phenotypes that can be reached via non-neutral point mutations from a given phenotype [11]. The second E_2 provides a more nuanced analysis of the potential to mutate from one phenotype to another [20]. Letting

$$f_{ij} = \begin{cases} \dfrac{v_{ij}}{\sum_{k \neq i} v_{ik}}, & \text{if } i \neq j \\ 0, & \text{if } i = j \end{cases} \tag{1}$$

denote the fraction of non-neutral point mutations to genotypes of phenotype i that result in genotypes of phenotype j, we define the evolvability E_2 of phenotype i as

$$E_{2,i} = 1 - \sum_j f_{ij}^2. \tag{2}$$

This captures the probability that two randomly chosen non-neutral point mutations to genotypes of phenotype i result in genotypes with distinct phenotypes. This metric takes on high values when a phenotype is adjacent to many other phenotypes and its non-neutral mutations are uniformly divided amongst these phenotypes. It takes on low values when a phenotype is adjacent to only a few other phenotypes and its non-neutral mutations are biased toward a subset of these phenotypes.

In addition to measuring the propensity to mutate away from a phenotype, we also measure phenotypic accessibility [20], which is formally defined as

$$A_i = \sum_j f_{ji}. \tag{3}$$

This metric takes on high values if phenotype i is relatively easy to access from other phenotypes, and low values otherwise.

3   Results

3.1   Statistical Characteristics of Genotype and Phenotype Space

To characterize the genotype and phenotype spaces of the two-input, one-output LGP system of $L = 4$ instructions, we exhaustively enumerated the size and structure of the genotype networks corresponding to each of the 2^{28} individual genotypes.

In Fig. 2, we depict the corresponding phenotype network. The network is fully connected, such that any phenotype can be reached directly from any other. However, the number of non-neutral point mutations between phenotypes, depicted by edge width, is heterogeneous. Some phenotypes are mutationally biased toward a small subset of phenotypes (e.g., Fig. 2, x), while others mutate nearly uniformly to all other phenotypes (e.g., Fig. 2, x XOR y). Phenotypic robustness, depicted by vertex size, is also heterogeneous and ranges from a minimum of 24,832 genotypes to a maximum of 60,393,728 genotypes (occupying $\ll 1\%$ and 23% of genotype space, respectively). Each phenotype comprises a single genotype network, as opposed to multiple independent genotype networks. Thus, any genotype yielding a specific phenotype is reachable through a series of neutral point mutations from any other genotype with the same phenotype.

In Figs. 3a,b, we depict the distributions of genotypic robustness and genotypic evolvability for the representative phenotype FALSE. Within this phenotype, and all others in this system, the distribution of genotypic robustness is bimodal (Fig. 3a), while the distribution of genotypic evolvability is unimodal (Fig. 3b). These properties are inversely related, such that genotypes of greater robustness are less evolvable (Fig. 3c).

The means of the distributions of genotypic robustness and evolvability vary as a function of phenotypic robustness. Specifically, average genotypic evolvability

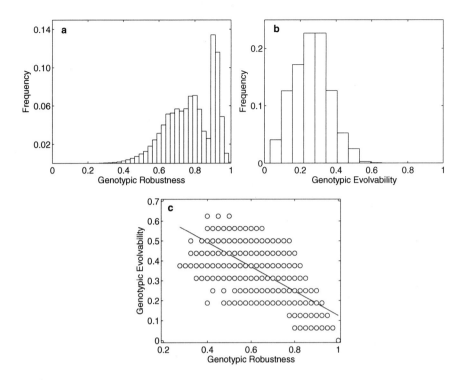

Fig. 3. Properties of genotypes within the phenotype `FALSE`. Distributions of (a) genotypic robustness and (b) genotypic evolvability for all ≈ 60 million genotypes. (c) Relationship between genotypic evolvability and genotypic robustness for 10,000 randomly sampled genotypes. The solid line represents the best linear fit to the data, and is provided as a guide for the eye.

decreases logarithmically as a function of phenotypic robustness (Fig. 4a, $R^2 = 0.95$). This intuitive observation implies that within robust phenotypes, most mutations are neutral and do not allow access to adjacent phenotypes. It follows that the individual genotypes that make up robust phenotypes are collectively more robust. Indeed, we observe that the average genotypic robustness increases logarithmically as a function of phenotypic robustness (Fig. 4b, $R^2 = 0.98$).

The relationship between phenotypic evolvability and phenotypic robustness is less intuitive. Because the phenotype network is fully connected, all phenotypes are equally and maximally evolvable according to E_1 (filled circles, Fig. 4c). In contrast, when mutational biases are taken into account with E_2, phenotypic evolvability exhibits a nonlinear relationship with phenotypic robustness (open circles, Fig. 4c). Phenotypic evolvability is lowest for phenotypes of intermediate robustness (`x AND !y`, `!x AND y`), and then increases logarithmically with increasing phenotypic robustness ($R^2 = 0.87$). The relationship is made non-monotonic by the high evolvability of the least robust phenotypes (`x XOR y`, `x == y`).

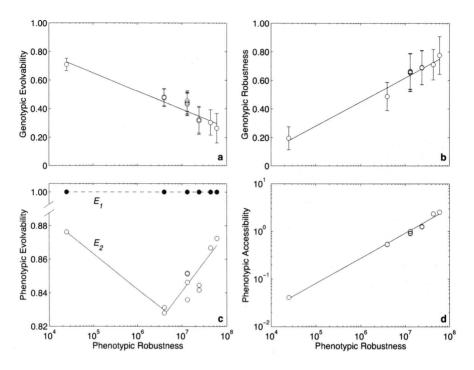

Fig. 4. Statistical properties of genotype and phenotype networks in linear genetic programming. Genotypic (a) evolvability and (b) robustness and phenotypic (c) evolvability and (d) accessibility as a function of phenotypic robustness. The data in (a,b) correspond to the average of all genotypes within a given phenotype and error bars denote their standard deviation. The solid lines correspond to the best (a,b) logarithmic, (c) piecewise logarithmic, and (d) power-law fit to the data, and are provided as a guide for the eye. In (c), two measures of phenotypic evolvability are shown.

Phenotypic accessibility increases monotonically as a function of phenotypic robustness, following a power-law (Fig. 4d, $R^2 = 0.99$). This implies that random mutations are more likely to lead to robust phenotypes than to non-robust phenotypes. Taken together, these results suggest that the most robust phenotypes are both easy to find (Fig. 4d) and highly evolvable (Fig. 4c), with the exception of the least robust phenotype, which is simultaneously the least accessible and the most evolvable of any of the phenotypes in this system.

3.2   Random Walks through Genotype and Phenotype Space

To begin to understand the implications of these observations for mutation-based search, we consider an ensemble of random walks. For each of the 16×16 possible combinations of phenotypes, we designate one phenotype as a source and the other as a target. We then perform 1000 random walks, starting from a randomly chosen genotype in the source phenotype and ending when the random

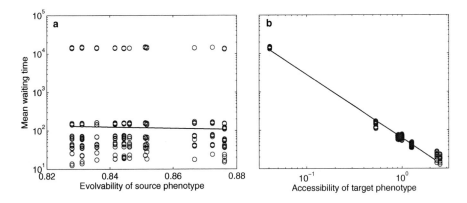

Fig. 5. Mean waiting time of a random walk as a function of (a) the source phenotype's evolvability and (b) the target phenotype's accessibility. The solid lines corresponds to the best (a) exponential and (b) power-law ($T \propto A^{-5/3}$) fit to the data. Both lines are provided as a guide for the eye.

walk reaches any genotype in the target phenotype. Each step in the random walk corresponds to a single point mutation. We record the average number of steps required to get from one phenotype to another, which we refer to as the mean waiting time.

In Fig. 5, we depict the mean waiting time of a random walk as a function of the source phenotype's evolvability (Fig. 5a) and the target phenotype's accessibility (Fig. 5b). While the mean waiting time is independent of the evolvability of the source phenotype ($R^2 = 0.001$), it is strongly correlated with the accessibility of the target phenotype ($R^2 = 0.99$). Specifically, the mean waiting time T decreases as a function of target accessibility according to the power-law $T \propto A^{-5/3}$. Thus, accessible phenotypes are found more rapidly by random mutation than less accessible phenotypes. As highly accessible phenotypes are also highly robust (Fig. 4d), random mutation leads to robust phenotypes.

4   Discussion

The results of this study have helped to clarify how the distributions of neutrality at the genotypic and phenotypic scales affect the relationship between robustness and evolvability in linear genetic programming (LGP). At the genotypic level, robustness and evolvability were found to be negatively correlated with one another (Fig. 3c), echoing previous results regarding RNA landscapes [11]. This intuitive observation suggests that robust genotypes are located far from the periphery of the genotype network, prohibiting direct mutational access to adjacent genotype networks. Further support for this claim could be obtained through a topological analysis of the genotype networks, which could also help to explain the observed bimodality in the distributions of genotypic robustness (Fig. 3a).

Within a given phenotype, the distribution of genotypic robustness was heterogeneous, where some genotypes were vastly more robust than others (Fig. 3a). In addition, all genotypes within a phenotype formed a single genotype network, such that any genotype was reachable from any other through a series of neutral point mutations. Taken together, these results suggest that genotypic robustness is an evolvable property in LGP. A caveat of this observation is that any selective pressure for genotypic robustness comes at the expense of reduced genotypic evolvability (Fig. 3c).

At the phenotypic level, the relationship between robustness and evolvability varied depending on how evolvability was defined (Fig. 4c). When defined by the total connectivity of a phenotype in the phenotype network (E_1), evolvability was independent of robustness; all phenotypes were maximally evolvable. In contrast, when mutational biases were taken into account (E_2), the relationship between evolvability and robustness was nonlinear, with phenotypes of intermediate robustness exhibiting the lowest evolvabilities. These results contrast with those made in RNA systems where E_1 was found to be positively correlated [11], and E_2 negatively correlated, with robustness [20]. However, accessibility and robustness were positively correlated (Fig. 4d), in line with observations made in RNA systems and supporting the intuitive notion that larger phenotypes are easier to access.

To explore the implications of these observations for evolutionary search, we considered an ensemble of random walks between source and target phenotypes. The mean waiting time of random mutation to reach a target phenotype was found to be uncorrelated with the evolvability of the source phenotype (Fig. 5a), a result that calls into question the utility of existing phenotypic evolvability measures. While these measures provide useful information concerning the immediate adjacency of phenotypes [11] and their mutational biases [20], they are too myopic to predict the length of an evolutionary trajectory from one phenotype to another. Consider, for example, that correlations may exist between the evolvabilities of adjacent phenotypes, such that high evolvability phenotypes are mutationally biased toward low evolvability phenotypes. As these correlations (a.k.a. mixing patterns [21]) are not taken into account, the applicability of current phenotypic evolvability measures are left severely constrained, at least for this system.

In contrast, the mean waiting time of random mutation to reach a target phenotype was strongly correlated with the target phenotype's accessibility (Fig. 5b). Our observations suggest that this relationship follows a power-law of the form $T \propto A^{-5/3}$. We believe this relationship may lend itself to analytical treatment, which could prove useful in generalizing our results to other phenotype networks. For instance, it remains to be seen whether target accessibility is predictive of mean waiting time in phenotype networks that are not fully connected. Recent analyses of random walks on weighted complex networks [22] may provide a useful starting point for this analysis.

There has been much debate regarding the benefit of neutrality in evolutionary computation. Using a simple set of experiments, recent analysis has demonstrated this benefit to be problem dependent [9]. Our results provide a more subtle picture. Using the notion of robustness to compartmentalize neutrality into specific phenotypes, we have shown that it is not only the presence of neutrality that affects mutation-based search, but also how this neutrality is distributed amongst phenotypes. The more robust the target phenotype, the easier it will be for evolution to identify it.

To fully understand how robustness, evolvability, and accessibility influence evolutionary search, future analyses will have to take fitness into consideration. In the Boolean search problem considered herein, for example, fitness could be assigned as a function of phenotypic distance from the target [19]. As only five distinct fitness values exist in our LGP system, this form of fitness assignment would add a new layer of robustness and evolvability to the analysis. Specifically, this layer would consist of a set of vertices that each comprise multiple phenotypes, and links between these vertices would capture all mutational events that lead from one fitness value to another. While fitnesss would therefore modify the transition probabilities between phenotypes, selection would modify the transition probabilities between fitness values. Understanding how these layers interact to influence evolutionary search is left for future research.

Future work will also consider population-based evolutionary search and the role of recombination. We are particularly interested in analyzing population diffusions throughout genotype and phenotype networks [23], under mutational and recombinative variation operators, to understand how robustness and evolvability change as a function of phenotypic distance to the target.

Acknowledgments

This work was partially supported by NIH grants R01-LM009012, R01-LM010098, and R01-AI59694. J.L.P. was supported by NIH grant K25-CA134286. W.B. acknowledges support from NSERC Discovery Grants, under RGPIN 283304-07.

References

1. Banzhaf, W.: Genotype-phenotype mapping and neutral variation - a case study in genetic programming. In: Davidor, Y., Männer, R., Schwefel, H.-P. (eds.) PPSN 1994. LNCS, vol. 866, pp. 322–332. Springer, Heidelberg (1994)
2. Smith, T., Husbands, P., O'Shea, M.: Neutral networks and evolvability with complex genotype-phenotype mapping. In: Kelemen, J., Sosík, P. (eds.) ECAL 2001. LNCS (LNAI), vol. 2159, pp. 272–282. Springer, Heidelberg (2001)
3. Smith, T., Husbands, P., O'Shea, M.: Neutral networks in an evolutionary robotics search space. In: Proceedings of the IEEE Congress on Evolutionary Computation, pp. 136–145 (2001)
4. Collins, M.: Finding needles in haystacks is harder with neutrality. Genetic Programming and Evolvable Machines 7, 131–144 (2006)
5. Yu, T., Miller, J.F.: Through the interaction of neutral and adaptive mutations, evolutionary search finds a way. Artificial Life 12, 525–551 (2006)

6. Soule, T.: Resilient individuals improve evolutionary search. Artificial Life 12, 17–34 (2006)
7. Harvey, I., Thompson, A.: Through the labyrinth evolution finds a way: A silicon ridge. In: Proceedings of the First International Conference on Evolvable Systems: From Biology to Hardware, pp. 406–422 (1996)
8. Ebner, M., Shackleton, M., Shipman, R.: How neutral networks influence evolvability. Complexity 7(2), 19–33 (2002)
9. Galván-López, E., Poli, R.: An empirical investigation of how and why neutrality affects evolutionary search. In: Proceedings of the Genetic and Evolutionary Computation Conference, pp. 1149–1156 (2006)
10. Wagner, A.: Neutralism and selectionism: a network-based reconciliation. Nature Reviews Genetics 9, 965–974 (2008)
11. Wagner, A.: Robustness and evolvability: a paradox resolved. Proceedings of the Royal Society London B 275, 91–100 (2008)
12. Bloom, J.D., Labthavikul, S.T., Otey, C.R., Arnold, F.H.: Protein stability promotes evolvability. Proceedings of the National Academy of the Sciences 103(15), 5869–5874 (2006)
13. Ferrada, E., Wagner, A.: Protein robustness promotes evolutionary innovations on large evolutionary time-scales. Proceedings of the Royal Society London B 275, 1595–1602 (2008)
14. Isalan, M., Lemerle, C., Michalodimitrakis, K., Horn, C., Beltrao, P., Raineri, E., Garriga-Canut, M., Serrano, L.: Evolvability and hierarchy in rewired bacterial gene networks. Nature 452, 840–846 (2008)
15. Draghi, J.A., Parsons, T.L., Wagner, G.P., Plotkin, J.B.: Mutational robustness can facilitate adaptation. Nature 463, 353–355 (2010)
16. Huynen, M.A., Stadler, P.F., Fontana, W.: Smoothness within ruggedness: The role of neutrality in adaptation. Proceedings of the National Academy of the Sciences 93, 397–401 (1996)
17. Newman, M., Engelhardt, R.: Effects of selective neutrality on the evolution of molecular species. Proceedings of the Royal Society London B 265, 1333–1338 (1998)
18. Whitacre, J., Bender, A.: Degeneracy: a design principle for achieving robustness and evolvability. Journal of Theoretical Biology 263, 143–153 (2010)
19. Banzhaf, W., Leier, A.: Genetic Programming Theory and Practice III. In: Genetic Programming Theory and Practice III, pp. 207–221. Springer, Heidelberg (2006)
20. Cowperthwaite, M.C., Economo, E.P., Harcombe, W.R., Miller, E.L., Meyers, L.A.: The ascent of the abundant: how mutational networks constrain evolution. PLoS Computational Biology 4(7), e10000110 (2008)
21. Newman, M.: Mixing patterns in networks. Physical Review E 67, 026126 (2003)
22. Noh, J.D., Rieger, H.: Random walks on complex networks. Physical Review Letters 92(11), 118701 (2004)
23. van Nimwegen, E., Crutchfield, J., Huynen, M.: Neutral evolution of mutational robustness. Proceedings of the National Academy of the Sciences 96, 9716–9720 (1999)

A Quantitative Study of Learning and Generalization in Genetic Programming

Mauro Castelli[1], Luca Manzoni[1], Sara Silva[2], and Leonardo Vanneschi[1,2]

[1] Dipartimento di Informatica, Sistemistica e Comunicazione (D.I.S.Co.)
University of Milano-Bicocca, Milan Italy
[2] INESC-ID Lisboa, KDBIO group, Lisbon Portugal

Abstract. The relationship between generalization and solutions functional complexity in genetic programming (GP) has been recently investigated. Three main contributions are contained in this paper: (1) a new measure of functional complexity for GP solutions, called Graph Based Complexity (GBC) is defined and we show that it has a higher correlation with GP performance on out-of-sample data than another complexity measure introduced in a recent publication. (2) A new measure is presented, called Graph Based Learning Ability (GBLA). It is inspired by the GBC and its goal is to quantify the ability of GP to learn "difficult" training points; we show that GBLA is negatively correlated with the performance of GP on out-of-sample data. (3) Finally, we use the ideas that have inspired the definition of GBC and GBLA to define a new fitness function, whose suitability is empirically demonstrated. The experimental results reported in this paper have been obtained using three real-life multidimensional regression problems.

1 Introduction

The issue of generalization in Genetic Programming (GP) [12] has received a growing attention in the last few years (see [9] for a survey and [11] and references therein for a more recent discussion). A common agreement of many researchers is the so called minimum description length principle (see for instance [13]), which states that the best model is the one that minimizes the amount of information needed to encode it. Simpler solutions are thought to be more robust and generalize better [14], while complex solutions are more likely to incorporate specific information from the training set, thus overfitting it. A superficial interpretation of the minimum description length may be that bloat (i.e. an excess of code growth without a corresponding improvement in fitness [10,15]) and overfitting should be two related phenomena: bloated programs could in fact use too much information, thus overfitting training data. Nevertheless, the observations in [16,18] seem to contradict this idea. In [17] Vanneschi and coworkers defined measures to quantify bloat, overfitting and functional complexity of GP solutions, and they showed that functional complexity, rather than bloat, seems to be related to generalization. But the experimental results reported in [17] show that the proposed complexity measure has a positive correlation with overfitting only for a subset of the studied test problems, while for other problems no clear relationship appeared between this measure and the ability of GP to find general solutions. We hint that one of the reasons why, in some cases, that complexity measure shows poor correlation with overfitting is that the

S. Silva et al. (Eds.): EuroGP 2011, LNCS 6621, pp. 25–36, 2011.

measure itself is not rotationally invariant, i.e. it is calculated on all the different dimensions separately, and it is obtained by averaging the partial calculations on the single dimensions. The present work has three different, but related, goals: (1) defining a new measure of functional complexity, that is rotationally invariant and that overcomes the limitations of the measure proposed in [17]; (2) defining a new measure to quantify the ability of GP to learn "difficult" points (our intuition of what a "difficult" point is will be explained in Section 3 where the measure will be defined) and studying its correlation with generalization; and (3) defining a new fitness function (inspired by the two previously defined measures) to improve GP generalization ability in those cases where standard GP (i.e. GP that calculates fitness using the root mean squared error between outputs and targets) has poor generalization ability.

The paper is structured as follows: in Section 2 we discuss previous contributions in which the issue of generalization in GP has been investigated, focusing in particular on those references where an attempt has been made to relate generalization with the functional complexity of the solutions. In Section 3 we present the measures proposed in this work. Section 4 introduces the real-life problems used as test cases. In Section 5 we present our experimental settings and we discuss the obtained results. Finally, Section 6 concludes the paper.

2   Previous and Related Work

The contributions on generalization in GP are so numerous that it is impossible to analyze all of them in the strict length limit imposed to this publication. Thus, we focus on those contributions that we consider more similar and related to the present work, referring the interested reader to [9] for a survey on the issue. In 1995 Zhang and Mühlenbein investigated the relationship between generalization and parsimony and proposed an adaptive learning method that automatically balances these two factors. One year later, in [5], a new GP system called Compiling GP System was introduced and in [3], Banzhaf and coworkers showed the positive effect of an extensive use of the mutation operator on generalization in GP using sparse data sets. More recently, in [6], Gagné and coworkers have investigated two methods to improve generalization in GP: the selection of individuals using a three data sets methodology, and the application of parsimony pressure to reduce the size of the solutions. In the last few years the idea of quantitatively studying the relationship between generalization and solution complexity was tackled in several contributions. For instance, in [1] the authors propose a theoretical analysis of GP from the perspective of statistical learning theory and prove the advantage of a parsimonious fitness using Vapnik-Chervonenkis theory. Another important contribution, even though not explicitly focused on GP, but on evolutionary algorithms in general, is represented by [7], where the authors measure behavioral complexity explicitly using compression, and use it as a separate objective to be optimized. In [19], the authors define two measures of complexity for the GP individuals: a genotypic measure and a phenotypic one. While the genotypic measure is related to counting the number of nodes of a tree and its subtrees, the phenotypic one, called *order of nonlinearity*, is related to functional complexity and consists in calculating the degree of the Chebyshev polynomial approximation of the function. The authors then use these two measures as criteria in a multi-objective system and they show that this system is able to counteract both bloat and overfitting. As already pointed out in the previous section, in [17] indicators of bloat, overfitting and complexity have been introduced and their mutual relationships have been investigated. The overfitting measure quantified the relationship between training and test fitness, normalizing it with the training and test fitness of the point with the best test fitness found by GP so far (the reader is referred to page 878 of [17] for the pseudo-code used to calculate this measure). The complexity measure was used to approximate the "degree of curvature" (or "ruggedness") of the function expressed by GP solutions and it was basically a weighted sum of the slopes of the segments joining the various training points in the single dimensions (the formal definition of this complexity measure can be found at page 880 in [17]). The measures of overfitting and complexity proposed in [17] will be experimentally compared with the measures introduced in this work (and presented in the next section). The experimental comparison will be presented in Section 5.

3   The Proposed Measures

The complexity measure proposed in this work, as the one introduced in [17], is inspired by the idea that complex functions should have a larger "degree of curvature" (or "ruggedness") than simple ones. But, contrarily to [17], in the present work we quantify the idea of "degree of curvature" using the following intuition: let g be a GP individual; the "degree of curvature" of g can be expressed by counting the number of pairs of "*close*" training points α and β (where both α and β are points that belong to the domain of g) for which the corresponding values $g(\alpha)$ and $g(\beta)$ are "*far*". In more formal terms, given a GP individual g, let $S = \{(\mathbf{x_1}, g(\mathbf{x_1})), (\mathbf{x_2}, g(\mathbf{x_2})), \ldots, (\mathbf{x_m}, g(\mathbf{x_m}))\}$ be the set of training points $\mathbf{x_1}, \mathbf{x_2}, \ldots, \mathbf{x_m}$ associated with the corresponding values assumed by g on them. Let $\mathbf{x_1}, \mathbf{x_2}, \ldots, \mathbf{x_m} \in \mathbf{X}$ and $g(\mathbf{x_1}), g(\mathbf{x_2}), \ldots, g(\mathbf{x_m}) \in \mathbf{Y}$, with both $\mathbf{X}$ and $\mathbf{Y}$ being metric spaces equipped with metrics $d_{\mathbf{X}}$ and $d_{\mathbf{Y}}$ respectively. For any $i = 1, 2, \ldots, m$, and for any prefixed constant value δ, let $B_\delta(\mathbf{x_i})$ be the open ball of radius δ centered on $\mathbf{x_i}$ in metric space $\mathbf{X}$, i.e. $B_\delta(\mathbf{x_i}) = \{\mathbf{x_j} \mid d_{\mathbf{X}}(\mathbf{x_i}, \mathbf{x_j}) < \delta\}$. Analogously, for any $i = 1, 2, \ldots, m$, and for any prefixed constant value ε, let $B_\varepsilon(g(\mathbf{x_i}))$ be the open ball of radius ε centered on $g(\mathbf{x_i})$ in metric space $\mathbf{Y}$. For every training point $\mathbf{x_i}$, we define the set: $V(\mathbf{x_i}) = \{\mathbf{x_j} \in B_\delta(\mathbf{x_i}) \mid g(\mathbf{x_j}) \notin B_\varepsilon(g(\mathbf{x_i})) \text{ and } \mathbf{x_j} \neq \mathbf{x_i}\}$. This set contains all the points of the sample set that are close (i.e. nearer than a given δ) to $\mathbf{x_i}$ in the $\mathbf{X}$ metric space, but whose values under g are not close (i.e. farther than a given ε) to the value of $g(\mathbf{x_i})$ in the $\mathbf{Y}$ metric space. We now consider the set $V = \cup_{i=1}^m V(\mathbf{x_i})$. V can be addressed as the set of points in $\mathbf{X}$ in which the function represented by the GP individual g is *rugged*, thus the fraction of training points that belong to set V can be used as a measure to quantify our intuition of "degree of curvature" (or "ruggedness"): $\frac{|V|}{m}$. Clearly, values near 1 denote a very rugged function, while values near 0 indicate flat (or straight) and thus less complex functions. It is interesting to consider not only the set $V(\mathbf{x_i})$ containing the *points* of ruggedness since, in the union $\cup_{i=1}^m V(\mathbf{x_i})$ one misses the information about which *pairs* of points that are close to each other have corresponding function values which are far apart. So, we introduce the set $E = \cup_{i=1}^m (\{\mathbf{x_i}\} \times V(\mathbf{x_i}))$, which is a relation that associates each $\mathbf{x_i}$ with all the corresponding points in $V(\mathbf{x_i})$. Now, if we define the set $E_{tot} = \{(\mathbf{x_i}, \mathbf{x_j}) \mid \mathbf{x_j} \in B_\delta(\mathbf{x_i}) \setminus \{\mathbf{x_i}\}\}$, we can define our complexity measure as:

$$\mathrm{GBC} = |E|/|E_{tot}|$$

We remark that, in case of symbolic regression problems, we usually have $\mathbf{X} \subseteq \mathbf{R}^n$ and $\mathbf{Y} \subseteq \mathbf{R}$. Thus, it is possible to calculate GBC, for instance, using the Euclidean distance as the $d_\mathbf{X}$ and $d_\mathbf{Y}$ metrics. Thus GBC is rotationally invariant, contrarily to what happens for the complexity measure defined in [17]. The acronym we have chosen as the name of this measure (GBC, which stands for Graph Based Complexity) depends on the fact that it is possible to represent it in terms of counting operations on a graph. Let $G = (\{\mathbf{x_1}, \ldots, \mathbf{x_m}\}, E_{tot})$ be a graph defined on the training points, where two vertices are connected if their distance on the metric space $\mathbf{X}$ is less than δ. Now consider the subgraph $G_\varepsilon = (\{\mathbf{x_1}, \ldots, \mathbf{x_m}\}, E)$ which only contains the edges $(\mathbf{x_i}, \mathbf{x_j})$ of G such that the distance between $g(\mathbf{x_i})$ and $g(\mathbf{x_j})$ in the $\mathbf{Y}$ metric space is greater or equal than ε. The GBC measure is clearly equal to the ratio between the number of connections in G_ε and the number of connections in G.

The GBC function can be used to quantify the complexity of GP individuals. However, it is also clear that the same calculation can be performed using the known target values (instead of using the values assumed by the learned function) on the different training points. In particular, if we indicate by $f(\mathbf{x_i})$ the target value on a training point $\mathbf{x_i}$, it is possible to define a set V' (analogous to the set V previously defined) as follows: $V'(\mathbf{x_i}) = \{\mathbf{x_j} \in B_\delta(\mathbf{x_i}) \mid f(\mathbf{x_j}) \notin B_\varepsilon(f(\mathbf{x_i})) \text{ and } \mathbf{x_j} \neq \mathbf{x_i}\}$. And thus, it is also possible to define a set E' as follows: $E' = \cup_{i=1}^m (\{\mathbf{x_i}\} \times V'(\mathbf{x_i}))$. Following the same idea that we have used to define the GBC measure, we can state that $\mathrm{GBC}_{target} = |E'|/|E_{tot}|$ is a measure that can be used to quantify the ruggedness of the target function.

Furthermore, we can also use *both* information coming from sets E and E' to quantify the ability of a GP individual to learn "difficult" points. For this aim we define the following measure, that we call GBLA (Graph Based Learning Ability):

$$\mathrm{GBLA} = (|E \triangle E'|)/|E_{tot}|$$

where $\triangle$ represents the operator of symmetrical difference between sets. GBLA quantifies the number of training points where the target function is rugged and the learned function is flat, plus the number of training points where the learned function is rugged and the target function is flat. For simplicity, we call these points "difficult" points.

Both the definitions of GBC and GBLA are based on the definition of the $V(\mathbf{x_i})$ set. The elements of $V(\mathbf{x_i})$ depend on the choice of the two parameters δ and ε. Consequently, also the values of GBC and GBLA depend on these two parameters. Nevertheless, a set of preliminary experiments (whose results are not reported here for lack of space) have indicated an interesting fact concerning these two parameters: if we consider many series composed by the values of the GBC of the best individual in the population for each iteration for several pairs of values of δ and ε, all these series have a positive value of their mutual cross correlation coefficient, with a magnitude of this coefficient approximately equal to 1. The same fact has also been observed for GBLA. Given that in the experimental study presented in Section 5 we are mainly interested in understanding the *correlation* of GBC and GBLA with other quantities during the GP runs (rather than the particular values of GBC and GBLA), we can assert that parameters δ and ε qualitatively affect neither the results of Section 5, nor the conclusions that we are able to draw from them. We have also repeated all the experiments reported in Section 5 for several other pairs of values of δ and ε and all the experimental results (not shown here for lack of space) have confirmed that the values of δ and ε do not affect the qualitative interpretation of the results.

4   Test Problems

The real-life applications considered in this paper to test the proposed measures are three multidimensional regression problems, whose goal is to predict the value of as many pharmacokinetic parameters of a set of candidate drug compounds on the basis of their molecular structure. The first pharmacokinetic parameter we consider is human oral bioavailability (indicated with %F from now on), the second one is median lethal dose (indicated with LD50 from now on), also informally called toxicity, and the third one is called plasma protein binding levels (indicated with %PPB from now on). %F measures the percentage of the initial orally submitted drug dose that effectively reaches the systemic blood circulation after the passage from the liver. LD50 refers to the amount of compound required to kill 50% of the considered test organisms (cavies). %PPB quantifies the percentage of the drug initial dose that reaches blood circulation and binds the proteins of plasma. For a detailed discussion of these three pharmacokinetic parameters the reader is referred to [2]. The datasets we have used are the same as in [2]: the %F dataset consists in a matrix composed by 260 rows (instances) and 242 columns (features). Each row is a vector of molecular descriptor values identifying a drug; each column represents a molecular descriptor, except the last one, that contains the known target values of %F. Both the LD50 and the %PPB datasets consist in a matrix composed by 234 rows (instances) and 627 columns (features). Also in this case, each row is a vector of molecular descriptors identifying a drug and each column represents a molecular descriptor except the last one, that contains the known values of the target. For all these datasets training and test sets have been obtained by random splitting: at each different GP run, 70% of the molecules have been randomly selected with uniform probability and inserted into the training set, while the remaining 30% formed the test set.

5   Experimental Study

Experimental setting. A total of 120 runs were performed to obtain the results reported in this section. All the runs used a population of 200 individuals. The number of generations that we have performed was equal to 100 for the LD50 and %F datasets and to 500 for %PPB (the reason why we have executed a larger number of generations for %PPB will be clear on page 33 in the current section). Tree initialization was performed with the Ramped Half-and-Half method [8] with a maximum initial depth of 6. The function set contained the four binary operators $+$, $-$, $\times$, and $/$, protected as in [8]. The terminal set contained 241 floating point variables for the %F dataset and 626 floating point variables for the LD50 and %PPB datasets. No random constants were added to the terminal set. Because the cardinalities of the function and terminal sets were so different, we have imposed a balanced choice between functions and terminals when selecting a random node. Unless where explicitly pointed out, fitness was calculated as the root mean squared error (RMSE) between outputs and targets. Tournament selection was used with size 10. The reproduction (replication) rate was 0.1, meaning that each selected parent has a 10% chance of being copied to the next generation instead of being engaged in breeding. Standard tree mutation and standard crossover (with uniform selection of crossover and mutation points) were used with probabilities of 0.1 and 0.9, respectively. Selection for survival used elitism (i.e. unchanged copy of the best individual in the next population). A fixed maximum depth equal to 17 was used for the trees in the population. We remark that these parameters are absolutely identical to the ones used in [17]. Given that, as pointed out in the previous section, the δ and ε parameters do not affect the qualitative interpretation of the results contained in this section, we report the results obtained for two arbitrary values, i.e.: $\delta = 0.06$ and $\varepsilon = 0.05$. All distance values have been normalized into the range [0,1] before comparing them with the values of δ and ε.

Experimental results: GBC and GBLA. In Figure 1 we report the median over 120 independent runs of the RMSE on the training set, the RMSE on the test set, the value of GBC, the value of 1-GBLA and finally the complexity and overfitting measures introduced in [17] for all the performed generations (first column: LD50; second column: %F; third column: %PPB). From now on we use the terms "complexity" and "overfitting" to indicate the complexity and overfitting measures introduced in [17] and with the terminology "RMSE on the test set" we indicate the RMSE on the test set of the individual with the best RMSE on the training set. Let us focus first on the relationship between RMSE on the training and test set for the studied problems. For LD50 the RMSE on the training set steadily decreases during the whole evolution, while the RMSE on the test set keeps increasing after generation 30, also showing an irregular and oscillating behaviour. For %F both the RMSE on training and test set are steadily decreasing during the studied 100 generations. For %PPB the RMSE on the training set steadily decreases during the whole evolution, while the RMSE on the test set is decreasing until generation 50, and then increasing until generation 500, also showing some oscillations. We conclude that GP has a worse generalization ability for LD50 and %PPB than for %F. Let us now focus on the curves representing GBC and 1-GBLA. Both GBC and 1-GBLA are increasing after generation 30 for LD50, steadily decreasing (except for the initial part of the run) for %F and increasing (except for the first 50 generations) for %PPB. These results hint a relationship between the trend of the RMSE on the test set and the GBC and GBLA measures for all the studied problems. In particular, GBC seems to have a positive correlation with the RMSE on the test set and GBLA seems to have a negative correlation with the RMSE on the test set. Before studying in details these correlations, let us first look at the trend of the complexity and overfitting measures. The complexity measure seems to have a less clear relationship with the RMSE on the test than GBC and GBLA for LD50 and %F. On the other hand, for %PPB the complexity measure seems to be growing with a higher speed than GBC and 1-GBLA, and thus it seems to have a stronger correlation with the RMSE on the test set. Finally,

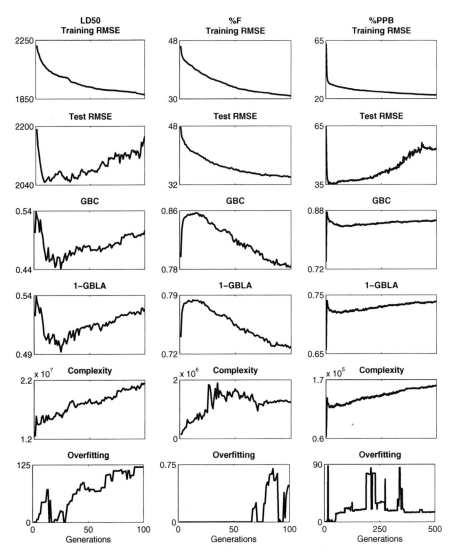

Fig. 1. The first (respectively second and third) column reports the results for LD50 (respectively %F and %PPB). For each column, from top to bottom, we report the RMSE on the training set, the RMSE on the test set, the GBC, 1-GBLA and the values of the complexity and overfitting measures introduced in [17]. All these results are reported against generations and they are medians of the value assumed by the best individual (i.e. the one with the best RMSE on the training set) over 120 independent runs.

we point out that the overfitting measure has a more oscillating and less regular behavior than the other measures. Nevertheless, its general trend seems related to GBC and to the complexity measure for LD50 and to GBC, GBLA and complexity for %PPB. On the other hand, no clear relationship appears between the overfitting measure and any of the other measures for %F.

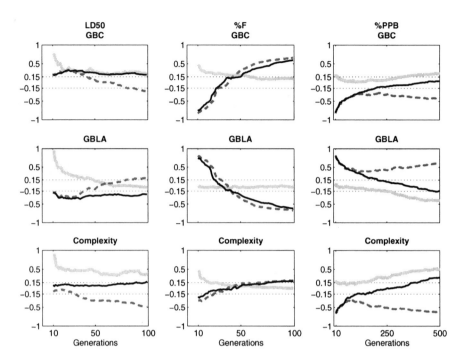

Fig. 2. The first (respectively second and third) column reports the results for LD50 (respectively %F and %PPB). For each column, from top to bottom, we report the cross correlation of GBC, GBLA and complexity with the RMSE on the training set (dashed dark grey line), the RMSE on the test set (black line) and the overfitting measure (solid light grey line). All these results are reported against generations and they are medians over 120 independent runs.

In order to better understand the mutual relationships between the quantities plotted in Figure 1, in Figure 2 we report the median values of the cross correlation at delay zero between them. Cross correlation is a standard method of estimating the degree to which two series $S_1 = \{x_1, x_2, ..., x_n\}$ and $S_2 = \{y_1, y_2, ..., y_n\}$ are correlated. It is defined by $r = (\sum_{i=d+1}^{n} |(x_i - \overline{S_1})(y_{i-d} - \overline{S_2})|)/(\sigma_{S_1}\sigma_{S_2})$, where $\overline{S_1}$ and $\overline{S_2}$ are the averages of the elements in the series S_1 and S_2 respectively, σ_{S_1} and σ_{S_2} are their respective standard deviations, and d is the delay (in this work, we have used $d = 0$). An introduction to cross correlation can be found, for instance, in [4]. Assume we want to calculate the correlation between GBC and RMSE on the test set, as it is the case, for instance, in the left-top plot of Figure 2. This can be done by considering as S_1 the series composed by the GBC values of the best individuals in the population at each generation and as S_2 the series of the respective RMSE values on the test set of the same individuals. Applying the same method, it is possible to calculate the cross correlation of any pair of measures reported in Figure 1. In Figure 2, we also draw two horizontal lines at the values -0.15 and 0.15, often empirically identified as thresholds between the presence of a correlation (either positive or negative) and the absence of it. For LD50: during the whole run, GBC has a positive cross correlation with the RMSE on the test set and overfitting and a negative cross correlation with the RMSE on the training set after generation 40. GBLA has a negative cross correlation with the RMSE on the test set after generation 50, a positive cross correlation with the RMSE on the training set and cross correlation approximately equal to zero with overfitting. Complexity has a positive cross correlation with overfitting and a negative cross correlation with the RMSE on the training set. The cross correlation between complexity and RMSE on the test set is approximately equal to zero, except for some small oscillations at the end of the run. For %F: GBC has a positive cross correlation with RMSE both on the training and test set (in particular after generation 50 and this value is steadily increasing during the run) and a cross correlation approximately equal to zero with overfitting. GBLA has a negative (and steadily decreasing) cross correlation both with the RMSE on the training and on the test set and a cross correlation approximately equal to zero with overfitting. Finally, complexity has a cross correlation approximately equal to zero with overfitting and a positive cross corrlation with both RMSE on the training and test set (even though in both cases the cross correlation becomes larger than 0.15 only in the final part of the run). One general interesting thing to be remarked is also that in some cases the the cross correlations seem, so to say, to "lag" the raw data; e.g., in Figure 1, for %F, GBC appears to be strongly correlated with test RMSE after generation 10-20, but in Figure 2 the cross correlation is not positive until generation 50. For this reason we are planning to investigate other measures instead of cross correlation in the future. For %PPB: the value of the cross correlation between GBC and the RMSE on the training and test set is negative, but steadily increasing, in the first 100 generations. For this reason, we have executed the simulations until generation 500, to see if some of these correlations became positive later in the evolution. We can see that the correlation between GBC and the RMSE on the test set becomes positive more or less at generation 350, and it keeps on growing, although without becoming larger than 0.15. Because of the steadily growing trend of the curve of the cross correlation between GBC and RMSE on the test set, we hypothesize that this cross corralation would become positive later in the run (this hypothesis will be verified in the future by performing runs for a larger number of generations than 500). On the other hand, the cross correlation between GBC and RMSE on the training set is clearly negative during the whole run. Finally, the cross correlation between GBC and overfitting is positive (it becomes larger than 0.15 around generation 280, and it remains larger than 0.15 until the end of the run). GBLA has a negative cross correlation with both overfitting and RMSE on the test set and a positive cross correlation with RMSE on the training set. Finally, complexity has a positive cross correlation with both overfitting and RMSE on the test set and a negative cross correlation with RMSE on the training set.

Summarizing: GBC is positively correlated with the RMSE on the test set and GBLA is negatively correlated with the RMSE on the test set. These facts seem independent on the generalization ability of GP (i.e. they hold for all the studied problems). Further- more, the magnitude of the correlation with the RMSE on the test set is larger for GBC and GBLA than for the complexity measure for all studied problems except %PPB. The negative values of the correlation between GBLA and RMSE on the test set can be in- terpreted as follows: GBLA quantifies the ability of GP to learn the "difficult" training points. It is intuitive that a good learning of those points leads GP to a poor gener- alization ability, because the solutions are too specialized on training data and thus overfit them. This can be caused, for instance, by the fact that those "difficult" points correspond to "noise", or even errors in the training data, or they are generally not useful to reliably reconstruct the target function. The existing relationship between GBC and GBLA with the RMSE on the test set seems to hint that the ideas used to define GBC and GBLA could be useful to build a new fitness function able to reduce the error on the test set. This is the goal of the next paragraph.

Experimental results: New Fitness Function. Using either GBC or GBLA as new fitness functions does not allow us to obtain interesting results. Consider, for instance, the case of GBLA: each function able to learn some particular points (the ones that are not considered as difficult) would have a good fitness, and thus it would receive a high probability of surviving and mating in the GP population, independently of the distance of that function from the target one. Besides our intuition, also a set of preliminary experimental results (not shown here for lack of space) confirm that using either GBC or GBLA as fitness functions does not allow us to obtain better results than standard GP (i.e. GP that uses the RMSE as fitness) on the test set. Nevertheless, the ideas used to define GBC and GBLA can be used to define a new fitness function, assuming to integrate them with the error between learned values and target ones. A possibility could be to use them together with RMSE in a multi-objective method. Even though the idea is interesting, and it definitely deserves to be investigated in the future, in this paper we want to define *one* new fitness function able to incorporate both the information derived from the RMSE and from the new measures. The idea is to give a *weight* to the error in each training point. For this reason, we call the new fitness function "weighted_fitness". The weight should depend on how rugged the learned function is in that point, reducing the weight of the rugged points. The new fitness measure is:

$$\text{weighted\_fitness}(g) = \sum_{i=0}^{m} \frac{(f(\mathbf{x_i}) - g(\mathbf{x_i}))^2}{1 + |V(\mathbf{x_i})|}, \quad \text{where } \mathbf{x_1}, \mathbf{x_2}, \ldots, \mathbf{x_m} \text{ are the training points,}$$

g is a GP individual, the values $f(\mathbf{x_1}), f(\mathbf{x_2}), \ldots, f(\mathbf{x_m})$ represent the targets on those points. The fact that the denominator in the equation of the new fitness is $1 + |V(\mathbf{x_i})|$ instead of $|V(\mathbf{x_i})|$ is due to the fact that the value of $|V(\mathbf{x_i})|$ could be equal to zero, and thus adding 1 prevents us from the eventuality of an error. In Figure 3 we report an experimental comparison between standard GP and GP that uses the new fitness function The same experimental settings and number of runs used for the initial analysis are used also here. We report the median of the RMSE on test data for each performed generation for both these models. For LD50, GP that uses the new fitness function is able to obtain better results. For both %F and %PPB, GP using the new fitness function seems to return very similar results than standard GP. We conclude that GP using the proposed fitness function is able to better generalize (compared to standard GP) for some problems where standard GP has a poor generalization ability (as it is the case of the LD50), while it behaves comparably to standard GP when standard GP itself has a good generalization ability (like for %F). Nevertheless, problems where standard GP has a poor generalization ability and the new fitness function is not able to improve it exist (it is the case of %PPB). But at least, we have shown that in this last case, the new fitness function does not worsen the results. These results suggest that the proposed fitness function could be a suitable one, given that in some cases it gives an advantage

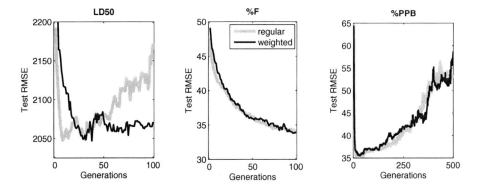

Fig. 3. RMSE on the test set of standard GP (i.e. GP that uses the RMSE as fitness, indicated by "regular" in figure) and of GP that uses the new fitness function (indicated by "weighted" in the figure). Left: LD50. Middle: %F. Right: %PPB.

when standard GP has poor generalization, and when it doesn't, at least, it does not give any disadvantage. Furthermore, the new fitness function is simple to implement and computationally reasonable (we do not report statistics about the execution time for lack of space).

6   Conclusions and Future Work

A study of Genetic Programming (GP) learning ability has been presented in this paper, offering the three following contributions: first, we have defined a new measure (GBC) to quantify the functional complexity of GP individuals. Compared to another complexity measure defined in [17], GBC is rotationally invariant and it has a higher correlation with the quality of GP solutions on out-of-sample data. Secondly, we have presented a new measure (GBLA) aimed at quantifying the ability of GP to learn "difficult" points and we have shown that this measure is negatively correlated with the quality of GP solutions on out-of-sample data. Based on these ideas, the third contribution consisted in defining a new fitness function for GP. Experimental results have shown that this new fitness function allows GP to better generalize for some problems where standard GP has a poor generalization, without worsening the results in all other cases. This seems to indicate the suitability of this fitness function in any possible case. The new fitness function has to be further studied in the future, and this is one of the main subjects of our current research. In particular, we are testing GP with the new fitness function on several other datasets of different complexities and also on a set of hand-tailored symbolic regression problems of various different difficulties. Furthermore, we are considering several possible extensions and improvements of the new fitness function.

Acknowledgments. This work was partially supported by FCT (INESC-ID multiannual funding) through the PIDDAC Program funds. The authors acknowledge project PTDC/EIA-CCO/103363/2008 from FCT, Portugal.

References

1. Amil, N.M., Bredeche, N., Gagné, C., Gelly, S., Schoenauer, M., Teytaud, O.: A statistical learning perspective of genetic programming. In: Vanneschi, L., Gustafson, S., Moraglio, A., De Falco, I., Ebner, M. (eds.) EuroGP 2009. LNCS, vol. 5481, pp. 327–338. Springer, Heidelberg (2009)
2. Archetti, F., Lanzeni, S., Messina, E., Vanneschi, L.: Genetic programming for human oral bioavailability of drugs. In: Keijzer, M., et al. (eds.) GECCO 2006, vol. 1, pp. 255–262. ACM Press, New York (2006)
3. Banzhaf, W., Francone, F.D., Nordin, P.: The effect of extensive use of the mutation operator on generalization in genetic programming using sparse data sets. In: Ebeling, W., Rechenberg, I., Voigt, H.-M., Schwefel, H.-P. (eds.) PPSN 1996. LNCS, vol. 1141, pp. 300–309. Springer, Heidelberg (1996)
4. Duda, R.O., Hart, P.E.: Pattern Classification and Scene Analysis. Wiley, Chichester (1973)
5. Francone, F.D., Nordin, P., Banzhaf, W.: Benchmarking the generalization capabilities of a compiling genetic programming system using sparse data sets. In: Koza, J.R., et al. (eds.) GP 1996, pp. 72–80. MIT Press, Cambridge (1996)
6. Gagné, C., Schoenauer, M., Parizeau, M., Tomassini, M.: Genetic programming, validation sets, and parsimony pressure. In: Collet, P., Tomassini, M., Ebner, M., Gustafson, S., Ekárt, A. (eds.) EuroGP 2006. LNCS, vol. 3905, pp. 109–120. Springer, Heidelberg (2006)
7. Gomez, F.J., Togelius, J., Schmidhuber, J.: Measuring and optimizing behavioral complexity for evolutionary reinforcement learning. In: Alippi, C., Polycarpou, M., Panayiotou, C., Ellinas, G. (eds.) ICANN 2009. LNCS, vol. 5769, pp. 765–774. Springer, Heidelberg (2009)
8. Koza, J.R.: Genetic Programming: On the Programming of Computers by Means of Natural Selection. MIT Press, Cambridge (1992)
9. Kushchu, I.: An evaluation of evolutionary generalization in genetic programming. Artificial Intelligence Review 18(1), 3–14 (2002)
10. Langdon, W.B., Poli, R.: Foundations of Genetic Programming. Springer, Heidelberg (2002)
11. O'Neill, M., Vanneschi, L., Gustafson, S., Banzhaf, W.: Open issues in genetic programming. Genetic Programming and Evolvable Machines 11(3-4), 339–363 (2010)
12. Poli, R., Langdon, W.B., McPhee, N.F.: A field guide to genetic programming (2008), Published via http://lulu.com and freely available at http://www.gp-field-guide.org.uk
13. Rissanen, J.: Modeling by shortest data description. Automatica 14, 465–471 (1978)
14. Rosca, J.: Generality versus size in genetic programming. In: Koza, J.R., et al. (eds.) GP 1996, pp. 381–387. MIT Press, Cambridge (1996)
15. Silva, S., Costa, E.: Dynamic limits for bloat control in genetic programming and a review of past and current bloat theories. Genetic Programming and Evolvable Machines 10(2), 141–179 (2009)
16. Silva, S., Vanneschi, L.: Operator equalisation, bloat and overfitting: a study on human oral bioavailability prediction. In: Raidl, G., et al. (eds.) GECCO 2009, pp. 1115–1122. ACM, New York (2009)
17. Vanneschi, L., Castelli, M., Silva, S.: Measuring bloat, overfitting and functional complexity in genetic programming. In: GECCO 2010, pp. 877–884. ACM, New York (2010)
18. Vanneschi, L., Silva, S.: Using operator equalisation for prediction of drug toxicity with genetic programming. In: Lopes, L.S., Lau, N., Mariano, P., Rocha, L.M. (eds.) EPIA 2009. LNCS, vol. 5816, pp. 65–76. Springer, Heidelberg (2009)
19. Vladislavleva, E.J., Smits, G.F., den Hertog, D.: Order of nonlinearity as a complexity measure for models generated by symbolic regression via pareto genetic programming. IEEE Transactions on Evolutionary Computation 13(2), 333–349 (2009)

GP-Based Electricity Price Forecasting

Alberto Bartoli, Giorgio Davanzo, Andrea De Lorenzo, and Eric Medvet

DIII, University of Trieste, Via Valerio, Trieste, Italy

Abstract. The electric power market is increasingly relying on competitive mechanisms taking the form of day-ahead auctions, in which buyers and sellers submit their bids in terms of prices and quantities for each hour of the next day. Methods for electricity price forecasting suitable for these contexts are crucial to the success of any bidding strategy. Such methods have thus become very important in practice, due to the economic relevance of electric power auctions.

In this work we propose a novel forecasting method based on Genetic Programming. Key feature of our proposal is the handling of outliers, i.e., regions of the input space rarely seen during the learning. Since a predictor generated with Genetic Programming can hardly provide acceptable performance in these regions, we use a classifier that attempts to determine whether the system is shifting toward a difficult-to-learn region. In those cases, we replace the prediction made by Genetic Programming by a constant value determined during learning and tailored to the specific subregion expected.

We evaluate the performance of our proposal against a challenging baseline representative of the state-of-the-art. The baseline analyzes a real-world dataset by means of a number of different methods, each calibrated separately for each hour of the day and recalibrated every day on a progressively growing learning set. Our proposal exhibits smaller prediction error, even though we construct one single model, valid for each hour of the day and used unmodified across the entire testing set. We believe that our results are highly promising and may open a broad range of novel solutions.

1 Introduction

The electric power industry has shifted from a centralized structure to a distributed and competitive one. In many countries of the world, electricity markets of several forms have been established that allow consumers to select among different providers according to reliability and cost metrics. Although the legal and technical features of such markets are regulated differently in each country, the presence of *auctions* in which buyers and sellers submit their bids in terms of prices and quantities is commonplace [13].

An important form of such auctions can be found in the *day-ahead* market, in which producers and consumers present price-sensitive supply offers and demands for each hour of the next day. Each day a coordinating authority determines the outcome of the auction in terms of electricity flows and final prices.

S. Silva et al. (Eds.): EuroGP 2011, LNCS 6621, pp. 37–48, 2011.
© Springer-Verlag Berlin Heidelberg 2011

The economic relevance of these auctions makes the ability to accurately predict next-day electricity prices very important in practice, for both producers and consumers: bidding strategies are based on price forecast information hence the actual benefit obviously depends heavily on the accuracy of such information. Not surprisingly, thus, many approaches to electricity price forecasting have been explored in the recent years [15,3,2,1,11,10,8,4].

In this work we examine the usefulness of Genetic Programming (GP) in the context of day-ahead electricity price forecasting. We propose two GP approaches that differ in the choice of the variables used as input of the evolutionary search and two hybrid approaches. The hybrid approaches augment the GP-built predictor with a classifier that predicts the interval to which the next prices will belong. When the predicted interval was rare in the learning set, the output of the GP-based predictor is replaced by the mean value that was observed, in the learning set, for the predicted interval. The rationale for this design is that the GP-generated predictor can hardly provide acceptable performance in regions of the input space that have been rarely seen during the learning. The classifier attempts to determine whether the system is shifting toward the difficult-to-learn region, in which case we simply predict a constant value tailored to the specific subregion expected.

We assess our results by comparing them to a very challenging baseline that, in our opinion, may be considered as representative of the state-of-the-art for the problem. The dataset consists of hourly market clearing prices set by the California Power Exchange (CalPX) from July 5, 1999 to June 11, 2000. This period includes an important market crisis, which started on May 1, 2000, that provoked significant volatility and large variations in prices, due to bankruptcy and strong financial problems of major players in the market [9]. The forecasting methods used as baseline are those discussed in [14], which evaluates the performance of 12 widely different approaches proposed in the literature. For each approach, 24 different models are constructed and carefully tuned, one for each hour of the next day. Each model is recalibrated every day, by shifting the end of the learning set to the current day—thereby growing the learning set every day. We also include in the comparison the results from [9], which apply 4 AI-based methods to the same dataset. Even in this case, each method is calibrated differently for each hour of the day and is recalibrated every day.

We evaluate the performance of our predictors with the same error index used in these works and obtain very interesting results. The GP-based approaches exhibit slightly worse performance than those of the traditional methods. The hybrid approaches, on the other hand, provide *better* performance. In fact, they even provide a performance better than a conceptual (not implementable) forecasting method obtained by selecting in each week of the testing set the best of all the other predictors for that week.

We remark that our results have been obtained in a scenario more challenging than the baseline: (i) we construct one single predictor, valid for every hour of each day; and (ii) we never recalibrate our predictor, i.e., we use the very same learning set used in [14,9] at the beginning of the simulation and then we leave the predictor unmodified across the entire testing set.

We believe our contribution is relevant for several reasons. First, we provide a novel solution to a problem highly relevant in practice that compares favorably to the current state-of-the-art. Second, we extend the set of application domains in which GP may outperform, or at least compete with, traditional approaches. Third, we show a simple yet effective way to cope with a dataset that do not cover the output space uniformly.

2   Our Approach

2.1   Overview

We propose two techniques for day-ahead electricity price forecasting. The first technique is entirely GP-based (Section 2.2), while the second one is a hybrid technique that combines the output of the GP-generated predictor with the output of a second simple predictor, to be used when the system is shifting toward regions that have been rarely seen during learning (Section 2.3).

We denote by P_h the observed price for hour h and by $\hat{P}_h$ the predicted price for that hour.

Every day at midnight the prediction machinery generates a forecast $\hat{P}_h$ for the following 24 hours, i.e., for each $h \in \{1, \ldots, 24\}$. This is the usual pattern used in the literature, although in practice prediction occurs around mid-day, not at midnight.

The variables potentially available for generating $\hat{P}_h$ are:

- $P_{h-24}, \ldots, P_{h-168}$, that represent the previously observed values for the price (e.g., P_{h-168} indicates the observed price one week before the generation of the prediction).
- $H_{h-24}, H_{h-48}, H_{h-72}, H_{h-96}, H_{h-120}, H_{h-144}$ and H_{h-168}, that represent the maximum value observed for the price in the corresponding day (e.g., H_{h-48} indicates the maximum price in the day that precedes the generation of the prediction).
- $I_{h-24}, I_{h-48}, I_{h-72}, I_{h-96}, I_{h-120}, I_{h-144}$ and I_{h-168}, that represent the minimum value observed for the price in the corresponding day.
- $N_h, N_{h-1}, \ldots, N_{h-168}$, a set of binary values that represent whether an hour corresponds to night-time, i.e., $N_k = 1$ if $1 \leq k \leq 5$ and $N_k = 0$ otherwise.
- $I_h, I_{h-1}, \ldots, I_{h-168}$, a set of binary values that represent whether an hour corresponds to holidays.
- An enumerated variable $h \in \{1, 2, \ldots, 24\}$ that represents the hour of the day for $\hat{P}_h$.
- An enumerated variable $d \in \{1, 2, \ldots, 7\}$ that represents the day of the week for $\hat{P}_h$ (from Sunday to Saturday).

We remark that we rely only on measured values for the variable to be predicted, i.e., we do not require any exogenous variable. Existing literature, in contrast, often assumes that the prediction machinery has some exogenous variables available, e.g., temperature, actual or forecasted load and alike. Indeed, 6 of the 12 models in [14] use load forecast as exogenous variable. We believe that our approach may be more practical, simpler to implement and less dependent on "magic" tuning numbers—e.g., if temperature were to be used, where and at which hour of the day itshould be taken?

We partition the dataset in three consecutive time intervals, as follows: the *training set*, for performing the GP-based evolutionary search; the *validation set*, for selecting the best solution amongst those found by GP; the *testing set*, for assessing the performance of the generated solution.

2.2 GP Approach

The set of variables potentially available to the prediction system is clearly too large to be handled by the GP search efficiently. We consider two configurations: one, that we call *GP-baseline*, in which the terminal set consists of the same variables used in the best-performing method of the baseline work (except for any exogenous variable, such as the load) [14]. The resulting terminal set is: $\{P_{h-24}, P_{h-48}, P_{h-168}, I_h, N_h, L_{h-24}\}$. The other configuration, that we call *GP-mutualInfo*, uses a terminal set that consists of variables selected by a feature selection procedure that we describe below.

The procedure is based on the notion of *mutual information* between pairs of random variables, which is a measure of how much knowing one of these variables reduces the uncertainty about the other [12]. The procedure consists of an iterative algorithm based on the training and validation portions of the dataset, as follows. Set S initially contains all the 498 variables potentially available to the prediction system. Set S_{out} is initially empty and contains the selected variables to be used for the GP search.

1. Compute the mutual information m_i between each variable $X_i \in S$ and the price variable Y.
2. For each pair of variables $X_i \in S, X_j \in S$, compute their mutual information m_{ij}.
3. Let $X_i \in S$ be the variable with highest m_i. Assign $S := S - X_i$ and $S_{out} := S_{out} + X_i$. For each variable $X_j \in S$, modify the corresponding mutual information m_j as $m_j := m_j - m_{ij}$.
4. Repeat the previous step until S_{out} contains a predefined number of elements.

We chose to execute this feature selection procedure for selecting 8 variables. The resulting terminal set to be submitted to GP is:

$$\{P_{h-24}, P_{h-168}, I_h, I_{h-24}, I_{h-168}, N_h, H_{h-24}, L_{h-24}, h, d\}$$

At this point we run the GP search on the training set, with parameters set as described in Section 3.2. Next, we compute the fitness of all individuals on the validation set. Finally, we select the individual that exhibits best performance on this set as predictor. This individual will be used along the entire testing set.

2.3   Hybrid Approach

Our hybrid approach generates a GP-based predictor exactly as described in the previous section, but introduces an additional component to be used in the testing phase. This component is synthesized using the training and validation portions of the dataset, as follows.

1. We define 10 equally-sized intervals for the observed price values in the training and validation set and define each such interval to be a class.
2. We compute the mean value for each class.
3. We execute a feature selection procedure [6] consisting of a genetic search algorithm [5] and select 95 of the 498 variables potentially available.
4. We train a classifier for the above classes based on the variables selected at the previous step. In other words, this classifier predicts the class to which the next price value will belong. In our experiments we have used a multilayer perceptron.

The choice of the specific algorithms used at step 3 and 4 has been influenced by the software tool used for this purpose (Weka [7]).

In the testing phase, the prediction is generated as follows. We denote by C_A the set of the 2 classes with more elements and by C_B the set of the other classes. Let $\hat{c}$ be the predicted class for P_i. If $\hat{c} \in C_A$ then the predicted value $\hat{P}_i$ is the value generated by the GP predictor, otherwise $\hat{P}_i$ is the mean value computed for $\hat{c}$ (step 2 above).

The rationale of this design is that the GP-generated predictor cannot be expected to perform well in regions of the input space that have been rarely seen during the learning. The classifier attempts to determine whether the system is shifting toward the difficult-to-learn region, in which case we simply predict a constant value determined during training and tailored to the specific subregion expected.

Figures 1(a) and 1(b) show the distributions of price values in the training set and in the testing set, respectively (in the validation set all values happen to belong to the first 2 classes). The percentage of elements in the 2 classes with more elements is 92% in the learning set (training and validation) and 82% in the testing set.

3   Experimental Evaluation

3.1   Dataset and Baseline

As clarified in the introduction, we believe the dataset and baseline that we have used are highly challenging and may be considered as representative of the state-of-the-art. The dataset consists of hourly market clearing prices set by the California Power Exchange (CalPX) from July 5, 1999 to June 11, 2000 (Figure 2). This period includes a market crisis period characterized by large price volatility, that started on May 1, 2000 and lasted beyond our dataset [9].

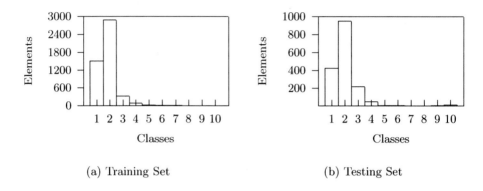

(a) Training Set   (b) Testing Set

Fig. 1. Distribution of price values in classes

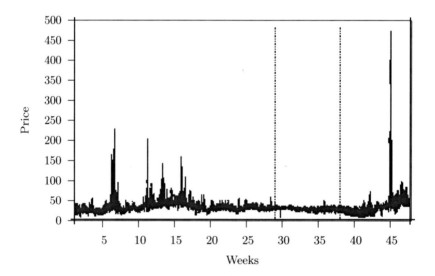

Fig. 2. Dataset used for evaluating the proposed methods. The vertical line at the right indicates the division between learning set and testing set. The vertical line at the left indicates the division between training set and validation set (used only in our approaches, see Section 3.2).

We use the results from [14] as main baseline. This work examines a set of widely differing approaches proposed earlier in the literature: basic autoregressive (AR), spike preprocessed (p-AR), regime switching (TAR), mean-reverting jump diffusion (MRJD), iterated Hsien-Manski estimator (IHMAR), smoothed non-parametric maximum likelihood (SNAR) (please refer to the cited work for full details). Each approach is applied with and without load forecast as exogenous variable. For each approach, 24 different models are constructed and carefully tuned, one for each hour of the next day. In the testing phase each model is recalibrated every day, by shifting the end of the learning set to the current day—thereby growing the learning set every day. The initial learning set contains the first 9 months, from July 5, 1999, to April 2, 2000. The next 10 weeks constitute the testing set, which thus includes the market crisis mentioned above[1].

We include in the comparison also the results from [9], which applies several AI-based approaches to the same dataset[2]: autoregressive neural network (ANN), local regression (LOCAL), linear regression tree (TREE), generalized additive model (GAM) (again, please refer to the cited work for full details). This work follows the same structuring as the previous one: it uses the same learning set, it models each hour of the day separately and recalibrates each model every day.

The performance index is the Weekly-weighted Mean Absolute Error (WMAE), defined as:

$$\text{WMAE} = \frac{\sum_{h=1}^{168} |P_h - \hat{P}_h|}{\sum_{h=1}^{168} P_h}$$

where P_h is the actual price for h and $\hat{P}_h$ is the predicted price for that hour.

3.2   Settings

We split the dataset as in [14,9]: the learning set contains the first 9 months, whereas the next 10 weeks constitute the *testing set*. We further split the learning data in two consecutive intervals used as described in Section 2.2: a *training set* from July 5, 1999, to January 30, 2000, is used for the GP search; a *validation set* from January 31, 2000 to April 2, 2000, is used for selecting the best individual produced by the GP search.

We used WMAE measured on the training set as fitness function. We could have used other fitness functions (e.g., squared error distances) and then assess the resulting performance on the testing set based on the performance index of interest in this work, i.e., WMAE. We have not explored this possibility in depth, but preliminary results suggest that there are no substantial differences between these two approaches. Indeed, this finding is in line with a similar assessment made in [14].

---

[1] The cited work analyzes also another dataset from the Nordic Power Exchange (http://www.nordpoolspot.com/) augmented with hourly temperatures in Sweden. We have not yet applied our approach to this dataset.

[2] This work actually considers a longer testing set. We include here the results for the same testing set used in [14].

We experimented with four configurations: GP-baseline, GP-mutualInfo, Hybrid-baseline (i.e., coupled with GP-baseline), Hybrid-mutualInfo. The GP searches have been made with the same set of parameters, except for the composition of the terminal set, that is different for the cases GP/Hybrid-baseline and GP/Hybrid-mutualInfo (see Section 2.2). The functions set includes only the four basic arithmetic operators and the terminal set always includes a few basic constants: 0.1, 1, 10. During our early tests we experimented with different combinations of population size and number of generations. Concerning the former, we swept the range between 500 and 1000 individuals, keeping fixed the number of generations, and found no significant differences in WMAE performance. However, we also found that a population with 1000 individuals triplicates the computation time required by one with 500 individuals, thus we decided to select 500 as population size. Concerning the number of generations, we decided to use 1200 generations after some exploratory experimentation. The full set of GP-related parameters is summarized in Table 1.

Table 1. GP parameters

Parameter	Settings
Populations size	500
Selection	Tournament of size 7
Initialization method	Ramped half-and-half
Initialization depths	1
Maximum depth	5
Elitism	1
Reproduction rate	5%
Crossover rate	80%
Mutation rate	15%
Number of generations	1200

For each configuration (i.e., GP-baseline, GP-mutualInfo, Hybrid-baseline, Hybrid-mutualInf): (i) we ran 128 GP executions, each with the parameters in Table 1; (ii) at the end of each execution we selected the individual with the best fitness on the training set, thereby obtaining a final population of 128 individuals; (iii) we evaluated the fitness of these individuals on the validation set and selected the one with best fitness as predictor to be used in the testing set. Concerning the hybrid approach, we used the Weka tool in the standard configuration [7] and experimented with several forms of classifier: Random Tree, Random Forest Tree, SVM, Multilayer Perceptron. The latter is the one that exhibited best performance and has been used for deriving the results presented here.

Finally, a few notes about execution time: each GP search took about 34 hours on 4 identical machines running in parallel, each machine being a quad-core Intel Xeon X3323 (2.53 GHz) with 2GB RAM; the training of the classifier took about 1 hour on a single core notebook (2 GHz), with 2GB RAM; the variable selection procedure (Section 2.2) took a few minutes on the same notebook.

3.3   Results

Table 2 presents the salient results. The first four rows contain the average WMAE along the testing set for each of the approaches that we have developed. To place these results in perspective, the next set of rows provides the same index, extracted from [14]. In particular, the first 12 rows correspond to the 6 approaches, each tested with and without predicted load as exogenous variable (models with the exogenous variable are denoted by the X suffix). Then, we provide the mean for the full set of 12 models, the mean for the 6 pure-price models only and the mean for the 6 models with exogenous variable. Finally, the row labeled Ideal gives the mean WMAE of an optimal (merely conceptual) model, constructed by selecting the best performing model in each week of the testing phase (the cited work provides the WMAE of each model in each week). The final set of rows provides the corresponding WMAE values from [9]. We excluded method LOCAL from the evaluation of mean values, as it is clearly an outlier. The row labeled Ideal has the same meaning as above whereas the row IdealBoth corresponds to selecting the best model in each week from the full set of 16 predictors provided by the cited works.

The key result is that both the hybrid methods perform better than all the other methods, including the "optimal" (and merely conceptual) predictors constructed by selecting the best predictor in each week. We believe this is a very promising result. The fact that our approaches construct one single model valid for every hour of the day and that we never recalibrate our models along the entire testing set, may only corroborate this claim.

Table 2. Mean WMAE results in the testing set. The upper portion of the left table corresponds to our approaches, the lower portion are results from [9] (the huge value for the LOCAL method is not a typing mistake), the right table are results from [14].

Method	Mean WMAE (%)
GP-mutualInfo	20.70
GP-baseline	16.17
Classifier-base	16.03
Hybrid-mutualInfo	11.84
Hybrid-baseline	12.32
ANN	13.11
LOCAL	154499.01
TREE	14.02
GAM	13.29
Mean	13.47
Ideal	12.83
Ideal both	12.42

Method	Mean WMAE (%)
AR	13.96
ARX	13.36
p-AR	13.44
p-ARX	12.96
TAR	13.99
TARX	13.31
MRJD	15.39
MRJDX	14.67
IHMAR	14.01
IHMARX	13.37
SNAR	13.87
SNARX	13.17
Mean	13.79
Mean pure-price only	14.11
Mean with load only	13.47
Ideal	12.64

Fig. 3. Distribution of mean WMAE performance for the final populations, with hybrid methods. Vertical lines indicate the WMAE for the three Ideal methods shown in Table 2.

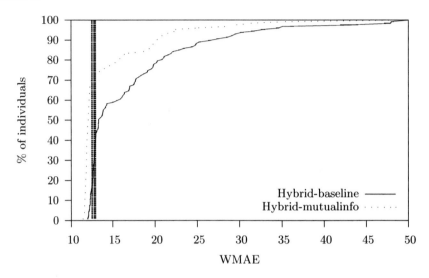

In order to gain further insights into the ability of our hybrid methods to effectively generate accurate predictors, we evaluated WMAE across the entire testing set for all the individuals of the final population. That is, rather than selecting one single individual based on its WMAE performance in the validation set, we take all the individuals. The distribution of the respective WMAE performance is shown in Figure 3. Vertical lines indicate the WMAE for the three Ideal methods shown in Table 2.

It can be seen that the better performance exhibited by the hybrid methods is not an occasional result provoked by a single "lucky" individual: these methods consistently tend to generate individuals whose performance is better than any of the baseline methods. Indeed, the baseline performance is improved by more than half of the final population (Hybrid-baseline) and by approximately three-quarters of the final population (Hybrid-mutualInfo).

For completeness of analysis, we assessed the impact of the classifier from several points of view. Concerning the prediction mistakes performed by the classifier in the testing set, it provoked a wrong replacement of the prediction by GP 2.5% of the times, and it provoked a wrong use of the prediction by GP 10.11% of the times. The performance (mean WMAE in the testing set) that one could obtain by our Hybrid approach implemented by a perfect classifier—i.e., one that never makes any prediction mistake in the testing set—is 10.50% (mutualInfo) and 10.96% (baseline). Finally, the performance that one could obtain by always using the mean value for the class predicted by a perfect classifier is 15.49%.

From these data we observe what follows. First, attempting to improve the prediction accuracy of the classifier further is probably not worthwhile (a perfect classifier does not deliver a substantial improvement over Hybrid-mutualInfo). Second, our hybrid approach indeed boosts performance of its building blocks—classifier-based prediction and GP-based prediction: the former is slightly better than the latter, but their combination is substantially better than either of them. Third, the simple classifier-based prediction exhibits performance that is better than GP-only methods and is only slightly worse than the 16 baseline methods.

4   Concluding Remarks

We have proposed novel GP-based methods for electricity price forecasting that are suitable for day-ahead auctions. We designed simple yet effective mechanisms for enabling GP to cope with the strong volatility that may be typical of this difficult application domain.

We assessed our proposal on a challenging real-world dataset including a period of market crisis, and compared our results against a baseline that is representative of the current state-of-the-art. Our hybrid methods performed better than all the 16 methods considered, and better than ideal (not implementable) predictors constructed by taking the best of those predictors in each week. We also showed that our methods tend to systematically generate predictors with good performance: we actually generated tens of predictors that exhibit better performance than those used as baseline.

Although our approach has certainly to be investigated further, in particular on other datasets, we believe that our results are significant and highly promising.

Acknowledgments

We are very grateful to Cyril Fillon for having initiated our group into the secrets (and power) of GP, for the development of Evolutionary Design (the API used in this research) and for his comments on this work. We are also grateful to Gabriele Del Prete for his hard work during the initial stage of this research.

References

1. Amjady, N., Keynia, F.: Day ahead price forecasting of electricity markets by a mixed data model and hybrid forecast method. International Journal of Electrical Power and Energy Systems 30(9), 533–546 (2008)
2. Areekul, P., Senjyu, T., Toyama, H., Yona, A.: A hybrid ARIMA and neural network model for Short-Term price forecasting in deregulated market. IEEE Transactions on Power Systems 25(1), 524–530 (2010)
3. Catalao, J.P.S., Pousinho, H.M.I., Mendes, V.M.F.: Hybrid Wavelet-PSO-ANFIS approach for Short-Term electricity prices forecasting. IEEE Transactions on Power Systems (99), 1–8 (2010)
4. Cuaresma, J.C., Hlouskova, J., Kossmeier, S., Obersteiner, M.: Forecasting electricity spot-prices using linear univariate time-series models. Applied Energy 77(1), 87–106 (2004)

5. Goldberg, D.E.: Genetic algorithms in search, optimization and machine learning (1989)
6. Hall, M.A.: Correlation-based feature selection for discrete and numeric class machine learning. In: Proc. 17th Intern. Conf. Machine Learning, pp. 359–366 (2000)
7. Hall, M., Frank, E., Holmes, G., Pfahringer, B., Reutemann, P., Witten, I.H.: The weka data mining software: An update. SIGKDD Explorations 11(1) (2009)
8. Koopman, S.J., Ooms, M.: Forecasting daily time series using periodic unobserved components time series models. Computational Statistics & Data Analysis 51(2), 885–903 (2006)
9. Mendes, E.F., Oxley, L., Reale, M.: Some new approaches to forecasting the price of electricity: a study of californian market, http://ir.canterbury.ac.nz/handle/10092/2069, RePEc Working Paper Series: No. 05/2008
10. Mount, T.D., Ning, Y., Cai, X.: Predicting price spikes in electricity markets using a regime-switching model with time-varying parameters. Energy Economics 28(1), 62–80 (2006)
11. Pedregal, D.J., Trapero, J.R.: Electricity prices forecasting by automatic dynamic harmonic regression models. Energy Conversion and Management 48(5), 1710–1719 (2007)
12. Peng, H., Long, F., Ding, C.: Feature selection based on mutual information criteria of max-dependency, max-relevance, and min-redundancy. IEEE Transactions on Pattern Analysis and Machine Intelligence 27(8), 1226–1238 (2005)
13. Sheblé, G.B.: Computational auction mechanisms for restructured power industry operation. Springer, Netherlands (1999)
14. Weron, R.: Misiorek: Forecasting spot electricity prices: A comparison of parametric and semiparametric time series models. International Journal of Forecasting, 744–763 (2008)
15. Wu, L., Shahidehpour, M.: A hybrid model for Day-Ahead price forecasting. IEEE Transactions on Power Systems 25(3), 1519–1530 (2010)

Novel Loop Structures and the Evolution of Mathematical Algorithms

Mingxu Wan, Thomas Weise, and Ke Tang

University of Science and Technology of China, Hefei, Anhui 230026, China

Abstract. In this paper, we analyze the capability of Genetic Programming (GP) to synthesize non-trivial, non-approximative, and deterministic mathematical algorithms with integer-valued results. Such algorithms usually involve loop structures. We raise the question which representation for loops would be most efficient. We define five tree-based program representations which realize the concept of loops in different ways, including two novel methods which use the convergence of variable values as implicit stopping criteria. Based on experiments on four problems under three fitness functions (error sum, hit rate, constant 1) we find that GP can statistically significantly outperform random walks. Still, evolving said algorithms seems to be hard for GP and the success rates are not high. Furthermore, we found that none of the program representations could consistently outperform the others, but the two novel methods with indirect stopping criteria are utilized to a much higher degree than the other three loop instructions.

1 Introduction

Genetic Programming (GP) [4, 15], today is used for many tasks such as symbolic regression, the synthesis of electronic circuits, or the design of distributed systems [16]. However, there exits little evidence that it is actually suitable to derive *programs* in the intuitive sense of the word, maybe in the form of C++ programs. More specifically, we consider the evolution of non-trivial, non-approximative, deterministic mathematical algorithms which compute distinct discrete results.

Such synthesis tasks are fundamentally different from the domains where GP excels. If an intermediate solution generates close-to-optimal outputs, this does not necessarily mean that its program structure is also similar to a good solution. In [14, 18], we pointed out that traditional imperative program representations (such as those commonly used for Standard or Linear Genetic Programming) exhibit strong epistatic effects [19]: It is not possible to modify one part of a program without affecting the behavior of the other parts [14, 17]. As a consequence, the fitness landscapes of the problems under consideration are often rugged and likely contain large neutral areas. Hence, Genetic Programming utilizing imperative program representations similar to high-level programming structures may not be efficient.

We first provide a comprehensive related work study in Section 2. Many of the works presented there are positive about the ability to synthesize programs

S. Silva et al. (Eds.): EuroGP 2011, LNCS 6621, pp. 49–60, 2011.

which effectively use loops with GP. However, their experimental setups are often specialized and their results cannot be generalized. With our work, we contribute evidence about the actual utility of loop instructions in Genetic Programming. We will show that some of the example problems we used in our experiments are hard for GP, although they are not too different from those listed in the related work.

We test five different loop representations which are specified in Section 3. Two novel loop representations, *conditional assignments* (CA) and *implicit loops* (IL) are defined. Both replace the traditional, explicit loop conditions with an *implicit* convergence condition based on side-effects. Whereas in CA, the whole programs are turned into single loops, IL allows the nesting of independent loops.

We conduct a large-scale experimental study in which we apply Genetic Programming using the five loop representations to four test problems (specified in Section 4) using different fitness functions (error sum, hit rate, and constant 1). In our experiments discussed in Section 5, we find that Genetic Programming is usually more efficient than random walks. However, its capability to solve the harder problems was not yet satisfying. We show that the new non-imperative and implicit loop structures CA and IL are utilized more efficiently by Genetic Programming than those presented in the related work. Although, from an absolute fitness point of view, the three other methods perform not significantly worse, Genetic Programming does more efficiently utilize our novel loop structures and with CA, even can solve one of the hardest benchmark tasks.

2   Related Work

The usage of loops in GP has a tradition of at least 15 years, but only few works considering it have actually been published. Teller [13] proved that (tree-based) Genetic Programming with indexed memory can be Turing complete and also proposed some methods to limit the runtime of programs [12]. These works do not contain any experimental validation and mainly introduce conceptual ideas.

Qi et al. [11] introduced a structure which uses a constant N as loop counter, where N is given *beforehand* by experience. Nesting of loops is forbidden. This method is used to solve the lawn mower problem for an 8-by-8 square and N is set to 8. Such high degree of incorporated a priori knowledge prevents drawing any general conclusions from this work.

Pioneering work in terms of loops in GP has been done by Koza [4]. In this paragraph, we outline the contributions of two of his students: A new syntax in which conditional loops and alternatives were used was established by Finkel [3]. She solved the factoring problem with this approach and applied a penalty in the fitness to keep the programs small. Lai [5] introduced a new method to solve the greatest common divisor (GCD) problem which we discuss in Section 4.3. 12 fitness cases were used and when number of hits was 12, additional test cases are used to test the success of the program. Lai [5] states that his results lack of generality because of the small number of test cases he used. Both, Finkel and Lai, use relatively large populations, up to 5000 individuals. With such large populations, high success rates could be achieved by sheer chance. Thus, in our experiments, we compare the performance of Genetic Programming also with random walks.

Ciesielski and Li [2] applied two forms of for-loop structures to solve a modified Santa Fe trail problem and a sorting task. We use a very similar loop structure, the *counter loop* CL (see Section 3), in our experiments. In [6], they used a similar approach to the visit-every-square problem and a modified Santa Fe trail task and showed that in these problems, loop structure are advantageous and lead to a reduction of program size. Both problems are, however, static by nature and cannot be compared with algorithm tasks with changing inputs (where issues such as overfitting arise).

Chen and Zhang [1] used a loop which is similar to our *while loop* (see Section 3) to solve the factorial problem which we discuss in Section 4.4 for the inputs from 1 to 15. Their GP method could find a perfect solution in 25 out of the 50 runs. However, the instruction set used was very small so such success rates maybe do not bear too much evidence and cannot be generalized.

Wijesinghe and Ciesielski [20] investigated how indexed loops can be implemented in Genetic Programming. He used them to train programs to produce regular patterns in bit strings but no results from this work are available.

In the past, various loop structures have been introduced and applied to several problems. Yet, no data is available on the comparison of different kinds of loop structures on different benchmark problems. Most works used much beforehand knowledge and utilized it to construct specialized loop structures. In our work, we use instruction sets which are not tailored for the problems we apply GP to, test them on different problems, and compare loop structures in a realistic way.

3   Loop Representations

For our experiments, we use tree-based Standard Genetic Programming with memory. The inner nodes of a (program) tree are expressions and the leaf nodes are terminals. We utilize the instruction set defined in Table 1 and combine it with one of the following five loop structures: the counter loop (CL), the memory loop (ML), the while loop (WL), conditional assignments (CA), and the implicit loop (IL).

The nodes which represent a *counter loop* (CL) have two children, say x and y, each of which is a subtree. x denotes an expression which, if evaluated, returns the number of times the loop body y should be executed. Fig. 1 (a) is an example for a program computing the GCD of two numbers in this representation.

Memory loop nodes ML also have two children x and y. x is a terminal node which identifies a variable and y is a subtree representing the loop body. The loop body (which can access and modify the variable x) is executed until x becomes less or equal 0. The variable x is also decreased by one in each loop iteration. An example of this representation is given in Fig. 1 (b).

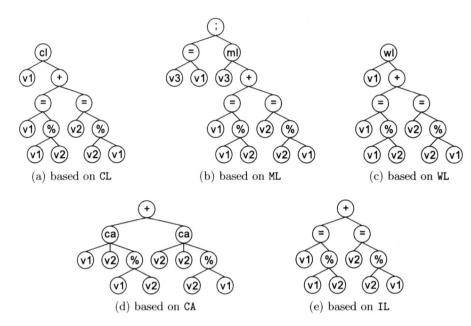

(a) based on CL   (b) based on ML   (c) based on WL

(d) based on CA   (e) based on IL

Fig. 1. Ideal solutions to the GCD problem

The *while loop* WL has the same structure as CL. Here, however, the subtree x represents a condition. The loop body subtree y is repeatedly evaluated until this condition becomes 0. Fig. 1 (c) illustrates a WL program for computing the GCD.

In our fourth loop representation, the new *conditional assignment* operator CA replaces the standard variable assignment =. Different from the = nodes used in the other program representations, CA has three children, x, y, and z. x is a condition expression subtree. y is a single terminal node identifying the target variable. The value of the expression subtree z will only be computed and assigned to the variable y if the condition x evaluates to a non-zero value. Like IL, the *complete* program tree will be evaluated repeatedly until no variable *changes* anymore. It thus takes on the function of a loop body. Again, we give an example for this notation in Fig. 1 (d).

Conditional assignments are different from the *Soft Assignment* method developed by McPhee and Poli [9, 10] where the strength of the assignment operator is weakened in order to allow for a more smooth change of the program fitness during the evolution. In the CA representation, y=z means that the variable y will actually take on the exact value of the expression z, but the assignment will only be performed if $x \neq 0$.

The *implicit loop* IL has only a single child x. x is a subtree and represents the loop body. The loop body may contain variable assignments and will be evaluated until there is no change in any variable. We sketch an IL program for computing the GCD in Fig. 1 (e).

Table 1. Basic operators in our experiments

Operator	Function
$+, -, *$	well-known arithmetic operators.
$a\%b$	protected modulo division, returns $a \bmod b$ if $b \neq 0$, and a otherwise
$a = b$	set the value of variable a to the result of the expression b (which is also returned)
$a; b$	concatenation of two expressions a and b, return value of b
$v_1, v_2, \ldots, v_n$	the value of the one of n memory variables
0, 1	the only two ephemeral random constants available

Different from the other three loop instructions, the novel loop representations (CA and IL) do not use a certain explicit expression (such as $x \leq 0$) to decide whether they should stop. Instead, they run until all variables have converged to stationary values. CA still retains some kind of explicit condition for the loop body, though decomposed into conditions for each assignment. In IL, *only* the side-effects determine the number of loop iterations.

4   Test Problems

We try to synthesize four different mathematical expressions, three of which can only be computed using a loop structure. For each problem, we use several training cases t_i. After executing an evolved program P, we expect its result $P(t_i)$ to be stored in its last memory cell v_n. We generally allowed the programs to utilize $n = 2$ two memory cells.

4.1   Polynomial Problem ($pr = 1$)

First, we propose a *trivial* polynomial problem $\phi_1(t_i) = t_i^3 + t_i^2 + 2 * t_i$ which does not require a loop in the instruction set at all. This test problem was designed to test whether the Genetic Programming system works correctly, can solve basic symbolic regression problem, and whether it utilizes some of the loops even in cases where they are not strictly necessary. We use $tc = 100$ training cases $t_i \in 1 \ldots 100$ and, at the beginning of each program execution, initialize all variables with t_i.

4.2   Sum Problem ($pr = 2$)

The second problem is to find the sum $\phi_2(t_i) = \sum_{j=1}^{t_i} j$ of the first t_i natural numbers. If a division operator is present in the instruction set, this problem can be solved without using loops. We omit the division operation from the instruction set, thus forcing Genetic Programming to synthesize suitable loops ($n(n + 1)/2$ *cannot* be discovered with the available instructions). We again use $tc = 100$ training cases $t_i \in 1 \ldots 100$ and, at the beginning of each program execution, initialize all variables with t_i except for the last one which is initialized with 0.

4.3   GCD Problem ($pr = 3$)

In the GCD problem [5], we try to find an algorithm which can compute the greatest common divisor $\phi_3(t_{i,1}, t_{i,2}) = \gcd(t_{i,1}, t_{i,2})$. A training case $t_i = (t_{i,1}, t_{i,2})$ this time consists of the two natural numbers $t_{i,1}$ and $t_{i,2}$. We randomly create $tc = 100$ training cases at the beginning of each generation. At the beginning of the program execution, the first two variables v_1 and v_2 are initialized with $t_{i,1}$ and $t_{i,2}$. In the *memory loop* representation ML, each possible solution requires at least three memory cells so we allowed this representation to utilize $n = 3$ variables (initially $v_3 = 0$). In Fig. 1, we sketch manually derived optimal solutions for the GCD task in each of the five program representations.

4.4   Factorial Problem ($pr = 4$)

The factorial problem [1], i.e., synthesizing a program which can compute $\phi_4(t_i) = t_i!$ of a natural number t_i, also requires at least one effective loop. We use $tc = 10$ training cases $t_i \in 1 \ldots 10$ and, at the beginning of each program execution, initialize the first variable with t_i and the other one with 1.

5   Experiments

5.1   Experimental Protocol

In our experiments, we used the Genetic Programming implementation of the ECJ framework [7] with a population size of 1000, a generation limit of 100, tournament selection with 7 contestants, 10% point mutation, 90% subtree exchange crossover, and a maximum tree depth of 17. For each configuration, we performed 100 independent runs in order to get statistically reliable results.

We used three different objective functions f_1 to f_3. f_1, defined in Equation 1, is the error sum and subject to minimization. For a given problem pr, it sums up the absolute difference of the problem-specific expected result $\phi_{pr}(t_i)$ for i^{th} training case (t_i) and the result produced by the evolved program P which, as stated in Section 4, is expected to occur in the last variable v_n after the execution of P.

$$f_1(P) = \sum_{i=1}^{tc} |\phi_{pr}(t_i) - P(t_i)| \tag{1}$$

$$f_2(P) = |\{(i \in 1..tc) \wedge (\phi_{pr}(t_i) = P(t_i))\}| \tag{2}$$

$$f_3(P) = 1 \tag{3}$$

Equation 2 specifies the hit rate f_2, an alternative for f_1 which counts the number of correctly solved training cases (and is subject to maximization). We also tested summing up the logarithms or the arc tangents of the error values in order to

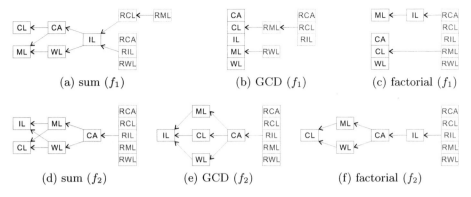

(a) sum (f_1) (b) GCD (f_1) (c) factorial (f_1)

(d) sum (f_2) (e) GCD (f_2) (f) factorial (f_2)

Fig. 2. Statistical comparisons of the loop representations in different problems

alleviate the different scales of the errors in different training cases (especially in the factorial problem). However, this did not lead to significantly different results and thus, was omitted here. The purpose of f_3 is to turn the Genetic Programming process into a parallel random walk by completely randomizing its (tournament) selection procedure. All other procedures such as crossover and mutation are left intact and retain their associated probabilities. Hence, we can fairly compare these runs with those using f_1 and f_2. We will prepend a R to the names of the random walk versions of the experiments.

5.2   Results in Terms of Fitness

As to be expected, in the polynomial problem, all the GP approaches found the solutions quickly. The parallel random walks were not able to discover even a single solution. A comparison shows that there is no statistically significant difference between the representations and all configurations using f_1 or f_2 are significantly better than those with f_3. Hence, the well-known utility of GP on symbolic regression tasks has been verified for the presented approaches and we will focus on the other three experiments in our evaluation.

In each of the sub-figures of Fig. 2, we illustrate the statistical evaluation of the performance of the different loop representations in 100 independent runs using either f_1 or f_2 on one problem. An arrow $a \leftarrow b$ from box b to box a means that approach b is statistically significantly worse than approach a according to a two-tailed Mann-Whitney U test [8] with a significance level of 2% based on the best f_1 (or f_2) value encountered in each run. The arrows are transitive, meaning that an arrow from $a \leftarrow b \leftarrow c$ also means $a \leftarrow c$. All approaches in a stack of boxes yield the same results in the tests with other methods whereas the tests amongst each others are inconclusive.

We find that for almost all program representations and problems, Genetic Programming is significantly more efficient than random walks. This adds evidence to the claim that GP may indeed be able to construct non-trivial

```
v1∈1..10; //input
v2=5; //output(init)
while(v1>3){
    v2=v2;
    v1--;
}
```
(a) Cat. A

```
v1∈1..10; //input
v2=5; //output(init)
while(v1<5){
    v2++;
    v1++;
}
```
(b) Cat. B

```
v1∈1..10; //input
v2=5; //output(init)
while(v1<5){
    v2=3;
    v1++;
}
```
(c) Cat. C

```
v1∈1..10; //input
v2=5; //output(init)
while(v1>0){
    v2=3;
    v1++;
}
```
(d) Cat. D

Fig. 3. Examples for different degrees of loop effectivity

non-approximative algorithms. Apart from this basic fact, the loop representations do not exhibit any consistent significant differences values.

5.3   Results in Terms of Loop Effectiveness

Besides a comparison in terms of fitness, we are interested in how often Genetic Programming could solve a given problem under the viewpoint of effective utilization of the loop instructions. In this respect, each evolved program can be categorized in exactly one of the following groups:

1. **Category A.** There is no loop instruction in the program that influences the result in any way. An example of this case is sketched in Fig. 3 (a).
2. **Category B.** At least one loop is present and is an effective loop (sketched in Fig. 3 (b)), i. e., its body is executed more than once and the result of the program depends on the number of executions. This can be verified by artificially limiting the maximum number of iterations to 1.
3. **Category C.** Otherwise, we check whether there is at least one loop which has degenerated to an alternative in the style of an `if-then` rule as illustrated in Fig. 3 (c). This can be verified by always executing the loop exactly once and checking whether the results differ from the normal execution.
4. **Category D.** One endless loop is present which exceeds maximum number of allowed iterations (1000 in our experiments), as sketched in Fig. 3 (d).

Notice that a loop may be effective or degenerated to an alternative but still can be endless at the same time. Categories **A** to **C** are exclusive, but programs of category **D** may additionally belong to any of the classes **A**, **B**, and **C**.

We now analyze the effectiveness of the loop representations in the different problems by categorizing the final results of each run into (at least) one of these four categories. In Table 2, we provide two numbers in the format "s/c" of each configuration and category, where c is the total number of programs of the given category returned by GP and s is the number of evolved programs of the category that actually solved the task perfectly.

What we can find in Table 2 is that the three non-trivial problems *cannot* be solved efficiently by Genetic Programming by utilizing the given population size and generation limit. Only the *memory loop* ML can solve the *sum* and *factorial* problem to some extent. This may be due to the fact that this representation is especially suitable for the underlying structure of these two problems.

Table 2. Evolved programs according to categories

Problem	A	B	C	D	Problem	A	B	C	D
CL f_1	60/60	0/0	40/40	0/0	CL f_1	0/2	0/97	0/1	0/96
CL f_2	33/72	1/28	0/0	1/25	CL f_2	0/53	1/47	0/0	0/42
RCL f_3	0/100	0/0	0/0	0/0	RCL f_3	0/99	0/1	0/0	0/0
ML f_1	90/90	10/10	0/0	0/0	ML f_1	0/4	26/96	0/0	0/64
ML f_2	22/76	3/24	0/0	0/9	ML f_2	0/46	27/54	0/0	0/14
RML f_3	0/91	0/9	0/0	0/0	RML f_3	0/95	0/5	0/0	0/0
WL f_1	92/90	0/1	0/0	0/1	WL f_1	0/77	0/22	0/1	0/23
WL f_2	91/91	0/9	0/0	0/9	WL f_2	0/85	0/15	0/0	0/15
RWL f_3	0/99	0/1	0/0	0/1	RWL f_3	0/98	0/2	0/0	0/2
CA f_1	0/0	100/100	0/0	0/0	CA f_1	0/0	0/100	0/0	0/0
CA f_2	59/59	41/41	0/0	0/0	CA f_2	0/16	1/84	0/0	0/0
RCA f_3	0/69	0/31	0/0	0/1	RCA f_3	0/70	0/30	0/0	0/5
IL f_1	56/56	44/44	0/0	13/13	IL f_1	0/5	0/95	0/0	0/0
IL f_2	100/100	0/0	0/0	0/0	IL f_2	0/26	0/74	0/0	0/1
RIL f_3	0/62	0/38	0/0	0/5	RIL f_3	0/70	0/30	0/0	0/5
CL f_1	0/29	0/46	0/25	0/46	CL f_1	0/2	0/92	0/6	0/90
CL f_2	29/83	0/17	0/0	0/17	CL f_2	0/35	0/45	0/0	0/39
RCL f_3	0/83	0/17	0/0	0/17	RCL f_3	0/99	0/1	0/0	0/0
ML f_1	0/26	0/74	0/0	0/74	ML f_1	0/25	14/75	0/0	0/45
ML f_2	29/89	1/11	0/0	1/11	ML f_2	0/37	0/63	0/0	0/40
RML f_3	0/93	0/7	0/0	0/7	RML f_3	0/92	0/8	0/0	0/0
WL f_1	0/31	0/69	0/0	0/69	WL f_1	0/77	0/23	0/0	0/23
WL f_2	30/85	5/15	0/0	0/10	WL f_2	0/94	0/6	0/0	0/6
RWL f_3	0/97	0/3	0/0	0/3	RWL f_3	0/100	0/0	0/0	0/0
CA f_1	0/0	9/100	0/0	0/7	CA f_1	0/0	0/100	0/0	0/1
CA f_2	0/0	7/100	0/0	0/7	CA f_2	0/14	0/86	0/0	0/1
RCA f_3	0/62	0/38	0/0	0/2	RCA f_3	0/63	0/37	0/0	0/3
IL f_1	0/0	1/100	0/0	0/13	IL f_1	0/13	0/87	0/0	0/0
IL f_2	0/1	86/99	0/0	0/0	IL f_2	0/72	0/28	0/0	0/1
RIL f_3	0/54	0/46	0/0	0/2	RIL f_3	0/59	0/41	0/0	0/4

Left-side top block row label: Polynomial ($pr = 1$). Left-side bottom block row label: GCD ($pr = 3$). Right-side top block row label: Sum ($pr = 2$). Right-side bottom block row label: Factorial ($pr = 4$).

The *GCD* problem was repeatedly solved especially by IL under f_2, but also by CA, and the WL approach using effective loops (cat. **B**). Most approaches here were able to find solutions without loops (cat. **A**) by creating sufficiently many modulo divisions and variable assignments in one program.

One interesting finding is Genetic Programming using f_1 and f_2 utilized the loop structures (created programs of category **B**) much more frequently than the random walks. We conclude that the reason for the better performance of GP in comparison with random walks is that it is able to discover the benefits of loops.

Even more interesting is that the new loop instructions CA and IL, which both replace explicit stopping criteria by side-effects, were utilized much more

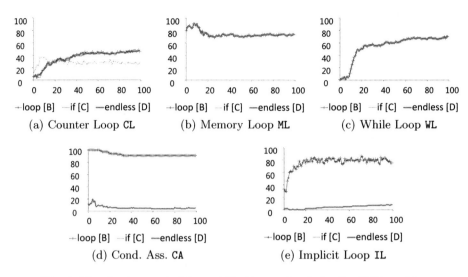

Fig. 4. The development of loop utilization under f_2 for the GCD problem

frequently than the traditional loop instructions. Also, the number of endless loops they produce is much lower. Although we could not find a consistent better performance, we find that it seems to be easier to discover and adapt the CA and IL structures for Genetic Programming. Together with the finding that the better performance of GP compared to random walks is rooted in its utilization of loops and that IL and CA can most effectively solve the *GCD* problem, this gives a clear indication that the convergence-based termination has a high utility. Yet, the simple memory loop ML, too, is very efficient.

5.4 The Evolution of the Loop Utilization

In Fig. 4, we illustrate the fractions of the programs in the population that belong the different categories for the GCD problem. Similar diagrams for the other two non-trivial problems look basically the same and have been omitted here. It can be seen that in CL, ML, and WL, the fraction of programs with endless loops (category **D**) is almost identical with the programs that have effective loops (category **B**). In other words, these approaches are able to find loops that somehow compute good results, but they are not able to discover suitable loop stopping criteria. In CA and IL, on the other hand, the fraction of category-**B** programs is always much higher than the proportion of category-**D** programs throughout the evolution (see also Table 2). It should be noted that runs that discover a solution before the 100 generations are elapsed will be terminated earlier, hence the number of programs with effective loops in the CA diagram goes down a bit after a few generations.

6   Conclusions

The goal of the research presented here was to *(1)* evaluate the capability of Genetic Programming approaches for synthesizing non-approximative, non-trivial, and deterministic mathematical algorithms with integer-valued results which involve loops and to *(2)* test whether novel loop structures with implicit stopping criteria can lead to better performance. We utilized five different program representations, two of which (CA and IL) possess features different from traditional loop instructions, and applied them to one basic and three non-trivial tasks.

We find that, on one hand, GP makes efficient use of the information provided by the objective functions, i. e., usually outperforms random walks. On the other hand, for non-trivial algorithm synthesis problems, Genetic Programming rarely finds suitable solutions. We thus have verified the "GP hardness" of the *sum*, *GCD*, and *factorial* problem on the presented instruction set and under the documented conditions.

We thus found both, evidence for the utility of GP and also evidence for the necessity of further research. It can be considered as a given fact that imperative Standard Genetic Programming in its current form does not perform well in the evolution of non-trivial and non-approximative algorithms. The positive impression promoted by the related work section could not be verified in our work here (under the parameters of experiments and validation).

With our experiments, we showed that implicit stopping criteria for loops, based on convergence of variables rather than explicit expressions, can much more easily be utilized by Genetic Programming. These representations are likely to have less epistatic effects since the stopping criteria do not "have to access" any variable. Hence, our future research will concentrate on further verifying low-epistatic control structures.

Acknowledgements. This work has been supported by China Postdoctoral Science Foundation Grant Number 20100470843.

References

[1] Chen, G., Zhang, M.: Evolving while-loop structures in genetic programming for factorial and ant problems. In: 18th Australian Joint Conference on Artificial Intelligence, Sydney, Australia, pp. 1079–1085 (2005)

[2] Ciesielski, V., Li, X.: Experiments with explicit for-loops in genetic programming. In: IEEE Congress on Evolutionary Computation, Portland, OR, USA, vol. 1, pp. 494–501 (2004)

[3] Finkel, J.R.: Using genetic programming to evolve an algorithm for factoring numbers. In: Genetic Algorithms and Genetic Programming at Stanford, pp. 52–60. Stanford Bookstore, Stanford (2003)

[4] Koza, J.R.: Genetic Programming: On the Programming of Computers by Means of Natural Selection. MIT Press, Cambridge (1992)

[5] Lai, T.: Discovery of understandable math formulas using genetic programming. In: Genetic Algorithms and Genetic Programming at Stanford, pp. 118–127. Stanford Bookstore, Stanford (2003)

[6] Li, X., Ciesielski, V.: An analysis of explicit loops in genetic programming. In: IEEE Congress on Evolutionary Computation, Edinburgh, UK, pp. 2522–2529 (2005)

[7] Luke, S., et al.: ECJ: A Java-based Evolutionary Computation Research System. George Mason University, Fairfax (2006)

[8] Mann, H.B., Whitney, D.R.: On a test of whether one of two random variables is stochastically larger than the other. The Annals of Mathematical Statistics 18(1), 50–60 (1947)

[9] McPhee, N.F., Poli, R.: Memory with memory: Soft assignment in genetic programming. In: Genetic and Evolutionary Computation Conference, Atlanta, GA, USA, pp. 1235–1242 (2008)

[10] Poli, R., McPhee, N.F., Citi, L., Crane, E.: Memory with memory in genetic programming. Journal of Artificial Evolution and Applications, Article ID 570606 (2009)

[11] Qi, Y., Wang, B., Kang, L.: Genetic programming with simple loops. Journal of Computer Science and Technology 14(4), 429–433 (1999)

[12] Teller, A.: Genetic programming, indexed memory, the halting problem, and other curiosities. In: 7th Florida Artificial Intelligence Research Symposium, Pensacola Beach, FL, USA, pp. 270–274 (1994)

[13] Teller, A.: Turing completeness in the language of genetic programming with indexed memory. In: 1st IEEE Conference on Evolutionary Computation, Orlando, FL, USA, pp. 136–141 (1994)

[14] Weise, T.: Evolving Distributed Algorithms with Genetic Programming. PhD thesis, University of Kassel, Kassel, Germany (2009)

[15] Weise, T.: Global Optimization Algorithms – Theory and Application (2009b), http://www.it-weise.de/

[16] Weise, T., Chiong, R.: Evolutionary approaches and their applications to distributed systems. In: Intelligent Systems for Automated Learning and Adaptation: Emerging Trends and Applications, pp. 114–149 (2009)

[17] Weise, T., Tang, K.: Evolving distributed algorithms with genetic programming. IEEE Transactions on Evolutionary Computation (2010) (accepted for publication)

[18] Weise, T., Zapf, M., Geihs, K.: Rule-based genetic programming. In: 2nd International Conference on Bio-Inspired Models of Network, Information, and Computing Systems, Budapest, Hungary, pp. 8–15 (2007)

[19] Weise, T., Zapf, M., Chiong, R., Nebro Urbaneja, A.J.: Why is optimization difficult? In: Chiong, R. (ed.) Nature-Inspired Algorithms for Optimisation. SCI, vol. 193, pp. 1–50. Springer, Heidelberg (2009)

[20] Wijesinghe, G., Ciesielski, V.: Experiments with indexed for-loops in genetic programming. In: Genetic and Evolutionary Computation Conference, Atlanta, GA, USA, pp. 1347–1348 (2008)

Maximum Margin Decision Surfaces for Increased Generalisation in Evolutionary Decision Tree Learning

Alexandros Agapitos[1], Michael O'Neill[1], Anthony Brabazon[1], and Theodoros Theodoridis[2]

[1] Financial Mathematics and Computation Research Cluster
Natural Computing Research and Applications Group
University College Dublin, Ireland
[2] School of Computer Science and Electronic Engineering
University of Essex, UK

Abstract. Decision tree learning is one of the most widely used and practical methods for inductive inference. We present a novel method that increases the generalisation of genetically-induced classification trees, which employ linear discriminants as the partitioning function at each internal node. Genetic Programming is employed to search the space of oblique decision trees. At the end of the evolutionary run, a (1+1) Evolution Strategy is used to geometrically optimise the boundaries in the decision space, which are represented by the linear discriminant functions. The evolutionary optimisation concerns maximising the decision-surface margin that is defined to be the smallest distance between the decision-surface and any of the samples. Initial empirical results of the application of our method to a series of datasets from the UCI repository suggest that model generalisation benefits from the margin maximisation, and that the new method is a very competent approach to pattern classification as compared to other learning algorithms.

1 Introduction

This paper introduces a novel, hybrid approach to the evolutionary learning of Decision Trees (DT), by means of Genetic Programming (GP) [1], that improves their generalisation performance. The evolutionary method is based on the concept of maximum margin linear discriminant functions to search among a number of potential decision surfaces for the one for which the margin is maximised. This concept is borrowed from the Support Vector Machine (SVM) [2] approach to overfitting avoidance. The goal of training a SVM is to find the separating hyperplane with the largest margin; it is expected that the larger the margin, the better generalisation of the classifier. Initially, oblique DTs are learned by means of GP using the classification accuracy as the fitness measure. At the end of the evolutionary run the best classifier is further optimised to improve its generalisation. The optimisation concerns maximising the margin, which is defined to be the smallest distance between the decision surface

S. Silva et al. (Eds.): EuroGP 2011, LNCS 6621, pp. 61–72, 2011.
© Springer-Verlag Berlin Heidelberg 2011

and any of the samples. Finding the maximally-separating decision hyperplane is a constrained optimisation problem that is tackled by means of quadratic programming in the case of SVMs. Our method is instead using an evolutionary optimisation algorithm (EA), namely a (1+1) Evolution Strategy (ES). The proposed methodology is compared in this study against the axis-parallel DT induction method C4.5, SVM (without any use of kernel functions), and naive Bayesian on a series of benchmark classification datasets from the UCI machine learning repository. Initial empirical results suggest that increased generalisation is accrued from margin maximisation, and that the new approach compares favourably, often outperforming other learning algorithms.

The rest of the paper is organised as follows. We first present relevant literature for inducing DTs using GP, and discuss the issue of model generalisation, motivating the need for maximum margin classifiers. The geometry of linear discriminant function is analysed in Section 2, which defines the main tool for calculating distances in the feature space. Section 3 presents the margin optimisation algorithm that is based on a novel distance-based fitness function. Section 4 outlines the experimental approach, describes the grammar-based GP system, and details the benchmark datasets. Section 5 presents the experimental results, and Section 6 draws our conclusions.

1.1   Model Generalisation in Genetically Programmed DTs

Genetic Programming, inherently adopting an expression-tree representation, has been an obvious choice for an application of stochastic search to the space of decision-tree structures, seeking to maximise some sort of classification performance metric. GP-induced partitioning functions can be composed of both linear and non-linear combinations of features, therefore oblique and non-linear splits can be represented as easily and efficiently as univariate splits. In the domain of axis-parallel decision trees the works of [3,4,5] evolve DTs using classification accuracy as the fitness function to drive the search. Oblique decision trees have been evolutionarily constructed in [6,7], whereas work on non-linear DTs has been reported in [8,9,10].

A crucial aspect of artificial learning systems is their ability to extract a precise underlying representation of the concept that is being inferred in a supervised learning task via a set of training instances, so as to be able to generalise well to unseen examples from that concept. Decision tree learning, which is based on adaptive tree-structures that are being iteratively refined on a set of training instances, suffers from the problem of model overfitting inherent in any process of data-driven modelling. The highly expressive representation offered by tree-structures often results in a close fit on the training instances that does not allow for effective generalisation if their complexity is not somehow kept under control. An approximate inductive bias of classical DT growing algorithms is that shorter trees are preferred over larger ones [11]. Dominant approaches to control the size of a DT are reduced-error pruning and rule post-pruning [11]. In GP-based DT induction, research on improving model generalisation has focused on controlling the size of the evolved structures mainly by modifying the fitness function to exhibit a bias towards smaller expression-trees [12,13,14,15,10]. A different approach to the problem of overfitting in genetically-induced DTs is presented in [16]. In that study, a statistical significance test of each program's performance is employed, and multi-objective fitness functions are designed to bias the evolutionary search towards better generalising individuals.

The generalisation performance of a learning machine can be studied by means of uniform convergence bounds, using a technique introduced by Vapnik and Chervonenkis [17]. The theoretical motivations behind this approach lie in the data-dependent *structural risk minimisation* (SRM) [18] principle. The central concept is the *effective capacity* of the class of hypotheses accessible by the machine: the richer such a class, the higher the risk of overfitting. This feature of a learning machine is referred to as its complexity or *capacity*. The statistical learning principle of SRM provides an upper bound to the generalisation error of the classifier in terms of its training error, the number of training examples and the model capacity. In this regard, SRM is just another way to express generalisation error as a tradeoff between training error and model complexity. In the case where a linear discriminant function in trained on a linearly-separable dataset, there exist an indefinite number of partitioning hyperplanes that attain a perfect split. However, there is no guarantee that any of these hyperplanes will generalise well on new patterns. Maximum margin classifiers approach the decision surface generalisation-potential through the concept of the *margin*, which is defined to be the smallest distance between the decision surface and any of the samples. In [2], it is shown that the capacity of a linear model is inversely related to its margin. Models with small margins have higher capacities because they are more flexible and can fit many training sets, unlike models with large margins. However, according to the SRM principle, as the capacity increases, the generalisation error bound will also increase. Therefore, it is desired to design linear discriminant functions that maximise the margins of their decision boundaries in order to ensure that their worst-case generalisation errors are minimised. The state-of-the-art maximum margin classifier is the support vector machine (SVM) [2].

2   Geometry of a Linear Discriminant Function

Oblique splits are essentially represented by linear discriminant functions of the form $y(x) = w^T x + w_0$, where w is called the weight vector, and w_0 is a bias. For this class of discriminant functions, the decision surfaces are $(D-1)$-dimensional hyperplanes within the D-dimensional feature space. An oblique DT can be therefore regarded as a collection of linear discriminants instrumented in such a way so as to provide a classification technique for both binary and multi-category pattern classification tasks.

The simplest representation of a linear discriminant function is obtained by taking a linear function of the input vector x so that $y(x) = w^T x + w_0 = \sum_{i=1}^{D} w_i x_i + w_0$, where w^T is the transpose of the weight vector, w_0 is a bias, and $w^T x$ represent the inner product of vectors w^T and x. Assuming a binary classification task, an input vector x is assigned to class C_1 if $y(x) > 0$, and to C_2 otherwise. The corresponding decision surface is defined by the expression $y(x) = w^T x + w_0 = 0$, which corresponds to a $(D - 1)$-dimensional hyperplane within the D-dimensional input space. Consider two points x_A and x_B that lie on the hyperplane. Because $y(x_A) = y(x_B) = 0$ we have:

$$w^T x_A + w_0 = w^T x_B + w_0 = 0 \Rightarrow$$
$$w^T (x_A - x_B) = 0 \tag{1}$$

and hence the vector w is perpendicular to every vector lying within the decision surface, given that their inner product is equal to zero. Thus, vector w determines the orientation of the decision surface. Similarly, if x is a point on the decision surface, then $y(x) = 0$, and so the normal distance from the origin to the decision surface is given by the following:

$$\frac{w^T x}{||w||} = -\frac{w_0}{||w||} \tag{2}$$

where $||w||$ represents the magnitude or Euclidean norm of vector w. We therefore see that the bias w_0 determines the displacement of the decision surface from the axis origin. These properties are illustrated for the case of $D = 2$ in Figure 1(a).

Furthermore, we note that the value of $y(x)$ gives a signed measure of the perpendicular distance r of the point x from the decision surface. To illustrate this, consider an arbitrary point x, and let $x_\perp$ be its orthogonal projection onto the decision surface. The perpendicular distance r of x from the decision surface is given as follows:

$$x = x_\perp + r\frac{w}{||x||} \Rightarrow$$
$$wx + w_0 = wx_\perp + r\frac{w}{||x||}w + w_0 \tag{3}$$

Given that $y(x) = w^T x + w_0 = 0$ and $y(x_\perp) = w^T x_\perp + w_0 = 0$, Equation 3 becomes:

$$y(x) = y(x_\perp) + r\frac{w}{||x||}w \Rightarrow$$
$$y(x) = r\frac{w}{||x||}w \tag{4}$$

Given that the inner product of two vectors a and b is given by $a \cdot b = ||a|| ||b|| cos\theta$, where θ is a measure of the angle between a and b, and that the angle between vector w an itself is zero, we have:

$$y(x) = \frac{r}{||w||}||w|| ||w|| cos(0) \Rightarrow$$
$$r = \frac{y(x)}{||w||} \tag{5}$$

This result is illustrated in Figure 1(a). For this case where $D = 2$, the absolute distance r of point $x = [x_1, x_2]$ from the decision hyperplane $y(x)$ with $w = [w_1, w_2]$ becomes:

$$r = \frac{|y(x)|}{||w||} = \frac{|w_1 x_1 + w_2 x_2 + w_0|}{\sqrt{w_1{}^2 + w_2{}^2}} \tag{6}$$

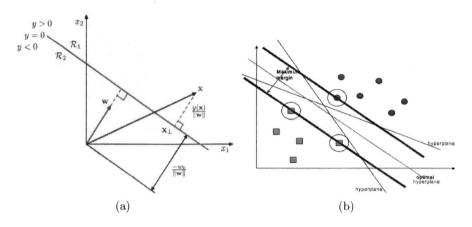

(a)   (b)

Fig. 1. (a) Illustration of the geometry of a linear discriminant function in two dimensions. The decision surface, shown in red, is perpendicular to w, thus its orientation depends on it, whereas its displacement from the axis origin is controlled by the bias parameter w_0. The signed orthogonal distance of a general point x from the decision surface is given by $y(x)/||w||$. Figure taken from [19] (page 182). (b) Possible decision surfaces for a linearly separable dataset. The margin is defined as the perpendicular distance between the decision surface and the closest data points. Maximising the margin leads to a particular choice of decision boundary, as shown with the red line. The location of the boundary is determined by a subset of the data points, which is indicated by the circles.

3 Evolutionary Optimisation of a Maximum Margin Decision Surface via an ES(1+1)

Decision boundaries with large margins tend to have lower generalisation errors than those with small margins. Intuitively, if the margin is small, then any slight perturbations to the decision boundary can have quite a significant impact on its classification. Classifiers that produce decision boundaries with small margins are therefore susceptible to model ovefitting. Figure 1(b) illustrate various decision hyperplanes for a linearly separable dataset. It should be clear that in this case the solution region incorporates a number of hyperplanes, with arbitrary orientation and displacement, that satisfy the requirement of providing a perfect classification on the training patterns. In the example of Figure 1(b), we illustrate three possible hyperplanes that would result in a 100% accurate classifier. One way to impose additional requirements in order to bias the selected hyperplane towards better generalisation is to introduce the concept of margin, which is defined as the perpendicular distance between the decision surface and the closest data points. Given this definition, we observe that if we wish to maximise this margin, the location of the required hyperplane is determined by a subset of the closest data points, which is indicated by the circles in Figure 1(b).

We adopted an evolutionary approach to optimise the margin of a decision surface using an ES(1+1) without self-adaptive mutations. Margin maximisation can be regarded as a constrained optimisation problem that requires the hyperplane to freely wander within the solution region of arbitrarily oriented hyperplanes for the one with optimal margin, without causing any drop in the classification accuracy measured originally on the training instances. We transform the constrained optimisation problem into a multi-objective problem by adding a penalty function that evaluates constraint violations realised through misclassifications. So far, we have assumed that the training data points are linearly separable in the feature space. In practice, however, the class-conditional distributions may overlap, in which case exact linear separation of the training data may not be possible, and if finally some kind of separation does succeed it usually leads to poor generalisation. We therefore need a way to allow some of the training points to be misclassified. Data points are allowed to be on the "wrong" side of the margin boundary, but with a penalty that increases proportionally to the distance from the boundary. A distinguishing characteristic of maximum margin classification approaches is that these learning algorithms have *sparse* solutions so that the location of the decision surface is determined by a subset of data points.

In order to quantify the size of the margin we start by determining the closest pair of data points (in terms of Euclidean distance), where each data point belongs to a different class and lies on the correct side of the margin. Then the distance of these points from the hyperplane is calculated using Equation 5; let us call these distances as cpd_{classA} and cpd_{classB}. We also use the same equation to calculate the distances between each correctly-classified data point and the hyperplane, for both classes; let us call the smallest of these distances as sv_{classA} and sv_{classB} for data points of classes A and B respectively. The distance-based objective function that quantifies a hyperplane's margin size is defined as follows:

$$f = \left[\frac{\mu_{cpd}}{\sigma_{cpd}} + sv_{classA} + sv_{classB} \right] - \frac{1}{N} \sum_{i=1}^{N} penalty(x_i) \qquad (7)$$

where μ_{cpd} is the mean distance of cpd_{classA} and cpd_{classB} distances, σ_{cpd} is their standard deviation, N is the total number of points for classes A and B , and penalty(x) is defined as:

$$penalty(x) = \begin{cases} \frac{|y(x)|}{||w||} & \text{if } predicted \neq actual \\ 0 & \text{otherwise} \end{cases} \qquad (8)$$

Maximisation of this objective function leads to maximisation of the hyperplane's margin. The first compartment of this function requires that the fraction

$\frac{\mu_{cpd}}{\sigma_{cpd}}$ is maximised along with the distances sv_{classA} and sv_{classB}, leading to a margin that maximises the minimum distance of the samples from the surface, while ensuring that the separating hyperplane is equally spaced in-between those closest samples. At the same time, the amount of misclassification penalties is required to be kept at a minimum by the second compartment.

The first step in optimising the margin of the decision boundaries represented by the linear discriminant functions serving as predicate nodes in a DT, is to analyse the expression-tree and map each of the linear discriminant functions to pairs of classes that they discriminate. For that purpose we instrument the fitness evaluation process in order to be able to track down the path that is being followed during a DT's execution with an input training pattern. For each fitness evaluation, we log the execution-trace by recording each linear discriminant that has been visited starting from the root node until we reach a final classification leaf-node. Having mapped the linear-discriminant functions to possible classifications, we then create all possible classification pairs that a linear discriminant attempts to partition. For every such classification pair we determine the size separating hyperplane's margin size using the objective function in Equation 7. The size of the margin for the overall DT is given by averaging the resulting margin sizes of all linear discriminants that compose the expression-tree.

We have chosen an ES(1+1), a simple but powerful random hill-climbing method to search the solution region of a separating hyperplane for the maximum possible margin. The candidate hyperplane is represented as a vector of coefficients that combines the components of vector w (determining the orientation) with the bias w_0 (determining the displacement from origin). The number of optimisation iterations are governed by a user-defined parameter. At each iteration a coefficient or a number of coefficients are being perturbed using a Gaussian mutation with known mean and standard deviation. Currently, we do not allow for self-adaptation of the mutation rates. The sequence in which coefficients are perturbed, as well as their number is either deterministic or random, with a real-valued parameter governing the probability of application of these two coefficient-examination schemes, set to 0.7 is favour of the deterministic sequential examination.

4   Methods

4.1   Experimental Approach

This empirical study attempts to hybridise the traditional method of constructing oblique DTs using GP, with the approach taken in maximum margin classification methodologies. The concept of learning revolves around the notion of generalisation, and it is highly desired that methods reinforcing model generalisation are extensively studied in the scope of evolutionary learning. We are contrasting four different learning algorithms in terms of their generalisation ability in a series of five benchmark classification datasets from the UCI

Table 1. Benchmarking Datasets

Dataset	Size	No. of Classes	No. of numerical pattern features
Wisconsin Breast Cancer (WBC)	699	2	9
BUPA liver disorders	345	2	6
PIMA diabetes	768	2	8
IRIS	150	3	4
Vehicle	846	4	18

machine learning repository [20]. These algorithms are: (a) GP hybridised with margin maximisation (GP-MM), (b) standard GP (GP-ST), (c) axis-parallel DT inductor C4.5, (d) SVM (without any use of kernel functions), and (e) naive Bayesian, all taken from WEKA software [21]. Performance statistics are evaluated through the process of 10-fold cross validation. Table 1 presents a description of the benchmarking datasets. Prior to classification all numeric features were normalised within the range of [0.0, 1.0].

4.2   Grammar-Based Genetic Programming

We employ a grammar-based GP system to evolve DTs. The context-free grammar that is used to type the DT representation language is presented bellow:

```
<DT> ::= <if>
<if> ::= <ldf> <expr> <expr>
<expr> ::= <if> | <classification>
<classification> ::= classA | classB | ... | classZ
<ldf> ::= <constant> * <feature> + <constant> * <feature> + <constant> > 0
<constant> ::= <constant> <op> <constant> | double in the range of [-1.0, 1.0]
<op> ::= + | - | * | /
<feature> ::= depending on the number of features in the problem
```

Using this grammar, a DT is represented as a set of two-dimensional linear discriminant functions. We deliberately constrained the system to the induction of two-dimensional linear discriminants in order to study the effectiveness of the proposed methodology on the simplest representation setting possible. Future extensions will employ a more expressive representation that allows the composition of multi-dimensional separating hyperplanes.

The GP algorithm employs a panmictic, generational, elitist genetic algorithm. The algorithm uses tournament selection with a tournament size of 12. The population size is set to 1,000 individuals. Ramped-half-and-half tree creation with a maximum depth of 5 is used to perform a random sampling of DTs during run initialisation. Throughout evolution, expression-trees are allowed to grow up to depth of 8. The evolutionary search employs a mutation-based variation scheme, where subtree mutation is combined with point-mutation; a probability governing the application of each, set to 0.6 in favour of subtree mutation. No reproduction nor recombination were used.

In the case where margin maximisation is included in the learning process (GP-MM), GP is run for 100 generations using classification accuracy as the fitness function (phase I), and optimisation proceeds for an additional 10,000 hill-climbing iterations (equally divided among the linear discriminants of a DT) using the objective function of Equation 7 (phase II), totalling 110,000 fitness evaluations. The Gaussian mutation in ES(1+1) has a mean of 0.0 and a standard deviation of 0.01. In the case of standard GP runs (GP-ST), evolution proceeds for 110 generations to allow for a comparison on the same number of fitness evaluations. Classification accuracy is similarly used as a fitness function.

5   Results

For the evolutionary learning algorithms we performed 50 independent runs to calculate performance statistics. Table 2 summarises the out-of-sample

Table 2. Comparison of test classification accuracy obtained by different learning algorithms in 10-fold cross validation. Standard errors in parentheses.

Dataset	Learning algorithm	Best accuracy	Average accuracy (phase I)	Average accuracy (phase II)	Generalisation increase
WBC	GP-MM	98.5%	94.6% (0.008)	97.5% (0.006)	2.8% (0.004)
	GP-ST	97.9%	95.1% (0.008)	-	-
	C4.5	96.1%	-	-	-
	SVM	96.6%	-	-	-
	Bayesian	97.4%	-	-	-
BUPA	GP-MM	81.3%	65.8% (0.01)	73.8% (0.01)	13.0% (0.02)
	GP-ST	71.8%	66.1% (0.01)	-	-
	C4.5	63.1%	-	-	-
	SVM	70.3%	-	-	-
	Bayesian	63.2%	-	-	-
PIMA	GP-MM	81.8%	75.3% (0.008)	78.7% (0.007)	4.8% (0.005)
	GP-ST	77.6%	74.9% (0.008)	-	-
	C4.5	75.5%	-	-	-
	SVM	78.1%	-	-	-
	Bayesian	75.0%	-	-	-
IRIS	GP-MM	100%	95.6% (0.01)	97.7% (0.008)	2.3% (0.004)
	GP-ST	98.2%	95.9% (0.01)	-	-
	C4.5	95.3%	-	-	-
	SVM	97.1%	-	-	-
	Bayesian	94.0%	-	-	-
VEHICLE	GP-MM	78.0%	66.7% (0.008)	71.2% (0.008)	6.7% (0.003)
	GP-ST	72.4%	66.3 % (0.007)	-	-
	C4.5	72.6%	-	-	-
	SVM	49.6%	-	-	-
	Bayesian	60.5%	-	-	-

classification performance accrued from the process of 10-fold cross validation. In the case of evolutionary algorithms we also present the average accuracy for phases I and II, corresponding to the values obtained during GP and ES(1+1) procedures respectively. Recall that for the case of GP-MM phase I lasts for 100 generations, whereas for GP-ST it lasts for 110 generations. Results suggest a superiority of evolutionary methods compared to other learning algorithms, which becomes more pronounced in the BUPA and VEHICLES datasets, with the EAs largely outperforming the rest of the classifiers. A comparison between GP-MM and GP-ST, further suggests that margin maximisation does indeed play an important role in increasing the model generalisation performance. The percentage of increase in model generalisation ranges from 2.8% in WBC dataset to 13.0%, suggesting that there are cases where margin maximisation is a very effective method to deal with the problem of model overfitting. The circumstances under which such an additional optimisation is fruitful remains to be seen from the application of our technique to a wider range of multi-category classification problems, by extending the expressiveness of linear discriminant function representation, allowing for separating hyperplanes in a much higher-dimensional feature space. An additional interesting observation in Table 2 is that we found no significant differences between the average classification accuracies in phase I of GP-MM and GP-ST, suggestive of no evident model overfitting from model overtraining. This is particularly interesting and surely warrants further study. Finally, we observe that for the four-class problem (VEHICLE), SVMs have a very low performance indicative of the inherent difficulty to tackle multi-class problems using this methodology. SVMs are traditionally binary classifiers, and practitioners need to rely on *one-versus-the-rest*, or *one-versus-one* binary problem decomposition strategies. On the other hand, tree-based structures can naturally represent multi-class problems.

6   Conclusion

The decision surface optimisation criterion revolves around the notion of a margin either side of a hyperplane that separates two data classes. Maximising the margin and thereby creating the largest possible distance between the separating hyperplane and the instances on either side of it has been theoretically shown and practically proven by SVMs to reduce the upper bound on the expected generalisation error. This study applies the methodology of maximum margin classifiers to evolutionary search, resulting in a hybrid method that learns oblique DTs via GP, and subsequently optimises the geometry of decision hyperplanes using an ES(1+1). Results are very encouraging, suggesting a superiority of the new approach as compared against other learning algorithms. An interesting point to note is that at the moment the ES optimises the geometry of lines in a two-dimensional feature space. In addition, the solution uses the support of all patterns in the training set. Further research is required to quantify the time-efficiency of the proposed approach when dealing with large datasets and high-dimensional feature spaces.

A substantial advantage of evolutionary DT induction is the inherent ability of the learning algorithm to perform feature construction and/or feature selection, while simultaneously searching for a good classification strategy. Therefore, with a careful design of an EA, the dimensionality of the feature space can be made invariant of the classification performance. Feature spaces that are linearly non-separable can take advantage of built-in or evolved kernels combined with margin-maximisation, providing a natural approach to a multi-category classification methodology that embodies the best practices of the state-of-the-art methods in pattern classification.

Acknowledgement

This publication has emanated from research conducted with the financial support of Science Foundation Ireland under Grant Number 08/SRC/FM1389.

References

1. Koza, J.R.: Genetic Programming: on the programming of computers by means of natural selection. MIT Press, Cambridge (1992)
2. Vladimir, V.: The nature of statistical learning theory, 2nd edn. Springer, Heidelberg (1999)
3. Koza, J.R.: Concept formation and decision tree induction using the genetic programming paradigm. In: Schwefel, H.-P., Männer, R. (eds.) PPSN 1990. LNCS, vol. 496, pp. 124–128. Springer, Heidelberg (1991)
4. Folino, G., Pizzuti, C., Spezzano, G.: Genetic Programming and Simulated Annealing: A Hybrid Method to Evolve Decision Trees. In: Poli, R., Banzhaf, W., Langdon, W.B., Miller, J., Nordin, P., Fogarty, T.C. (eds.) EuroGP 2000. LNCS, vol. 1802, pp. 294–303. Springer, Heidelberg (2000)
5. Eggermont, J.: Evolving Fuzzy Decision Trees with Genetic Programming and Clustering. In: Foster, J.A., Lutton, E., Miller, J., Ryan, C., Tettamanzi, A.G.B. (eds.) EuroGP 2002. LNCS, vol. 2278, pp. 71–82. Springer, Heidelberg (2002)
6. Rouwhorst, S.E., Engelbrecht, A.P.: Searching the forest: Using decision trees as building blocks for evolutionary search in classification databases. In: Proceedings of the, Congress on Evolutionary Computation CEC 2000, vol. 1, pp. 633–638 (2000)
7. Bot, M., Langdon, W.B.: Application of genetic programming to induction of linear classification trees. In: Proceedings of the Eleventh Belgium/Netherlands Conference on Artificial Intelligence, BNAIC 1999 (1999)
8. Marmelstein, R.E., Lamont, G.B.: Pattern classification using a hybrid genetic program decision tree approach. In: Genetic Programming 1998: Proceedings of the Third Annual Conference (1998)
9. Tsakonas, A.: A comparison of classification accuracy of four genetic programming-evolved intelligent structures. Information Sciences 176(6), 691–724 (2006)
10. Mugambi, E.M., Hunter, A., Oatley, G., Kennedy, L.: Polynomial-fuzzy decision tree structures for classifying medical data. Knowledge-Based Systems 17(2-4), 81–87 (2004)
11. Mitchel, T.: Machine Learning. McGraw-Hill, New York (1997)

12. Estrada-Gil, J.K., Fernandez-Lopez, J.C., Hernandez-Lemus, E., Silva-Zolezzi, I., Hidalgo-Miranda, A., Jimenez-Sanchez, G., Vallejo-Clemente, E.E.: GPDTI: A genetic programming decision tree induction method to find epistatic effects in common complex diseases. Bioinformatics 13(13), i167–i174 (2007)
13. Kuo, C.-S., Hong, T.-P., Chen, C.-L.: Applying genetic programming technique in classification trees. Soft Computing 11(12), 1165–1172 (2007)
14. Haruyama, S., Zhao, Q.: Designing smaller decision trees using multiple objective optimization based gps. In: IEEE International Conference on Systems, Man and Cybernetics, vol. 6, p. 5 (2002)
15. Folino, G., Pizzuti, C., Spezzano, G.: Improving induction decision trees with parallel genetic programming. In: Proceedings 10th Euromicro Workshop on Parallel, Distributed and Network-based Processing, Canary Islands, January 9-11, pp. 181–187. IEEE, Los Alamitos (2002)
16. Agapitos, A., O'Neill, M., Brabazon, A.: Evolutionary Learning of Technical Trading Rules without Data-Mining Bias. In: Schaefer, R., Cotta, C., Kołodziej, J., Rudolph, G. (eds.) PPSN XI. LNCS, vol. 6238, pp. 294–303. Springer, Heidelberg (2010)
17. Vapnik, V.N., Chervonenkis, A.Y.: On the uniform convergence of relative frequencies of events to their probabilities. Theory of Probability and its Applications 16(2), 264–280 (1971)
18. Shawe-Taylor, J., Bartlett, P.L., Williamson, R.C., Anthony, M.: Structural risk minimization over data-dependent hierarchies. IEEE Transactions on Information Theory 44(5) (1998)
19. Bishop, C.M.: Pattern Recognition and Machine Learning. Springer, Heidelberg (2006)
20. Newman, D.J., Hettich, S., Blake, C.L., Merz, C.J.: UCI repository of machine learning databases (1998)
21. Hall, M., Frank, E., Holmes, G., Pfahringer, B., Reutemann, P., Witten, I.H.: The weka data mining software: an update. SIGKDD Explor. Newsl. 11, 10–18 (2009)

Evolving Cell Array Configurations Using CGP

Paul Bremner[1], Mohammad Samie[1], Gabriel Dragffy[1],
Anthony G. Pipe[1], and Yang Liu[2]

[1] Bristol Robotics Laboratory, University of the West of England, Bristol, BS16 1QY
[2] Intelligent Systems Group, Department of Electronics, University of York,
Heslington, York, YO10 5DD

Abstract. A cell array is a proposed type of custom FPGA, where digital circuits can be formed from interconnected configurable cells. In this paper we have presented a means by which CGP might be adapted to evolve configurations of a proposed cell array. As part of doing so, we have suggested an additional genetic operator that exploits modularity by copying sections of the genome within a solution, and investigated its efficacy. Additionally, we have investigated applying selection pressure for parsimony during functional evolution, rather than in a subsequent stage as proposed in other work. Our results show that solutions to benchmark problems can be evolved with a good degree of efficiency, and that compact solutions can be found with no significant impact on the required number of circuit evaluations.

1 Introduction

A potential design paradigm for fault tolerant digital systems is a cell array. This sort of system is made up of an array of identical cells, capable of being confgured to perform a variety of functions [1]; such an array can be designed so that, when a cell is detected as being faulty, it is excised from the system, and its functionality taken over by a spare cell. Thus, we have chosen a granularity of cells in order to best facilitate this process, i.e., balancing cell size (of functional and testing components), with functionality appropriate for decomposition of typical digital designs. Such a cell array is somewhat analogous to an FPGA, i.e., it is composed of configurable elements. However, there currently exists no synthesis tool to automatically produce digital circuits from functional descriptions for this *custom* FPGA, thus, the cell functions and the configuration bits must be painstakingly specified by hand. The powerful functionality of the cells makes performing this task, for all but the simplest of programs, a non-trivial process. An alternative approach, presented here, is to use Cartesian Genetic Programming (CGP) to generate the cell functions and the data flow between cells. Cartesian Genetic Programming, is an algorithm developed by Miller and Thomson [2] to facilitate the evolution of digital circuits. It uses a directed graph representation, with nodes representing digital functions, and edges the connections between them. To use CGP for cell array evolution the node functions are cell congfigurations; in order to do this, we have adapted the structure of

S. Silva et al. (Eds.): EuroGP 2011, LNCS 6621, pp. 73–84, 2011.

the genotype and the method by which mutation is performed. We have shown that, despite the increase in complexity, fewer solution evaluations are necessary (than CGP with logic gates) for a series of benchmark problems. Additionally, we have investigated two techniques for improving the evolutionary process. Firstly, a genetic operator that seeks to exploit modularity in solutions by copying cell configurations, from one part of a solution, to another. Secondly, a method of applying selection pressure for parsimony throughout functional evolution. We have found that by applying these two methods in concert, significantly smaller, more regular solutions are produced with no significant effect on the number of solution evaluations.

2   Motivation and Related Work

In their original work on CGP Miller and Thomson demonstrated its ability to evolve solutions to 2-bit and 3-bit multipliers, and the even 4 parity problem [2]. In their work, on what we shall term standard CGP, the digital functions were a small set of boolean logic gates. Beyond this standard implementation, it has been demonstrated that the algorithm can be successfully applied to evolution of circuits using a wider range of node functions. Sekanina [3] demonstrated its use with a mixture of normal and polymorphic logic gates. Bremner et al. [4] demonstrated its efficacy when using complex building blocks similar to the cells described here. The work presented here builds on the ideas in [4], making the node functions more powerful still and, as a consequence, also changing how the mutation operator functions. We investigate how this further increase in complexity affects the efficiency of evolution (i.e., the number of individuals that must be processed to find a solution) for several benchmark problems. A common method for improving the efficiency of genetic programming (GP) techniques is to exploit the modularity inherent in many problems. Koza and Rice [5] suggested doing so using automatically defined functions (ADFs), subroutines that are simultaneously evolved with the main program, and can be reused within it. Since the inception of ADFs, other researchers have used this concept to apply such ideas to their own work [6][7]. A suggested method of doing so for CGP is embedded CGP (ECGP) proposed by Walker and Miller [8]. In ECGP groups of nodes are compressed into modules that can be reused within a solution. This module creation is performed at random during evolution, and the evolutionary algorithm is structured to produce a selection pressure for useful modules. We used this concept to develop a genetic operator that exploits modularity at a lower level than ECGP by utilising the structure of the genome. When using CGP for evolution of the proposed cell array, the function of each node is controlled by a bit-string of genes that is subject to mutation as part of the evolutionary process. The proposed operator, node-copy, allows the copying of a node's functional genes to another node elsewhere within the solution. The efficacy of this operator in reducing the number of generations required to find a solution has been investigated for several benchmark problems. In addition to the efficiency of a GP technique in finding a solution to a particular digital circuit problem, the solution itself is also relevant. The smaller a solution is, the more useful it is from an implementation point of view. Previously proposed methods for finding smaller, i.e., more parsimonious, solutions have used a two stage evolutionary process, requiring a very high number of generations [3][9][10][11]. In the first stage of evolution functional correctness is evolved, then, in the second stage, functionally correct solutions that are evolved are selected between for size. The second stage is left to run for a long time (several orders of magnitude, in generations, longer than the first stage) to ensure stag- nation occurs and the smallest possible solution is found. We propose a method by which selection pressure for parsimony can be applied during evolution for functionality, resulting in solutions of a comparable size to those produced using a 2 stage technique; thus, a long second stage is no longer required.

3   Description of Cell Array

In the cell array targeted for the work presented here, each cell consists of two nearly identical halves termed slices. Each slice has two functional outputs that can be configured to operate in different ways depending on the required functionality of the cell. The principle functions that the slices can be configured for are: routing data to and from a communication bus, each output is used to route a piece of data on one of the input lines; as a full or half adder, one output produces sum the other carry; as a 3 input look-up table (LUT), the second output is used to either produce a fixed binary value or route one of the inputs. An additional configuration is if both slices are used as LUTs, then their LUT outputs can be selected between using an additional input for each slice, if the inputs to both slices are identical then this can result in a 4 input LUT; if they are not, the input to output relationship is somewhat complicated, how they are combined is shown in Fig. 1. The second output of both slices in this cooperative configuration can either produce a fixed binary value, or route one of the input lines. However, in the work presented here only functional evolution is considered, i.e., for the purposes of the evolution it is assumed that the data is successfully routed between cells. Thus, input routing on the second output

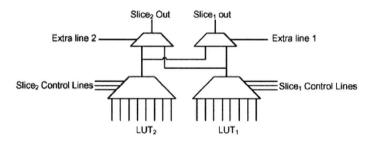

Fig. 1. Function of the combination of two 3-input LUT slices. If the control lines for the two slices are from the same sources then the two LUTs can be combined to form a 4-input LUT.

in the LUT configurations is not implemented here, nor are routing slices; it is intended that the routing will be solved separately in future work. The method by which the slice configurations are implemented within CGP is described in the following section.

4   CGP Overview

In CGP a digital circuit is represented as a directed graph, each node representing some form of digital processing element. It is described as Cartesian because the nodes are laid out in a grid, so the Cartesian coordinates of a node are used to identify the connections of the edges of the graph. A benefit of this type of representation is that the outputs of a given node can be connected to any other, allowing implicit reuse of the processing performed (Fig. 2).

CGP has been shown to be most efficient when only a single column (or row) of nodes is used, rather than a grid of nodes as suggested in the original implementation [12]. Additionally, the graph is acyclic as all the functions to be evolved are feed-forward, combinational logic. Thus, a node may only have input connections from preceding nodes in the graph and program inputs; the outputs are further restricted in that they may not be connected (directly) to program inputs.

Fig. 2. Acyclic Directed Graph, 3 nodes each with 2 inputs and 1 output, 2 program inputs (A,B) and one program output (C)

The genotype in CGP, is made up of a number of sets of integers, one set for each node in the graph. The genotype length is fixed, a specified number of nodes is defined for every member of the population. Typically, CGP uses only mutation as a genetic operator. Depending on where in the genome mutation acts, it either changes one of the input sources, or the functionality, of a node. However, the genotype-phenotype mapping is such that each node need not necessarily contribute to the value produced at any of the outputs. Thus, although the genotype is bounded, the phenotype is of variable length. The unconnected nodes represent redundant genetic information that may be expressed if a mutation results in their inclusion in an input to output path. Therefore the effect of a single point mutation on the genotype can have a dramatic effect on the phenotype, an example of this is shown in Fig 3. This phenomenon is often referred to as neutrality, as mutations in the redundant sections of the genome have no (immediate) effect on the fitness of the individual; it has been shown to

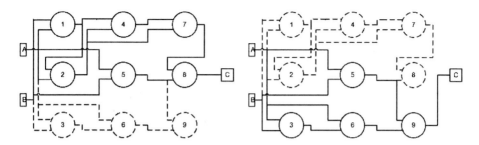

Fig. 3. Point mutation occurs changing which node the output C is connected to. Redundant nodes are indicated by dotted lines.

be beneficial to the operation of CGP [2][12][13]. A degree of redundancy as high as 95% is suggested by Miller and Smith [13] as providing optimum increase in performance. In order to ensure these neutral mutations influence the evolution, new population members that have the same fitness as the parent are deemed fitter than parents.

5   Adaptation of CGP for Cell Array Evolution

5.1   Basic Operation Changes

When using CGP for cell array evolution the nodes in the graph represent slice configurations. There are three possible configurations that the nodes can instantiate, 1-bit adder, 3 input LUT, and two cooperative 3 input LUTs (Fig. 1). The genotype is structured to enable this functionality, the group of numbers that represents each node is made up of: a node type identifier, which is used to define node features; function genes, for LUT nodes these are the contents of the LUT(s) and the value of the fixed outputs(s), for adder nodes they are not used; a set of input conections, how many is determined by the arity of the node type, each connection has two values, identifying the node the input comes from and which output of said node. The features of the node types are summarised in Table 1.

Table 1. Summary of node types

Node Type	Node ID	Inputs	Outputs	Functional Genes
3 input LUT	1	3 LUT contol bits	LUT out Fixed value	8-bit LUT contents 1-bit fixed value output
1-bit adder	2	2 adding bits 1 carry-in bit	Sum Carry	Not used
2 cooperative LUTs	3	8 LUT control bits	2 LUT outs 2 fixed values	2*8-bit LUT contents 2*1-bit fixed value output

As a consequence of the genome structure used here, mutation operates differently from in standard CGP. If a node input is selected to be mutated, then a new legal node is generated as in normal CGP, however which output of the selected node must also be generated. If the function of a node is selected to be mutated this can occur in two ways: firstly, the type can change, resulting in generation of a new set of function genes as appropriate, and adding or removing inputs if the arity changes; secondly, if the node is of type 1 or 3, the functional genes can be mutated, this is done by flipping a random number of bits in the functional gene string (comprised of the LUT contents and the fixed output value).

5.2   Additional Genetic Operator, Node-Copy

In many problem domains, including digital circuit design, there exists modularity within solutions. As previously stated, CGP begins to address expoitation of this modularity by allowing implicit functional reuse, due to the structure of the representation. A further method of exploitation is suggested here, in the form of an additional genetic operator, *node-copy*, which enables the copying of the functionality of nodes present in the genome. Part of the evolutionary process in the variant of CGP presented here is the evolution of functions by means of LUT contents, as well as what data is operated on by these functions. Clearly the two different types of LUT nodes represent a far greater functional range than is present in standard CGP (that typically uses a short predefined list of logic gates), and some LUT configurations are likely to be more useful processing elements for a given problem. Thus, the purpose of this operator is to exploit useful LUT contents present within a solution, allowing more than one set of inputs to be operated on in the same way.

Node-copy operates by first selecting (with an even chance) whether to copy a node from a list of previously copied nodes, or to copy from a new node not on the list (only node types 1 and 3 can be copied from), and add it to the list; the first time it operates the list is empty so it automatically selects to copy a new node. The node type and functional genes are then assigned to the node being copied to, and, as with the type change mutation, if the arity of the function changes then the inputs are adjusted accordingly. Nodes that are copied from can be selected from anywhere within the genome, thus, the operator can both reuse nodes already expressed in the phenotype, and explore useage of nodes from neutral sections of the genome. In order to try and prevent the copied nodes list becoming bloated with useless nodes, at the start of each generation it is pruned of all nodes not phenotypically expressed in individuals selected to be parents.

5.3   Applying Selection Pressure for Parsimony

It is desirous to have functional designs that use a minimum number of cells. However, the standard CGP approach evolves only for functionality, and the number of nodes used is not taken into account in fitness evaluations. We propose a method by which parsimony can be included in the fitness evaluations, while still ensuring fully functional solutions are found. Each generation the population is first sorted according to fitness, then, if there is more than one solution that has the best fitness (i.e., they have the same best fitness value) these solutions are further sorted by size, thus, the smallest of the fittest solutions will be selected as parents. In order to promote genetic drift (shown to be benficial to CGP operation [13]), parents are deemed to be less fit than offspring. As a result the fitness may not change from one generation to the next, but the size may grow if all fittest offspring use more nodes than the parents; however, the trend over many generations will still be for smaller solutions.

6   Performance Experiments

In order to test the performance of the proposed algorithm it has been used to evolve a series of benchmark problems, 2 and 3 bit multipliers, even 4 and even 8 bit parity problems. These are commonly used as benchmarks for digital circuit evolution techniques, and thus, allow some degree of comparrisson between the work presented here and other approaches. For all tested problems, the affects of node-copy and parsimony selection pressure have been investigated, thus there are four configurations to test their efficacy individually, and in combination.

Most previous work on CGP has employed a $1 + \lambda$ evoltionary strategy, typically with a λ of 4 (1 parent and 4 offspring), and that approach is followed here. For each problem 50 runs have been performed, and each run is continued until a successful solution is found. A successful solution is deemed to be one that is fully functionally correct as defined by the truth-table for the given problem. Thus, the fitness function is purely the total hamming distance between an individuals outputs and those specified on the truth-table for all input combinations.

Mutation is the only genetic operator used and operates on a fixed number of nodes within the genome, for each mutation a node is selected at random, it is then randomly determined whether the node function, or an input connection is mutated; if the function is mutated then there is a uniform chance of type change, gene string mutation, or node-copy operating. A mutation rate of 3% of the nodes in the genome has been determined to provide good performance, through testing of various mutation rates.

The other parameter that must be correctly selected for efficient operation is genome length. Although previous findings with CGP indicate that a large percentage of the genome being inactive provides optimal evolution in terms of computational effort [13], it has been observed that this is both difficult to acheive with our implementation, and detrimental to the finding of parsimoneous solutions. This is the case as the evolution tends to exploit the available resources; in experiments evolving a 3-bit multiplier, without selection pressure for parsimony, it was observed that average node expression of around 65% occured with 100, 200 and 300 nodes in the genome. Further, repeating the experiments with selection pressue for parsimony included, the solutions significantly increased in size as more nodes were added to the genome, i.e., the perfomance of the parsimony selection method was significantly impaired. An additional issue is that, although increasing the nodes in the genotype offered improvements to the number of evaluations required to find a solution, the actual computation time was significantly longer. Thus, it is important to select the correct genotype length for a given problem to ensure good performance, and evolution time; whereas in standard CGP a longer genome is typically better, here trial and error must be used to find the optimum length. As a consequence of our findings, we have used 50 nodes for the 2-bit multiplier and even parity problems, and 100 nodes for the 3-bit multiplier problem.

In order to fully analyse the efficacy of the parsimony selection pressure algorithm we have implemented a 2 stage evolution similar to that used in [3], that is used for comparison. In the first stage of evolution, fitness is purely functional correctness, once it has been achieved the second stage commences, whereby if more than one functionally correct individual exists in a generation the smallest (fewest phenotypically expressed nodes) is selected to be the next parent; the evolution continues until a predefined evolution limit of 2 million generations is reached, to ensure stagnation. It should be noted that the increased evolution time has meant that only 5 runs were performed for each of the problems for the 2 stage evolution.

6.1   Results and Discussion

The principle metric used for analysis of the proposed algorithm is number of solutions evaluated, Mann-Whitney U tests have been used to test the results for significance; further statistical analysis is presented in the form of median evaluations (ME), median absolute deviation (MAD), and inter-quartile range (IQR). These are the metrics used in [8]; they give a clear indication of performance, and allow some comparison between their results and our own. In the results presented in Tables 2-5 the first four rows are the work described here, configured to allow testing of the efficacy of node-copy and size-sort. The U values (from the Mann-Whitney tests) relate to the two rows they straddle, the number of stars indicates the level of significance[1]. The other 3 rows are results from related works [4][8].

In addition, in order to analyse the efficacy of the parsimony selection algorithm the mean and minimum number of expressed nodes over the 50 runs is also presented. Values are given both with and without its application. The U values (from the Mann-Whitney tests) compare the results for each problem with and without size-sort, the number of stars indicates the level of significance[1]. The results for the two stage evolution are presented for comparison.

6.2   Comparison to Similar Techniques

In order to give an overall indication of the efficiency of the proposed algorithm, we have compared it with other CGP implementations used to evolve the presented benchmark problems. We are not trying to show that one implementation

---

[1] * = slightly significant, $p < 0.05$. ** = significant, $p < 0.01$. *** = highly significant, $p < 0.001$.

Table 2. Efficiency statistics for 2-bit multiplier evolution

Evolution Algorithm	ME	MAD	IQR	U
Node-Copy, No Size-sort	38822	19020	35399	1366
No Node-Copy, No Size-sort	39062	18726	35558	1882***
No Node-Copy, Size-sort	95760	56218	156637	1261
Node-Copy, Size-sort	114740	68444	140959	
MolCGP	36250	16598	36568	-
CGP	6197	4130	8489	-
ECGP	7485	3304	9875	-

Table 3. Efficiency statistics for 3-bit multiplier evolution

Evolution Algorithm	ME	MAD	IQR	U
Node-Copy, No Size-sort	589010	289620	516000	1368
No Node-Copy, No Size-sort	473940	185882	382232	1450
No Node-Copy, Size-sort	973642	440642	897283	1638**
Node-Copy, Size-sort	545228	244096	838832	
MolCGP	1127692	515744	1342780	
CGP	4030201	2181656	6110863	
ECGP	854579	257354	773789	

Table 4. Efficiency statistics for even 4 parity evolution

Evolution Algorithm	ME	MAD	IQR	U
Node-Copy, No Size-sort	2768	1238	2405	1283
No Node-Copy, No Size-sort	2612	1506	4730	1335.5
No Node-Copy, Size-sort	4302	3118	7003	1402
Node-Copy, Size-sort	2216	1730	10580	
MolCGP	3584	2122	6107	
CGP	30589	12942	25438	
ECGP	37961	21124	49552	

Table 5. Efficiency statistics for even 8 parity evolution

Evolution Algorithm	ME	MAD	IQR	U
Node-Copy, No Size-sort	73854	62792	161980	1479
No Node-Copy, No Size-sort	59718	47324	122875	1474
No Node-Copy, Size-sort	65470	56352	140254	1555*
Node-Copy, Size-sort	57572	47052	247029	
MolCGP	21656	8982	19552	
CGP	7166369	3210224	6363465	
ECGP	745549	500924	1108934	

Table 6. The number of nodes that are phenotypically expressed in the final solutions for the benchmark problems

Problem	Without SS		With SS		U	2 Stage Evo	
	Mean	Min	Mean	Min		Mean	Min
2m2	35	16	19	4	2314***	7	5
3m3	70	43	48	25	2172***	25	20
e4p	25	2	13	2	1786***	3	2
e8p	26	8	14	4	2101***	4	4

is better than another (as different function sets make it a moot comparison), but rather try to establish whether the increased design space results in significant differences in efficiency; i.e., is CGP a practical means to evolve a solution using cells. It is apparent from comparison of the presented ME figures that using CGP with cells, is more efficient than standard CGP and ECGP for more complex problems such as even parity and the three-bit multiplier. Although the increase in search space has resulted in less efficiency for the parity problems than molCGP, the results are still in the same order of magnitude. These findings indicate that using cells within CGP is a practical means of evolving digital circuits.

6.3 Node-Copy

It can be observed that node-copy has no statistically signifficant effect on the efficiency of evolution for any of the problems tested when it is implemented without size-sort (Tables 2-6). A likely reason for this is the presence of junk nodes within a solution, i.e., nodes that are phenotypically expressed, but do not contribute to the fitness. Their presence is apparent from observation of the number of expressed nodes during evolution; while fitness values remain unchanged, the number increases and deceases by tens of nodes over the course of many generations. Thus, junk nodes are likely to be copied, and remain on the copyable list (and thus be recopied), despite not being beneficial.

However, when size-sort is implemented, parsimony selection pressure means that junk nodes will tend not to be expressed in the solution. Thus, nodes that do contribute to the fitness of an individual are far more likely to be copied and be maintained on the copyable list, when size-sort is used in conjunction with node-copy. The significant difference in evaluations between when node-copy is and is not used in conjunction with size-sort, for the 3-bit multiplier (Table 3, $p < 0.001$) and even 8 parity (Table 5, $p < 0.05$) problems, demonstrates this fact. Further, studying several of the evolved solutions to both problems it was observed that not only are there several repeated nodes, but also nodes that have very similar genes (hamming distance less than 25% of the functional genes). This implies that nodes contributing to the fitness of the solution have been copied during evolution, and then slightly altered to further improve fitness; a possible further reason for the observed benefit of node-copy.

By contrast, node-copy has no significant effect on the evolution of the 2-bit multiplier (Table 2), and even 4 parity problems (Table 4). We suggest that this is because there is insufficient modularity in the small number of functional nodes that they require.

6.4   Size-Sort

The U values reported in Table 6 show that size-sort has a highly significant effect on the number of nodes expressed in a solution for all problems ($p < 0.001$). The critical value, from an implementation point of view, is the minimum number of nodes. Although, in the case of the even 4 parity problem an equal minimum size was found without size-sort (likely due to the simplicity of the problem for our algorithm), it is far more likely to be found with size-sort, indicated by the difference in mean size. Despite the fact that size-sort did not match the performance of the 2 stage evolution, the minimum solution sizes are close in value, with far fewer evaluations. We suggest that the massive reduction in the number of solutions that must be evaluated compensates for the difference in performance, and makes it a useful alternative to 2 stage evolution.

There is however, one minor detrimental effect of size-sort in that, as can be observed from the IQR figures (Tables 2-6), there is a marked increase in variance; i.e., the accuracy of the evolutionary technique is impaired. However, we do not believe this is a significant issue.

7   Conclusions and Further Work

In this paper we have shown that solutions to benchmark digital circuit problems can be evolved, with a practical degree of efficiency, using our cells as the nodes in CGP. We have shown that using our size-sort algorithm, parsimony selection pressure can be applied throughout evolution to result in significantly smaller solutions. While the performance of size-sort is not quite as good as the two stage evolution method, it is still a useful method given the massive reduction in the number of solutions that must be evaluated. In future work we intend to apply the two methods consecutively to investigate whether even smaller solutions might be evolved, and convergence times of the second stage might be reduced.

We have shown that our suggested node-copy operator can negate the undesirable effect of size-sort on the number of solution evaluations. An additional benefit of using node-copy is that the similar cells within a solution are potentially useful from the point of view of correlated redundancy, a fault correction mechanism suggested for use in cell arrays by Samie et al. [1] This mechanism uses the similarities between genes to form colonies of similar cells; within these colonies a damaged configuration string can be recovered from any member of the colony.

In future work we intend to investigate how the method might scale to larger digital circuit problems. In order to do so it may be necessary to use ECGP with our cells, to constrain the number of evaluations required (as it is shown to do for standard CGP).

Acknowledgments. This research work is supported by the Engineering and Physical Sciences Research Council of the United Kingdom under Grant Number EP/F062192/1.

References

1. Samie, M., Dragffy, G., Popescu, A., Pipe, T., Melhuish, C.: Prokaryotic bio-inspired model for embryonics. In: NASA/ESA Conference on Adaptive Hardware and Systems, pp. 163–170 (2009)
2. Miller, J.F., Thomson, P.: Cartesian genetic programming. In: Poli, R., Banzhaf, W., Langdon, W.B., Miller, J., Nordin, P., Fogarty, T.C. (eds.) EuroGP 2000. LNCS, vol. 1802, pp. 121–132. Springer, Heidelberg (2000)
3. Sekanina, L.: Evolutionary design of gate-level polymorphic digital circuits. In: Rothlauf, F., Branke, J., Cagnoni, S., Corne, D.W., Drechsler, R., Jin, Y., Machado, P., Marchiori, E., Romero, J., Smith, G.D., Squillero, G. (eds.) EvoWorkshops 2005. LNCS, vol. 3449, pp. 185–194. Springer, Heidelberg (2005)
4. Bremner, P., Samie, M., Dragffy, G., Pipe, T., Walker, J., Tyrrell, A.: Evolving digital circuits using complex building blocks. In: Tempesti, G., Tyrrell, A.M., Miller, J.F. (eds.) ICES 2010. LNCS, vol. 6274, pp. 37–48. Springer, Heidelberg (2010)
5. Koza, J.: Genetic programming: on the programming of computers by means of natural selection. MIT Press, Cambridge (1996)
6. Dessi, A., Giani, A., Starita, A.: An analysis of automatic subroutine discovery in genetic programming. In: GECCO, pp. 996–1001 (1999)
7. Rosca, J.: Genetic programming exploratory power and the discovery of functions. In: Proceedings of the Fourth Annual Conference on Evolutionary Programming, pp. 719–736 (1995)
8. Walker, J.A., Miller, J.F.: The automatic acquisition, evolution and reuse of modules in cartesian genetic programming. IEEE Trans. Evolutionary Computation (4), 397–417 (2008)
9. Vassilev, V.K., Job, D., Miller, J.F.: Towards the automatic design of more efficient digital circuits. In: EH 2000: Proceedings of the 2nd NASA/DoD workshop on Evolvable Hardware, p. 151 (2000)
10. Hilder, J., Walker, J., Tyrrell, A.: Use of a multi-objective fitness function to improve cartesian genetic programming circuits. In: 2010 NASA/ESA Conference on Adaptive Hardware and Systems (AHS), pp. 179–185 (2010)
11. Gajda, Z., Sekanina, L.: An efficient selection strategy for digital circuit evolution. In: Tempesti, G., Tyrrell, A.M., Miller, J.F. (eds.) ICES 2010. LNCS, vol. 6274, pp. 13–24. Springer, Heidelberg (2010)
12. Yu, T., Miller, J.F.: Neutrality and the evolvability of boolean function landscape. In: Miller, J., Tomassini, M., Lanzi, P.L., Ryan, C., Tetamanzi, A.G.B., Langdon, W.B. (eds.) EuroGP 2001. LNCS, vol. 2038, pp. 204–217. Springer, Heidelberg (2001)
13. Miller, J.F., Smith, S.L.: Redundancy and computational efficiency in cartesian genetic programming. IEEE Transactions on Evolutionary Computation (2), 167–174 (2006)

Designing Pheromone Update Strategies with Strongly Typed Genetic Programming

Jorge Tavares[1] and Francisco B. Pereira[1,2]

[1] CISUC, Department of Informatics Engineering, University of Coimbra
Polo II - Pinhal de Marrocos, 3030 Coimbra, Portugal
[2] ISEC, Quinta da Nora, 3030 Coimbra, Portugal
jorge.tavares@ieee.org, xico@dei.uc.pt

Abstract. Ant Colony algorithms are population-based methods widely used in combinatorial optimization problems. We propose a strongly typed genetic programming approach to automatically evolve the communication mechanism that allows ants to cooperatively solve a given problem. Results obtained with several TSP instances show that the evolved pheromone update strategies are effective, exhibit a good generalization capability and are competitive with human designed variants.

1 Introduction

The application of a global optimization algorithm to a specific problem usually requires some customization of the method. Tailoring ranges from simple parameter specification to more delicate problem-specific modifications, such as the design of operators or search strategy tuning. This is a crucial step for the success of the optimization, as one aims to maintain the ability of the algorithm to perform a robust exploration of the search space, while granting it some specific information that helps to efficiently discover good quality solutions for that particular problem. Usually, the design and adaptation of the algorithms is carried out manually, even though there are some recent proposals that discuss the possibility to automate the design of optimization methods [8,4,9].

Ant Colony Optimization (ACO) is a population-based method loosely inspired by pheromone-based strategies of ant foraging. It was originally proposed by Dorigo in 1992 and, since then, it has been successfully applied to difficult optimization problems [5]. Actually, there are several variants of ACO algorithms with differences, e.g., in the way pheromone levels are updated throughout the optimization process. Adjusting the key components of an ACO algorithm may allow its application to new situations and/or enhance its effectiveness on problems that it usually addresses [1]. However, performing the right modifications is far from trivial and requires a deep understanding of both the algorithm's behavior and the properties of the problem to solve.

In this paper we propose a framework to automatically discover effective pheromone update strategies for an ACO algorithm. In our approach, a Genetic Programming algorithm (GP) [10] evolves a population of strategies, seeking for

S. Silva et al. (Eds.): EuroGP 2011, LNCS 6621, pp. 85–96, 2011.

candidates that can be used by an ACO method applied to the optimization of the Traveling Salesperson Problem (TSP). An accurate evaluation of individuals generated by the GP algorithm is a crucial step in the evolution and it must reflect how well they help ACO algorithms to discover TSP short tours. In concrete, the evaluation of each individual is done by inserting it in an ACO algorithm and verifying how well the encoded strategy behaves in the optimization of a simple TSP instance.

In a recent work, we reported a preliminary application of standard GP to evolve pheromone update methods [13]. Results obtained with the TSP showed that the evolved strategies performed well when compared to three simple ACO architectures: Ant System, Elitist Ant System and the Rank-based Ant System. Here we adopt a strongly typed GP variant (STGP) [7] and consider several combinations of functions and terminals sets used to build solutions. A comprehensive analysis of the GP algorithm behavior is accomplished by investigating how the evolved strategies perform when compared to the competitive Max-Min Ant System (MMAS) [12,5]. Different instances of the TSP will be used to assess the effectiveness of the proposed approach and we focus our analysis on three key issues: i) Study how the composition of the function and terminal sets impact the performance of the GP algorithm; ii) Compare the effectiveness of strategies that were evolved by the GP against the MMAS human developed strategies; iii) Verify the generalization ability of the evolved strategies. Results presented in this paper show that, even though the GP algorithm relies on a single TSP instance, evolved strategies maintain its effectiveness when applied to other instances.

The paper is structured as follows: in section 2 we present a general description of ACO algorithms. Section 3 comprises a detailed presentation of the system used to evolve pheromone update strategies. Section 4 contains the experimentation and analysis. Finally, in section 5 we summarize the conclusions and highlight directions for future work.

2   ACO Algorithms

The first ACO algorithm, Ant System (AS), was conceived to find the shortest path for the well-known TSP, but soon it was applied to several different types of combinatorial optimization problems [5]. To apply an ACO algorithm to a given problem, one must first define the solution components. A connected graph is then created by associating each component with a vertex and by creating edges to link vertices. Ants build solutions by starting at a random vertex and iteratively selecting edges to add new components. From a specific vertex, ants make a probabilistic choice of the new edge to cross. The probability of choosing an edge depends on the heuristic information and pheromone level of that specific path. Higher pheromone levels signal components that tend to appear in the best solutions already found by the colony. After completing a solution, ants provide feedback by depositing pheromone in the edges they just crossed. The amount of pheromone is proportional to the quality of the solution. To avoid stagnation, pheromone trail levels are periodically decreased by a certain factor. Following these simple rules until a termination criterion is met, a solution to the problem will emerge from the interaction and cooperation made by the ants.

MMAS is an ACO variant proposed by Stützle and Hoos [12]. It focuses on the exploitation of recent search history since only the best ant is allowed to update the pheromone trail according to the following rule:

$$\tau_{ij}(t+1) = \rho \times \tau_{ij}(t) + \frac{1}{f(s^{best})} \qquad (1)$$

where τ_{ij} is the pheromone level on edge joining solution components i and j, ρ is the evaporation rate and $f(s^{best})$ is the cost of the solution of the best ant. The selected ant might be the one that found the best solution in the current iteration or the one that found the best solution since the beginning of the run. Additionally, MMAS has a mechanism to limit the range of possible pheromone trail levels and it may perform a restart when no improvement is seen in a given number of iterations.

3   Evolving Pheromone Trail Update Methods

The framework used to evolve pheromone update strategies contains two components: a STGP engine and an AS algorithm. The main task of GP is to evolve individuals that encode effective trail update strategies, i.e., it aims to evolve a rule that replaces equation 1. It starts with a population of random strategies and iteratively seeks for enhanced solutions. The job of the AS algorithm is to assign fitness to each solution generated by GP: whenever an evolved pheromone update strategy needs to be evaluated, GP executes the AS algorithm to solve a given TSP instance (using the encoded strategy as the update policy). The result of the optimization is assigned as the fitness value of that individual.

3.1   Strongly Typed Genetic Programming Engine

The GP engine adopts a standard architecture: individuals are encoded as trees and ramped half-and-half initialization is used for creating the initial population. The algorithm follows a steady-state model, tournament selection chooses parents and standard genetic operators for manipulating trees are used to generate descendants. STGP is a variant of GP that enforces data type constraints in the evolved programs [7]. In STGP, each terminal has an assigned type and every function has a return type and a specified type for each of its arguments. Restrictions enforced by STGP provide an advantage over standard GP when dealing with situations that consider multiples data types. This is what happens with the problem addressed in this paper, as the GP algorithm must deal with, e.g., real values or sets of ants. In a previous work [13] we assumed the closure principle [10], but the analysis of results showed that forcing all functions to handle any possible data type hindered the performance of the optimization framework. Therefore, in this paper, we adopt the strong typing principle and hence assure that GP only generates trees satisfying type constraints (see [7] for details about the initialization procedure and the application of genetic operators).

A key decision in the development of the framework is the definition of the function and terminal sets used by GP, as they will determine which components can be used in the design of pheromone update strategies. Our aim is to show that the proposed approach is able to evolve such strategies. Therefore, to validate our ideas we keep the definition of the function and terminal sets as simple as possible. We consider three different function and terminals sets. The first set allows the replication of standard AS and EAS methods and is composed by:

- *(prog2 p1 p2)* and *(prog3 p1 p2 p3)*: Sequential execution of two or three functions/terminals. The last result is returned; all types are generic, i.e., any type is accepted.
- *(evaporate rate)*: Standard evaporation formula with a given *rate*. The rate is of type real and the return type is generic.
- *(deposit ants amount)*: ants deposit a given *amount* of pheromone (type integer). The parameter *ants* can be an array of ants or a single one (type ant). The return type is generic.
- *(all-ants)*, *(best-ant)*, *(rho)*: Return respectively an array with all ants (type ant), the best ant found so far or a fixed evaporation rate (type real).
- *(integer)* and *(real)*: Ephemeral integer and real constants.

The second and third sets differ from the first one in two ways. We remove an explicit way for all the ants to make a deposit, i.e., *(all-ants)* does not exist. In the second set, we add the function *(rank number)* that receives an integer and returns an array of ants. This function sorts the ants by decreasing quality and returns the best *number* of them. In the third set, this function is transformed into a terminal that always returns the 10% best ants. A comparison between the results obtained by the three different sets will help us to gain insight into the key features that help to evolve successful update strategies.

3.2   Related Work

There are several efforts for granting bio-inspired approaches the ability to self adapt their strategies. On-the-fly adaptation may occur just on the parameter settings or be extended to the algorithmic components. One pioneer example of self-adaptation is the well-known 1/5 success rule used to control the mutation strength for the (1+1)-ES. Hyper-heuristics [2] and multimeme strategies [6] deal with the development of the best combination of methods for a given problem. The concept of hyper-heuristics identifies a framework composed by an EA that evolves a combination of specific heuristics for effectively solving a problem. Multimeme strategies are memetic algorithms that learn on-the-fly which local search component should be used. In what concerns the adaptation of the optimization algorithm, Diosan and Oltean proposed an evolutionary framework that aims to evolve a full-featured Evolutionary Algorithm (EA) [4].

As for the Swarm Intelligence area, there are some reports describing the self-adaptation of parameter settings (see, e.g., [1,14]). Additionally, a couple of approaches resemble the framework proposed in this paper. Poli et. al [9] use GP to evolve the equation that controls particle movement in Particle Swarm Optimization (PSO). Diosan and Oltean also did some work with PSO structures [3]. Finally, Runka [11] applies GP to evolve the probabilistic rule used by an ACO variant to select the solution components in the construction phase.

4   Experiments and Analysis

Selected instances from the TSPLIB[1] are used in the experiments. For all tests, the GP settings are: Population size: 100; Maximum tree depth: 5; Crossover rate: 0.9; Mutation rate: 0.05; Tourney size: 3. For the AS algorithm used to evaluate GP individuals we adopt the standard parameters found in the literature [5]: number of ants is the number of cities, $\alpha = 1$, $\beta = 2$, $\rho = 0.5$. The number of runs for all experiments is 30.

4.1   Evolution of the Update Strategies

In the first set of experiments we aim to establish the ability of GP to evolve update strategies and to identify performance differences between the three function/terminal sets defined in the previous section. The GP algorithm relies on a single TSP instance to evolve the strategies (eil51 – a small instance with 51 cities). The main reason for this choice is that the evaluation of each evolved strategy requires the execution of an AS algorithm. Two parameters, the number of runs and the number of iterations per run, define the optimization effort of the AS. On the one hand, if we grant the AS a small optimization period to evaluate a GP individual, then it might not be enough to correctly estimate the quality of an update strategy. On the other hand, a longer evaluation period can increase computational costs to unbearable levels. For the eil51 instance, AS finds near-optimal solutions within approximately 100 iterations. Extensive testing allows us to conclude that a single AS run is enough to evaluate a GP individual. The fitness value sent back to the GP engine is given by the best solution found (see [13] for a discussion of other evaluation configurations).

Table 1. Results from GP evolution for 30 runs with 25 generations

	Hits	Best	MBF	Deviation	Average Depth	Average Nodes
STGP 1	0	427	430.03	4.90	4.10	13.77
STGP 2	23	426	439.17	26.77	4.07	9.64
STGP 3	30	426	426.00	0.0	3.77	10.06
GP 1	1	426	430.63	4.14	n/a	n/a
GP 2	6	426	446.33	43.38	n/a	n/a

---

[1] http://comopt.ifi.uni-heidelberg.de/software/TSPLIB95/

Table 1 contains the results of the STGP evolution using the three sets (STGP 1, STGP 2 and STGP 3). For completeness, we also present the evolution outcomes of standard GP, i.e., the non strongly typed GP described in [13]: the function/terminal sets of GP1 and GP2 are equivalent to STGP1 and STGP2, respectively. The column *Hits* displays the number of runs where evolved strategies were able to discover the optimal solution for eil51. A brief overview of these results reveals that there are noteworthy differences between the STGP sets. Whilst set 1 was unable to find a tree that helped AS to discover the optimum, the remaining sets performed well. Set 2 evolved trees that found the optimum in 23 runs and set 3 was successful on all the 30 runs.

As expected, the components used by STGP are relevant to the quality of the evolved strategies. The terminal/function sets available to STGP1 do not contain high level components granting immediate access to a set of good quality ants, thus complicating the evolutionary optimization. Also, poor results are probably amplified by the sharp experimental conditions (low number of generations performed by the GP and small maximum tree depth). On the contrary, ranking components used by STGP2 and STGP3 allow a simplified evolution of greedy update strategies. Results presented in table 1 confirm that greedy elements are important to enhance the effectiveness of ACO algorithms. The difference in performance between STGP2 and STGP3, particularly visible in the Mean Best Fitness (MBF) values, reveals that fixing the proportion of high

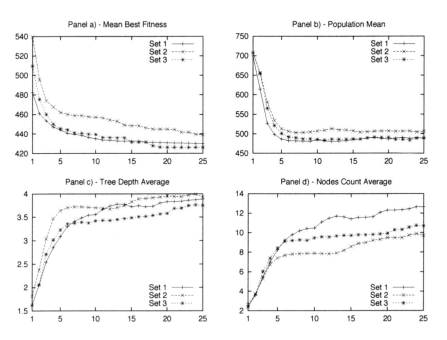

Fig. 1. STGP evolution plots with the 3 sets. Panels: a) Mean Best Fitness; b) Population Fitness; c) Tree Depth; d) Number of Nodes. Generations are on the x axis and fitness values are on the y axis in panel a) and b). Optimum solution has fitness 426.

quality ants further simplifies the discovery of effective strategies. Since GP does not have to learn an appropriate proportion of ants used to update pheromone levels. It is nevertheless important to notice that there is an obvious drawback in providing a fixed proportion for the rank function, as this might compromise the robustness of the evolved strategies. This effect is not visible in the results displayed in table 1, but it might be perceived when harder TSP instances are used. If we compare the results of STGP to the original GP version, one can see that the evolutionary pattern of STGP1 and GP1 is identical. On the contrary, STGP2 clearly outperforms GP2, both when we consider the number of hits or MBF. These results give evidence of the advantage provided by the addition of specific type constraints, particularly when combined with components that promote greedy update strategies.

The plots in Figure 1 summarize the evolutionary behavior of the GP engine. Panels a) and b) display, for the three sets, the evolution of the MBF and of the mean fitness of the population. The gradual discovery of increasingly effective update strategies is evident, confirming the usefulness of the GP exploration. The evolution of the tree depth is comparable in the three sets (panel c)), but there are some noteworthy differences in the average number of nodes (panel d)). As expected, strategies evolved by STGP1 use more nodes than those of other GP configurations, because only the best ant or all the ants are allowed to deposit pheromone. AS algorithms are greedy and, to explore this property, trees created by STGP1 require more nodes to reinforce the best solution trails. This situation does not happen in the other sets because the rank-based components allow a direct access to subsets of high quality ants.

4.2   Validation and Comparison with Max-Min Ant System

To confirm the effectiveness of the best evolved solutions we will describe a set of additional experiments. Specifically, we aim to: i) verify how update strategies behave in larger TSP instances; ii) assess the absolute optimization performance of evolved strategies, by comparing them with MMAS. To focus our analysis, 10 representative trees were selected from the set of the 53 evolved solutions that were able to discover the optimal solution for eil51. Trees are identified with a numerical ID: solutions obtained by STGP2 start with digit 2, whereas solutions discovered by STGP3 begin with digit 3. In figure 2 we present two examples of trees that were evolved by STGP3. To allow for a fair comparison, all

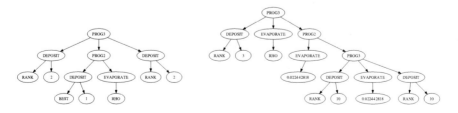

Fig. 2. Tree examples from set 3: on the left tree 326 and on the right tree 327

Table 2. Overview of the best trees, with 10000 iterations for 30 runs

Instance	Tree ID	Best	Dist	MBF	Deviation	Dist	Branching
	222	426	0.00	429.20	1.92	0.75	2.05
	224	426	0.00	426.17	0.37	0.04	3.15
	226	426	0.00	428.70	2.24	0.63	2.03
	230	426	0.00	429.00	2.16	0.70	2.09
eil51	31	426	0.00	426.97	1.11	0.23	2.34
	321	426	0.00	427.37	1.30	0.32	2.37
	326	426	0.00	430.90	3.24	1.15	2.07
	327	426	0.00	432.13	4.29	1.44	2.01
	329	426	0.00	426.50	0.56	0.12	2.36
	330	426	0.00	427.83	2.00	0.43	2.15
	Mean	426	0.00	428.48	1.92	0.58	2.26
	222	21282	0.00	21470.97	265.82	0.89	2.11
	224	21282	0.00	21315.30	28.68	0.16	3.06
	226	21282	0.00	21444.00	196.77	0.76	2.08
	230	21282	0.00	21308.20	57.59	0.12	2.19
kroA100	31	21282	0.00	21289.17	17.39	0.03	2.50
	321	21282	0.00	21293.20	21.90	0.05	2.50
	326	21282	0.00	21372.47	112.86	0.43	2.13
	327	21282	0.00	21582.87	317.85	1.41	2.03
	329	21282	0.00	21300.70	16.28	0.09	2.45
	330	21282	0.00	21312.87	55.29	0.15	2.30
	Mean	21282	0.00	21368.97	109.04	0.41	2.33
	222	16118	2.14	16340.97	114.86	3.55	3.02
	224	16075	1.87	16180.97	62.24	2.54	3.61
	226	15942	1.03	16061.13	70.23	1.78	2.61
	230	15901	0.77	16011.23	61.23	1.47	2.57
d198	31	16032	1.60	16197.67	72.98	2.65	3.27
	321	16009	1.45	16122.00	51.13	2.17	3.00
	326	15955	1.11	16031.80	63.22	1.60	2.45
	327	15849	0.44	15924.90	53.49	0.92	2.19
	329	16031	1.59	16199.87	74.02	2.66	2.82
	330	15944	1.04	16032.90	57.77	1.60	2.80
	Mean	15985.6	1.30	16110.34	68.12	2.09	2.83
	222	42468	1.04	42859.90	206.69	1.98	2.28
	224	43052	2.43	43388.27	211.70	3.23	3.28
	226	42223	0.46	42771.57	194.80	1.77	2.25
	230	42455	1.01	42792.23	199.48	1.82	2.46
lin318	31	42660	1.50	42919.60	159.26	2.12	2.73
	321	42507	1.14	43090.90	227.63	2.53	2.76
	326	42372	0.82	42817.07	214.48	1.88	2.35
	327	42203	0.41	43081.30	333.34	2.50	2.16
	329	42990	2.29	43514.30	271.44	3.53	2.68
	330	42231	0.48	42773.90	163.68	1.77	2.60
	Mean	42516.1	1.16	43000.90	218.25	2.31	2.56
	222	55452	9.20	56733.43	627.16	11.73	3.48
	224	56263	10.80	57386.23	498.18	13.01	3.71
	226	56199	10.68	56995.47	461.99	12.24	3.33
	230	51917	2.24	52755.30	382.93	3.89	2.85
pcb442	31	56108	10.50	57321.90	581.36	12.89	3.51
	321	54662	7.65	55754.27	645.12	9.80	3.51
	326	51804	2.02	52408.70	415.42	3.21	2.52
	327	51408	1.24	51870.83	274.58	2.15	2.22
	329	52815	4.01	53777.63	365.61	5.91	3.12
	330	54418	7.17	55788.73	724.40	9.87	3.43
	Mean	54104.6	6.55	55079.25	497.67	8.47	3.17

experiments presented in this section were performed in a MMAS environment. Therefore, evolved strategies were run in conjunction with a restart mechanism and a lower-upper bound update of the pheromone limits. Since the evolutionary process did not use these mechanisms, this is an additional robustness test for the evolved strategies. Five TSP instances were selected to assess the performance of the evolved methods: eil51, kroA100, d198, lin318 and pcb442. The size of each instance is given by the number in the name (e.g., lin318 has 318 cities). The AS maintains the same parameter settings used before, with two exceptions: ρ is set to 0.02 (the recommended value for MMAS) and the number of iterations is increased to 10000 to allow a correct optimization period for larger instances.

Table 2 contains the results for the 10 selected trees on the five TSP instances. This table displays the identification of the strategy (column *Tree ID*), the best solutions found, the MBF and standard deviation, with the respective percentage distance to the optimum for all these values (columns *Dist*). The last column is the average of the evolution of the branching factor (with $\lambda = 0.05$), a measure to determine the convergence of the pheromone matrix. For completeness, we add a row containing the average values for each instance. In general, results show that evolved strategies perform well across the instances, and thus, are able to generalize. For the two smaller instances, all strategies are able to find the optimum and the MBF is close to the optimal solution. For instances d198 and lin318, the best tour was not found but the solutions obtained by different trees are very close (between 0.41% and 2.43%). MBF values follow the same pattern, with distances still close to the optimum. As for the larger instance, the performance of the evolved strategies is not so effective. Nevertheless, some of the trees are able to generate good quality solutions. The most remarkable example is tree 327 with distances of 1.24% (best solution) and 2.15% (MBF). Two factors help to explain the performance differences visible in the larger instance. First, a visual inspection reveals that the most effective trees are the ones that include more evaporation instructions. This, coupled with the addition of the restart mechanism, prevents the pheromone matrix from saturating in certain positions and allows evolved strategies to avoid local minima. The difference in the composition of the function/terminal sets is another relevant issue. In the largest TSP instance, update strategies that were evolved by STGP3 tend to

Table 3. Comparison with MMAS, with 10000 iterations for 30 runs

Instance	AS	Best	Dist	MBF	Deviation	Dist	Branching
eil51	Tree 224	426	0.00	426.17	0.37	0.04	3.15
kroA100	Tree 31	21282	0.00	21289.17	17.39	0.03	2.50
d198	Tree 327	15849	0.44	15924.90	53.49	0.92	2.19
lin318	Tree 226	42223	0.46	42771.57	194.80	1.77	2.25
pcb442	Tree 327	51408	1.24	51870.83	274.58	2.15	2.22
eil51	MMAS	426	0.00	427.23	0.67	0.29	8.33
kroA100	MMAS	21431	0.70	21553.97	63.82	1.28	7.61
d198	MMAS	16141	2.29	16276.23	61.27	3.14	5.54
lin318	MMAS	45243	7.65	45912.77	335.12	9.24	6.11
pcb442	MMAS	58211	14.64	60009.20	714.16	18.18	4.40

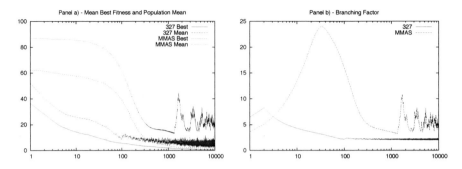

Fig. 3. Plots showing the evolution of the best solution, mean solutions and the branching factor for 10000 iterations (x axis) on the d198 instance

obtain better results. When compared to STGP2, the optimization task of STGP3 is simpler as it does not need to discover the proportion of high quality ants that update the pheromone matrix. For efficiency reasons, we selected a small TSP instance to evolve update strategies. Possibly this instance is too simple and it does not create enough evolutionary pressure to push STGP2 to discover the correct proportion of ants involved in pheromone level update. This inefficiency is not visible in easy instances, but as the number of cities increases, the performance of strategies evolved by STGP2 gradually decreases. The impact of the training instances on the quality and robustness of the evolved update strategies is a topic that we will address in future research.

A fundamental issue in the analysis is the comparison of the evolved solutions against human developed strategies. The top lines of table 3 contain the best results obtained by the evolved strategies on the selected TSP instances. The bottom half of the table shows the outcomes of our MMAS implementation without local search using standard parameter settings (see [5] for a complete description). Results clearly show that the evolved strategies outperform MMAS in every single instance and in every performance criteria (best solution found and MBF). With the exception of eil51, even the averages of all evolved trees from table 2 are better than MMAS results. The only weak point of the evolved solutions is the branching factor. MMAS consistently maintains a higher diversity level, suggesting that the evolved strategies might suffer from excessive greediness. In Figure 3 we can see the comparison between an evolved strategy and MMAS in instance d198. This is representative of all trees on all instances. Panel a) displays the evolution of the MBF and of the population mean (measured in % distance to the optimum). The standard behavior of MMAS is clearly seen: a slow improvement in the first 100 iterations, followed by a quick optimization until 1000 iterations and finally a slow improvement until the end. The evolved strategy performs a much quicker optimization attaining good quality solutions earlier. The restart mechanism is activated around iteration 100 with the evolved strategy, whereas for MMAS it only appears after 1000 iterations. The impact of restarts is higher on MMAS, as it seems to reintroduce a larger diversity in the

Table 4. Comparison with MMAS+LS [12], with n x 100 iterations for 30 runs

Variant	kroA100		d198		lin318		pcb442	
	MBF	Dist	MBF	Dist	MBF	Dist	MBF	Dist
Best Evolved	21289.17	0.03	15907.40	0.81	42586.73	1.33	51363.9	1.15
MMAS+LS	21481.00	0.94	16056.00	1.75	42934.00	2.15	52357.00	3.11
10+all+LS	21502.00	1.03	16197.00	2.64	43667.00	3.90	53993.00	6.33
10+best+LS	21427.00	0.68	15856.00	0.48	42426.00	0.94	51794.00	2.00

population. This is confirmed by the evolution of the branching factor, shown in panel b). The convergence of the pheromone matrix is stronger with the evolved strategy, thereby confirming its greedy behavior.

We also address the issue of using Local Search (LS). It is well-known that AS algorithms in general, and MMAS in particular, become more effective with the addition of a LS step. Our goal in this paper is to focus on the influence of pheromone update methods and not on the whole AS architecture. In spite of that, in table 4 we provide a simple performance comparison between the evolved strategies and MMAS with LS. The outcomes of the MMAS with LS were obtained in [12]. In that paper, the number of iterations performed is set to $n \times steps$, where n is the number of cites in the instance and $steps$ is a constant. Table 4 presents results when $steps$ is set to 100. The first row (*Best Evolved*) contains the best results obtained by the evolved strategies. The other rows display results from several MMAS+LS variants (see [12] for details). A brief perusal of the results reveals that the evolved strategies without LS are competitive with MMAS architectures with LS. They discover better solutions in instance kroA100 and, in the remaining instances, obtain tours close to those discovered by the most effective MMAS+LS variant. The indications given by this preliminary comparison are encouraging. Still, it is important to refer that the LS used is the 2-opt operator and more effective operators exist, e.g., the Lin-Kernighan heuristic.

5   Conclusions

We proposed a STGP framework to accomplish the automatic evolution of update pheromone strategies, a key component in the design of ACO algorithms. Results reveal that update strategies are effective for solving the TSP instance for which they were evolved, while they also exhibit a good generalization capability. Moreover, they are competitive in performance with MMAS and preliminary results also indicate that the evolved strategies without LS can be competitive with the standard MMAS using a basic LS operator. It must be pointed out that the system uses high-level function and terminal sets, resembling existing ACO methods. This choice undermines the possibility of evolving strategies that strongly deviate from standard ones. Moreover, the evolution is not too difficult as the key components are provided and GP just needs to find the proper arrangement. Nevertheless, the results show that evolution did not converge to the standard methods and found different designs in structure and in parameters.

The study presented here raises important questions that we intend to pursue in the near future. The evolved strategies show a greedy behavior, even more than standard AS architectures. We will study alternatives that help to reduce greediness, as we believe this will help to evolve more robust and scalable strategies. Also, the selection of the instance used to evaluate GP solutions plays an important role and we intend to investigate the impact of using instances of different size and/or distinct properties. These are all important aspects and this paper is a step forward in the effort of developing Self Ant Systems.

References

1. Botee, H.M., Bonabeau, E.: Evolving ant colony optimization. Advanced Complex Systems 1, 149–159 (1998)
2. Burke, E.K., Hyde, M., Kendall, G., Ochoa, G., Ozcan, E., Qu, R.: Hyper-heuristics: A survey of the state of the art. Tech. Rep. NOTTCS-TR-SUB-0906241418-2747, University of Nottingham (2010)
3. Diosan, L., Oltean, M.: Evolving the structure of the particle swarm optimization algorithms. In: Gottlieb, J., Raidl, G.R. (eds.) EvoCOP 2006. LNCS, vol. 3906, pp. 25–36. Springer, Heidelberg (2006)
4. Diosan, L., Oltean, M.: Evolutionary design of evolutionary algorithms. Genetic Programming and Evolvable Machines 10(3), 263–306 (2009)
5. Dorigo, M., Stützle, T.: Ant Colony Optimization. MIT Press, Cambridge (2004)
6. Krasnogor, N., Blackburnem, B., Hirst, J., Burke, E.: Multimeme algorithms for protein structure prediction. In: Guervós, J.J.M., Adamidis, P.A., Beyer, H.-G., Fernández-Villacañas, J.-L., Schwefel, H.-P. (eds.) PPSN 2002. LNCS, vol. 2439, pp. 769–778. Springer, Heidelberg (2002)
7. Montana, D.J.: Strongly typed genetic programming. Evolutionary Computation Journal 3(2), 199–230 (1995)
8. Oltean, M.: Evolving evolutionary algorithms using linear genetic programming. Evolutionary Computation Journal 13, 387–410 (2005)
9. Poli, R., Langdon, W.B., Holland, O.: Extending particle swarm optimisation via genetic programming. In: Keijzer, M., Tettamanzi, A.G.B., Collet, P., van Hemert, J., Tomassini, M. (eds.) EuroGP 2005. LNCS, vol. 3447, pp. 291–300. Springer, Heidelberg (2005)
10. Poli, R., Langdon, W.B., McPhee, N.F.: A field guide to genetic programming (2008); Published via http://lulu.com and freely available at http://www.gp-field-guide.org.uk (With contributions by J. R. Koza)
11. Runka, A.: Evolving an edge selection formula for ant colony optimization. In: GECCO 2009 Proceedings, pp. 1075–1082 (2009)
12. Stutzle, T., Hoos, H.: Max-min ant system and local search for the traveling salesman problem. In: ICEC Proceedings, pp. 309–314. IEEE Press, Los Alamitos (1997)
13. Tavares, J., Pereira, F.B.: Evolving strategies for updating pheromone trails: A case study with the tsp. In: Schaefer, R., Cotta, C., Kołodziej, J., Rudolph, G. (eds.) PPSN XI. LNCS, vol. 6239, pp. 523–532. Springer, Heidelberg (2010)
14. White, T., Pagurek, B., Oppacher, F.: ASGA: Improving the ant system by integration with genetic algorithms. In: Proceedings of the 3rd Genetic Programming Conference, pp. 610–617. Morgan Kaufmann, San Francisco (1998)

Random Lines: A Novel Population Set-Based Evolutionary Global Optimization Algorithm

İsmet Şahin[*]

Department of Materials Science and Engineering
Room 2123, Chemical and Nuclear Engineering Building (Bldg 90)
University of Maryland, College Park, MD 20742-2115 USA
isahin@gmail.com

Abstract. In this paper, we present a new population set-based evolutionary optimization algorithm which aims to find global minima of cost functions. This algorithm creates random lines passing through pairs of points (vectors) in population, fits a quadratic function based on three points on each line, and then applies the crossover operation to extrema of these quadratic functions, and lastly performs the selection operation. We refer to the points determining random lines as parent points and the extremum of a quadratic model as the descendant or mutated point under some conditions. In the crossover operation, some entries of a descendant vector are randomly replaced with the corresponding entries of one parent vector and some other entries of the descendant vector are replaced with the corresponding entries of the other parent vector based on the crossover constant. The above crossover and mutation operations make this algorithm robust and fast converging. One important property of this algorithm is that its robustness in general increases with increasing population size which may become useful when more processing units are available. This algorithm achieves comparable results with the well-known Differential Evolution (DE) algorithm over a wide range of cost functions.

Keywords: Global Optimization, Continuous Variable Optimization, Direct Search Methods, Evolutionary Computation, Random Lines, Differential Evolution.

1 Introduction

Global extrema of a function play an important role in design and analysis of many science applications [1]. Without loss of generality, we discuss only minimization of functions; therefore, we refer to these functions as cost functions. Direct search algorithms are global optimization algorithms which rely on only cost function evaluations

---

[*] The author works in direct collaboration with the NIST Center for Neutron Research, National Institute of Standards and Technology, 100 Bureau Drive, MS 6100, Gaithersburg, MD 20899-6100 USA. This work is supported in part by the US National Science Foundation under grant DMR-0520547.

S. Silva et al. (Eds.): EuroGP 2011, LNCS 6621, pp. 97–107, 2011.

for achieving minimization task [2]. As a result, these algorithms are very useful whenever derivative information (gradient or Hessian) about a cost function is not available due to its complexity, discontinuity, or nonnumeric structure. Multiple direct search algorithms [2] have been proposed and successfully used in literature including the Hooke and Jeeves algorithm [3] and the Downhill Simplex algorithm [4].

Population set-based algorithms are direct search algorithms which use multiple points simultaneously for creating variations and then selecting the variations yielding smaller function values than points of the current generation [5-11]. Points of the current generation are called target points and the points representing variations for this generation are called trial points. Variations are usually achieved by means of the mutation and crossover operations. One important example of population set-based algorithms is the DE algorithm [12,13] which is robust and efficient and therefore has been studied extensively and used successfully in many different applications [14-24]. The DE algorithm creates a mutated vector by choosing three different vectors, scaling the difference between two of these vectors, and then adding the scaled difference vector to the third vector [12]. In order to create a trial vector, the DE algorithm replaces some entries of the mutated vector with the entries of one parent vector in the crossover operation based on the crossover constant.

Quadratic functions have been used in many deterministic and stochastic optimization algorithms. The quasi-Newton algorithm in [25] finds a descent direction based on the BFGS (Broyden–Fletcher–Goldfarb–Shanno) Hessian update and performs an inexact search for the minimum along the line in this direction by using quadratic and cubic modals. The line search algorithm in [26] also makes use of quadratic interpolations. The stochastic algorithm Controlled Random Search [27] has a mutation operation which randomly chooses three points in population and fits a quadratic to these three points. A similar mutation operation for DE is also used in [28].

The proposed algorithm inherits main characteristics of population set-based direct search algorithms by using the mutation, crossover, and selection operations but it has major differences in using these operations compared to other population set-based algorithms. The most important difference is in the mechanism of creating variations. The proposed algorithm tries to learn cost function surface by fitting quadratic models and uses extrema of these quadratic models as mutated vectors under some conditions. This allows the algorithm to quickly locate the regions of search space with highly promising points and therefore to achieve fast convergence. In order to increase robustness, this algorithm uses a crossover operation for replacing some entries of a mutated vector with entries from two parent vectors rather than one vector as is usually performed in DE. Comparison of our algorithm with DE demonstrates its high efficiency and robustness over multiple cost functions.

2 Formulation

Consider a set of N-dimensional real vectors $X = \{x_1, x_2, ..., x_{N_P}\}$ with N_P elements. For each target point x_i in X, we randomly choose another point x_j in X and construct the line passing through x_i and x_j which are called parent points. This line can be represented by $x_i + \mu(x_j - x_i)$ where μ is a real number. In Fig. 1, a generation with five points are demonstrated where each pair of square points represent a specific x_i and the corresponding x_j. Next, we evaluate the function value at a randomly sampled point x_k from the line passing through x_i and x_j:

$$x_k = x_i + \mu_k(x_j - x_i) = x_i + \mu_k p \tag{1}$$

where $p = (x_j - x_i)$ and μ_k is a randomly chosen real number representing the step size constant, $\mu_k \neq 0$, $\mu_k \neq 1$. A step $\mu_k p$ from x_i toward x_j is taken if $\mu_k > 0$, otherwise a step in the opposite direction along the line is taken. We also note that the sampled point x_k is between x_i and x_j if $0 < \mu_k < 1$. Clearly, the point x_k will have larger distances from x_i for larger step sizes, therefore the local quadratic model may have larger mismatches with the underlying cost function surface. For this reason, we choose relatively small step sizes μ from the union of two intervals: [-0.95 -0.05] U [0.05 0.95] uniformly in this paper. Note that step sizes chosen from this distribution satisfy the conditions $\mu_k \neq 0$ and $\mu_k \neq 1$. The sampled points for the generation in Fig. 1 are represented by circle signs.

We consider one-dimensional real-valued function $\phi(\mu) = f(x_i + \mu p)$ which represents one-dimensional cross section of $f(x)$ along the line passing through x_i and x_j for the mutation operation. Since function values $\phi(0) = f(x_i)$, $\phi(1) = f(x_j)$, and $\phi(\mu_k) = f(x_i + \mu_k p)$ are known at three different points, a quadratic function $\hat{\phi}(\mu) = a\mu^2 + b\mu + c$ can be fit to $\phi(\mu)$. After some algebraic operations, we can show that the constraints $\hat{\phi}(0) = \phi(0)$, $\hat{\phi}(1) = \phi(1)$, and $\hat{\phi}(\mu_k) = \phi(\mu_k)$ uniquely determine the following coefficients of this quadratic model

$$a = \phi(1) - \phi(0) - b = -\frac{1}{\mu_k - 1}\phi(1) + \frac{1}{\mu_k}\phi(0) + \frac{1}{\mu_k(\mu_k - 1)}\phi(\mu_k)$$
$$b = \frac{\mu_k}{\mu_k - 1}\phi(1) - \frac{\mu_k + 1}{\mu_k}\phi(0) - \frac{1}{\mu_k(\mu_k - 1)}\phi(\mu_k) \tag{2}$$
$$c = \phi(0)$$

where $\mu_k \neq 0$ and $\mu_k \neq 1$. The critical point μ_* of this quadratic model is $\mu_* = -b/(2a)$, $a \neq 0$, and the corresponding point $x_* = x_i + \mu_* p$ represents a descendant or mutated point under some conditions. If $a > 0$, the quadratic model is convex with a unique minimum, and therefore x_* becomes the descendant point. Since all quadratic models in Fig. 1 are convex, their minima are the descendant points which are demonstrated with star signs. When $a < 0$, the model is concave with a maximum. In this case, we also use x_* as the descendant point if the function value at the sampled point is smaller than the function value of at least one parent point. This condition can be written as $a < 0$ and $[f(x_k) < f(x_i)$ or $f(x_k) < f(x_j)]$. If these conditions are not satisfied, x_* is not considered to be a descendant point and the target vector x_i

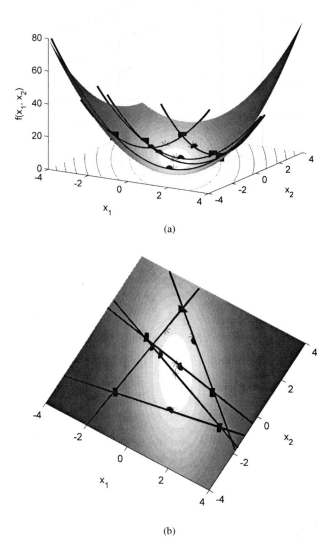

(a)

(b)

Fig. 1. Surface of the cost function Schwefel12 [21] is shown from side and top views in (a) and (b) respectively. Five square points represent the current generation. For each square point x_i, another square point x_j is chosen randomly in order to draw a line passing through both points. From each line a point x_k is randomly sampled and the function value at this point is evaluated. The sampled points are demonstrated with circle signs. Since function values at three different points on each line are known, the quadratic function passing through these three points is determined as shown in (a) and its minimum is marked with a star sign in the above figures. Star points constitute descendant points which are clearly located at lower altitudes of the cost function compared to the parent (square) points.

1	Create an initial generation $X_0 = \{x_1, x_2, ..., x_{N_P}\}$ and $X_c = X_0$
2	while (convergence is not achieved)
3	for each $x_i \in X_c$, $i = 1, 2, ..., N_P$
4	randomly choose $x_j \in X_c$, $x_i \neq x_j$ and evaluate $p = x_j - x_i$
5	choose a step size μ_k uniformly from the interval [-0.95 -0.05] U [0.05 0.95]
6	evaluate $f(x_k) = f(x_i + \mu_k p) = \phi(\mu_k)$
7	calculate a, b, and c by using equation (2)
8	if $a > 0$ or $[a < 0$ and $(f(x_k) < f(x_i)$ or $f(x_k) < f(x_j))]$
9	evaluate $\mu_* = -b/(2a)$, $x_* = x_i + \mu_* p$, and $f(x_*)$
10	calculate $\hat{x}_i$ by using the crossover operation in equation (3)
11	if $f(\hat{x}_i) < f(x_i)$
12	$\hat{x}_i \in X_+$
13	else
14	$x_i \in X_+$
15	end if
16	else
17	$x_i \in X_+$
18	end if
19	end for
20	$X_c = X_+$
21	end while

Fig. 2. The Random Lines (RL) Algorithm. The sign 'U' on line 5 denotes the union operation.

remains in the next generation. In addition to convex quadratic models, using concave quadratic models creates further diversity in search and therefore increases robustness of the proposed algorithm over multiple cost functions. When $a = 0$ we have a linear model rather than a quadratic model; also in this case, x_i remains in the next generation.

The crossover operation in our algorithm involves two parent points x_i and x_j, and the descendant point x_*. The trial vector $\hat{x}_i$ is determined based on the following rule:

$$\hat{\alpha}_k = \begin{cases} \alpha_k & \text{if } r_k \leq 0.5(1 - C_R) \\ \beta_k & \text{if } r_k \geq 0.5(1 + C_R) \\ \gamma_k & \text{if } 0.5(1 - C_R) < r_k < 0.5(1 + C_R) \end{cases} \quad (3)$$

where $\hat{\alpha}_k, \alpha_k, \beta_k, \gamma_k$, and r_k are k^{th} entries of vectors $\hat{x}_i, x_i, x_j, x_*$, and r respectively and C_R is the crossover constant chosen between 0 and 1. The vector r contains N entries, each of which is drawn randomly from the uniform distribution $U[0,1]$.

This rule means that $100 \cdot C_R$ percent of the trial vector $\hat{x}_i$ is determined by the descendant vector x_*, half of the remaining entries of the trial vector is determined by the first parent x_i and the other half is determined by the second parent x_j. For instance, if $C_R = 0.7$, the contributions of x_*, x_i, and x_j vectors to the trial vector $\hat{x}_i$ are 70, 15, and 15 percents on average respectively.

We evaluate function value $f(\hat{x}_i)$ at the trial point for performing the selection operation. If the trial point $\hat{x}_i$ achieves a smaller function value than the target point x_i, the trial point replaces the target point in the next generation, otherwise the target point remains in the next generation. In order to summarize this algorithm, let X_C and X_+ denote the current and next generations respectively. In Fig. 2, we summarize the steps of the proposed Random Lines (RL) algorithm.

3 Performance Evaluation

In this section, we compare performance of the RL algorithm with the well-known global optimization algorithm DE [12]. Since the DE/rand/1/bin variant of the DE algorithm is robust and fast convergent for different cost functions, this variant is often used in performance analysis in literature [5,8,12,21]. We also use this variant through this section. We use 30 cost functions listed in Table 1 including unimodal, multimodal, separable, non-separable, and badly scaled functions. Each algorithm minimizes each of these cost functions 100 times and its total number of successes N_S and average number of function evaluations N_F are recorded for comparing robustness and efficiency of these algorithms. The average number of function evaluations is calculated based on only successful runs. In order to compare number of function evaluations of the algorithms, we use the acceleration rate (AR) defined as the ratio of average number of function evaluations for RL to average number of function evaluations for DE [21]. The acceleration rate is only defined for the cost functions for which both algorithms achieve at least 30 successes since the average number of function evaluations may deviate largely for smaller number of successful runs.

Since both algorithms use parameters C_R and N_P, we choose the same parameter values $C_R = 0.9$ and $N_P = 10N$ for both algorithms. The DE algorithm also uses $F = 0.5$ for scaling difference vectors. Multiple studies [5, 21, 29, 30] also use similar DE parameters as the DE algorithm usually achieves larger number of successes with smaller number of function evaluations with these parameter values. Both algorithms use the Mersenne Twister random number generator [13,35]. We say that an algorithm is successful or it achieves convergence when the algorithm finds an acceptable point x_{best} satisfying $f(x_{best}) < f(x_{opt}) + \varepsilon$ where $\varepsilon = 10^{-5}$ and x_{opt} is the global minimizer. The algorithm is unsuccessful if it cannot find an acceptable point over 3000000 function evaluations. The algorithm is also unsuccessful if it cannot reduce the best function value over 500 generations.

Table 1. Cost functions and their limits and references. Here [a,b] means that each component of a vector in the initial generation is in interval [a,b]. Dimensions of functions are specified in paranthesis in first column.

Cost Functions	Limits	References
Ackley(N)	[-30, 30]	[21]
Alpine(N)	[-10, 10]	[21]
Beale(2)	[-10, 10]	[21]
Branin(2)	$x_1 \in$[-5, 10], $x_2 \in$[0, 15]	[21]
Brown Badly Scaled(2)	[-1e7, 1e7]	[32]
Camel6(2)	[-5, 5]	[21]
Colville(4)	[-10, 10]	[21]
Cubic Valley(2)	[-100, 100]	[34]
Dejong4(N)	[-1.28, 1.28]	[21]
GoldsteinPrice(2)	[-2, 2]	[33]
Griewangk(N)	[-600, 600]	[21]
Hartman3(3)	[0, 1]	[21]
Hartman6(6)	[0, 1]	[21]
Hyperellipsoid(N)	[-5.12, 5.12]	[21]
Kowalik(4)	[-5, 5]	[21]
Matyas(2)	[-10, 10]	[21]
Powell Badly Scaled(2)	[-10, 10]	[32]
Rastrigin(N)	[-5.12, 5.12]	[21]
Rosenbrock(N)	[-2.048, 2.048]	[21]
Schwefel12(N)	[-65, 65]	[21]
Schwefel221(N)	[-100, 100]	[21]
Schwefel222(N)	[-10, 10]	[21]
Shekel5(4)	[0, 10]	[21]
Shekel7(4)	[0, 10]	[21]
Shekel10(4)	[0, 10]	[21]
Shekel's Foxhole(5)	[0, 10]	[5]
Sphere(N)	[-5.12, 5.12]	[21]
Step(N)	[-100, 100]	[21]
Sum.Diff.Powers(N)	[-1, 1]	[21]
Zakharov(N)	[-5, 10]	[21]

Table 2 lists the number of successes and average number of function evaluations for DE and RL. For these results, we use two dimensional forms of the cost functions which can be extended to higher dimensions such as the Ackley and Rosenbrock's functions. Comparison of the results under the population size $N_P = 10N$ shows that RL achieves 2824 successes and DE achieves 2525 successes over 3000 runs. Both algorithms have the same number of successes over 16 functions. RL has larger number of successes than DE over 10 functions and DE has larger number of successes than RL over 4 cost functions. Both algorithms have at least 30 successes over 27 functions and DE is slightly more efficient than RL by requiring smaller number of function evaluations for 15 of these cost functions.

When the population size increases from $N_P = 10N$ to $N_P = 40N$, both algorithms have larger number of successes and function evaluations. RL converges successfully for 2904 runs and DE converges for 2659 runs over 3000 runs. They have the same number of successes for 26 cost functions. RL achieves larger number of successes than DE for the Brown, Hartman 6, and Powell cost functions, and DE has larger number of successes than RL for the Shekel's Foxholes function. From the last AR column,

Table 2. The number of successes and average number of function evaluations for DE and RL. Hyphens signify that there is no successful run for the corresponding cost function and therefore N_F and **AR** are not defined. Extended functions such as Rosenbrock are 2-dimensional.

Cost Functions	DE $N_P = 10N$		RL $N_P = 10N$		AR	DE $N_P = 40N$		RL $N_P = 40N$		AR
	N_S	N_F	N_S	N_F		N_S	N_F	N_S	N_F	
Ackley	100	1165	100	2472	0.47	100	4479	100	6719	0.67
Alpine	100	1269	100	2752	0.46	100	5502	100	8672	0.63
Beale	99	691	100	733	0.94	100	2456	100	1819	1.35
Branin	100	799	100	415	1.93	100	3495	100	1211	2.89
Brown	--	--	100	2259	--	2	15240	100	9099	--
Camel	100	673	100	508	1.32	100	2690	100	1652	1.63
Colville	70	4840	100	8336	0.58	100	20259	100	28361	0.71
Cube	58	1443	100	4350	0.33	100	5487	100	10198	0.54
De Jong 4	100	149	100	114	1.31	100	387	100	313	1.24
Golds. Price	100	685	100	659	1.04	100	2458	100	1926	1.28
Griewangk	92	1765	57	2587	0.68	100	7115	100	6375	1.12
Hartman3	100	990	100	1142	0.87	100	3604	100	3454	1.04
Hartman6	35	5929	100	6146	0.96	36	29753	100	17075	1.74
Hyperellips.	100	488	100	183	2.67	100	1739	100	523	3.33
Kowalik	99	3318	100	1297	2.56	100	4265	100	2915	1.46
Matyas	100	450	100	138	3.26	100	1612	100	396	4.07
Powell	--	--	83	5980	--	--	--	100	23934	--
Rastrigin	95	1123	96	1556	0.72	100	4494	100	2808	1.60
Rosenbrock	83	670	100	1339	0.50	100	2517	100	4202	0.60
Schw. 1.2	100	695	100	238	2.92	100	2546	100	697	3.65
Schw. 2.21	100	1178	100	3136	0.38	100	4536	100	7141	0.64
Schw. 2.22	100	1000	100	1862	0.54	100	3816	100	4850	0.79
Shekel 5	90	3468	95	7811	0.44	100	13852	100	20348	0.68
Shekel 7	100	3242	96	7148	0.45	100	12787	100	19839	0.64
Shekel 10	100	3289	97	6647	0.49	100	12931	100	18995	0.68
Shek. Foxh.	4	5987	--	--	--	21	27219	4	134205	--
Sphere	100	475	100	179	2.65	100	1686	100	507	3.33
Step	100	289	100	137	2.11	100	950	100	375	2.53
Sum of pow.	100	264	100	140	1.89	100	820	100	374	2.19
Zakharov	100	534	100	297	1.80	100	1906	100	963	1.98

notice that RL requires smaller number of function evaluations than DE over 17 functions and DE requires smaller number of evaluations than RL over 10 functions.

The number of successes and average number of function evaluations for 20 dimensional cost functions are given in Table 3. Since DE performance with $N_P = 30N$ and $N_P = 45N$ are usually worse than $N_P = 15N$ for these cost functions, only the DE results under $N_P = 15N$ are listed. Comparison of the results with $N_P = 15N$ shows that RL achieves slightly larger total number of successes than RL as they converge over 1219 and 1200 runs respectively out of 1400 runs. RL converges over 100 and 76 runs for the Alpine and Rastrigin cost functions respectively as DE does not converge for these cost functions. However, DE achieves 100 successes for the Step function while RL has 7 successes for this function. The sixth column specifies the acceleration rates for the results under $N_P = 15N$. From this column, notice that

Table 3. The number of successes and average number of function evaluations for DE and RL with 20-dimensional extended cost functions. Hyphens have the same meaning with hyphens in Table 2.

Cost Functions	DE $N_P = 15N$		RL $N_P = 15N$		AR	RL $N_P = 30N$		RL $N_P = 45N$	
	N_S	N_F	N_S	N_F		N_S	N_F	N_S	N_F
Ackley	100	294489	94	196893	1.50	100	292128	100	402248
Alpine	--	--	100	120937	--	100	208987	100	299383
Dejong4	100	82614	100	29052	2.84	100	28932	100	32616
Griewangk	100	436881	42	149311	2.93	68	289330	64	330154
Hyperellips.	100	169137	100	97157	1.74	100	161616	100	222191
Rastrigin	--	--	76	136360	--	100	222571	100	309498
Rosenbrock	100	500196	100	877172	0.57	100	1412765	100	1931361
Schwefel12	100	609909	100	294425	2.07	100	458028	100	613619
Schwefel221	100	578682	100	690687	0.84	100	747657	100	892602
Schwefel222	100	344670	100	137122	2.51	100	241600	100	343211
Sphere	100	148458	100	79955	1.86	100	134016	100	183455
Step	100	101364	7	73548	--	40	96062	71	121155
Sum of pow.	100	42546	100	3720	11.44	100	7092	100	10332
Zakharov	100	534828	100	216276	2.47	100	348900	100	477540

RL is more efficient over 9 functions and DE is more efficient over 2 cost functions. In particular, notice that DE requires approximately 11.4 times more function evaluations for the Sum of Powers function, 2.8 times more function evaluations for the Dejonk4 and Griewangk functions, and 2.5 times more function evaluations for the Schwefel222 and Zakharov functions than the RL algorithm.

Robustness of the RL algorithm in general increases with increasing population size. The number of successes for RL increases for the Ackley, Rastrigin, and Step functions with increasing population size in Table 3. Even though the number of function evaluations increases with increasing population size, the RL algorithm simultaneously achieves higher efficiency and robustness than DE. The total number of successes increases to 1308 and 1335 when population size is increased to $N_P = 30N$ and $N_P = 45N$ respectively for RL. For the population size $N_P = 30N$, RL is more efficient over 10 cost functions than DE and DE is more efficient than RL over 2 functions.

4 Conclusion

In this paper, we presented a new population set-based global optimization algorithm. The mutation operation of this algorithm makes an effort to learn cost function surface by constructing random lines passing through pairs of points in the current generation and then fitting a quadratic function by using three points on each line. The extrema of these quadratic models constitute descendant points which are subject to the crossover operation. The mutation operation with quadratic fit quickly finds regions of the search space with highly promising points and therefore allows this algorithm to achieve fast convergence. The crossover operation randomly selects components from two parents and replaces corresponding components of the descendant point. Using components from both parents increases diversity in search and therefore increases robustness. Minimization of 30 cost functions demonstrates that this algorithm achieves very promising results compared with the well-known DE algorithm. In particular, the robustness of this algorithm in general increases with increasing population size which becomes important when there are more processing units available.

Acknowledgements. I would like to thank Paul Kienzle, Florencia McAllister, Bulent Akgun, and Nuri Yilmazer for their useful comments.

References

1. Törn, A., Zilinskas, A.: Global Optimization. Springer, New York (1989)
2. Kolda, T.G., Lewis, R.M., Torczon, V.: Optimization by direct search: New perspectives on some classical and modern methods. SIAM Review 45(3), 385–482 (2003)
3. Hooke, R., Jeeves, T.A.: Direct Search Solution of Numerical and Statistical Problems. Journal of the ACM 8(2), 212–229 (1961)
4. Nelder, J.A., Mead, R.: A Simplex Method for Function Minimization. Computer Journal 7(4), 308–313 (1965)
5. Ali, M.M., Törn, A.: Population Set-based Global Optimization Algorithms: Some Modifications and Numerical Studies. Computers and Operations Research 31(10), 1703–1725 (2004)
6. Back, T., Schewefel, H.P.: An Overview of Evolutionary Algorithms for parameter optimization. Evolutionary Computation 1(1), 1–23 (1993)
7. Eiben, A.E., Smith, J.E.: Introduction to Evolutionary Computing. Springer, New York (2003)
8. Yao, X., Liu, Y., Lin, G.: Evolutionary Programming Made Faster. IEEE Transactions on Evolutionary Computation 3(2), 82–102 (1999)
9. Back, T., Hammel, U., Schwefel, H.P.: Evolutionary Computation: Comments on the History and Current State. IEEE Transactions on Evolutionary Computation 1(1), 3–17 (1997)
10. Fogel, D.B.: What Is Evolutionary Computation? IEEE Spectrum 37(2), 28–32 (2000)
11. Hinterding, R., Michalewicz, Z., Eiben, A.E.: Adaptation in Evolutionary Computation: A Survey. In: IEEE International Conference on Evolutionary Computation, pp. 65–69. IEEE Press, Los Alamitos (1997)
12. Storn, R., Price, K.: Differential Evolution – A Simple and Efficient Heuristic for Global Optimization over Continuous Spaces. Journal of Global Optimization 11(4), 341–359 (1997)
13. Price, K.V., Storn, R.M., Lampinen, J.A.: Differential Evolution: A Practical Approach to Global Optimization. Springer, Heidelberg (2005)
14. Veenhuis, C.B.: Tree Based Differential Evolution. In: Vanneschi, L., Gustafson, S., Moraglio, A., De Falco, I., Ebner, M. (eds.) EuroGP 2009. LNCS, vol. 5481, pp. 208–219. Springer, Heidelberg (2009)
15. Brest, J., Greiner, S., Boskovic, B., Mernik, M., Zumer, V.: Self-Adapting Control Parameters in Differential Evolution: A Comparative Study on Numerical Benchmark Problems. IEEE Transactions on Evolutionary Computation 10(6), 646–657 (2006)
16. Montgomery, J., Chen, S.: An Analysis of the Operation of Differential Evolution at High And Low Crossover Rates. In: IEEE Congress on Evolutionary Computation, pp. 1–8. IEEE Press, Los Alamitos (2010)

17. Caponio, A., Neri, F.: Differential Evolution with Noise Analyzer. In: Giacobini, M., Brabazon, A., Cagnoni, S., Di Caro, G.A., Ekárt, A., Esparcia-Alcázar, A., Farooq, M., Fink, A., Machado, P. (eds.) EvoWorkshops 2009. LNCS, vol. 5484, pp. 715–724. Springer, Heidelberg (2009)
18. Liu, G., Li, Y.X., He, G.L.: Design of Digital FIR Filters Using Differential Evolution Algorithm Based on Reserved Genes. In: IEEE Congress on Evolutionary Computation, pp. 1–7. IEEE Press, Los Alamitos (2010)
19. Das, S., Konar, A.: Automatic Image Pixel Clustering with an Improved Differential Evolution. Applied Soft Computing 9(1), 226–236 (2009)
20. Shi, Y., Teng, H., Li, Z.: Cooperative Co-evolutionary Differential Evolution for Function Optimization. In: Wang, L., Chen, K., Ong, Y.S. (eds.) ICNC 2005. LNCS, vol. 3611, pp. 1080–1088. Springer, Heidelberg (2005)
21. Rahnamayan, S., Tizhoosh, H.R., Salama, M.M.A.: Opposition-Based Differential Evolution. IEEE Transactions on Evolutionary Computation 12(1), 64–79 (2008)
22. Kundu, D., Suresh, K., Ghosh, S., Das, S., Abraham, A., Badr, Y.: Automatic Clustering Using a Synergy of Genetic Algorithm and Multi-objective Differential Evolution. In: Corchado, E., Wu, X., Oja, E., Herrero, Á., Baruque, B. (eds.) HAIS 2009. LNCS, vol. 5572, pp. 177–186. Springer, Heidelberg (2009)
23. Abbass, H.A., Sarker, R., Newton, C.: PDE: a Pareto-frontier Differential Evolution Approach for Multi-objective Optimization Problems. In: Proceedings of the 2001 Congress on Evolutionary Computation, Seoul, South Korea, vol. 2, pp. 971–978. IEEE Press, Los Alamitos (2001)
24. Vesterstroem, J., Thomsen, R.: A Comparative Study of Differential Evolution, Particle Swarm Optimization, and Evolutionary Algorithms on Numerical Benchmark Problems. Proc. Congr. Evol. Comput. 2, 1980–1987 (2004)
25. Dennis, J.E., Schnabel, R.B.: Numerical Methods for Unconstrained Optimization and Nonlinear Equations. Prentice-Hall, Inc., Englewood Cliffs (1983)
26. More, J.J., Thuente, D.J.: Line Search Algorithms with Guaranteed Sufficient Decrease. ACM Transactions on Mathematical Software 20, 286–307 (1992)
27. Mohan, C., Shanker, K.: A Controlled Random Search Technique for Global Optimization Using Quadratic Approximation. Asia-Pacific Journal of Operational Research 11, 93–101 (1994)
28. Thangaraj, R., Pant, M., Abraham, A.: New Mutation Schemes for Differential Algorithm and Their Application to the Optimization of Directional Over-current Relay Settings. Applied Mathematics and Computation 216, 532–544 (2010)
29. Liu, J., Lampinen, J.: A Fuzzy Adaptive Differential Evolution Algorithm. Soft Computing-A Fusion of Foundations, Methodologies and Applications 9(6), 448–462 (2005)
30. Brest, J., Greiner, S., Boskovic, B., Mernik, M., Zumer, V.: Selfadapting Control Parameters in Differential Evolution: A Comparative Study on Numerical Benchmark Problems. IEEE Transactions on Evolutionary Computation 10(6), 646–657 (2006)
31. Storn, R.: On the Usage of Differential Evolution for Function Optimization. In: Proc. Biennial Conf. North Amer. Fuzzy Inf. Process. Soc., pp. 519–523 (1996)
32. More, J.J., Garbow, B.S., Hillstrom, K.E.: Testing Unconstrained Optimization Software. ACM Transactions on Mathematical Software 7(1), 17–41 (1981)
33. Ali, M.M., Khompatraporn, C., Zabinsky, Z.B.: A Numerical Evaluation of Several Stochastic Algorithms on Selected Continuous Global Optimization Test Problems. Journal of Global Optimization 31, 635–672 (2005)
34. Pierre, D.A.: Optimization Theory with Applications. Dover Publications Inc., Mineola (1969)
35. Matsumoto, M., Nishimura, T.: Mersenne twister: a 623-dimensionally equidistributed uniform pseudo-random number generator. ACM Transactions on Modeling and Computer Simulation 8(1), 3–30 (1998)

A Peer-to-Peer Approach to Genetic Programming

Juan Luis Jiménez Laredo[1], Daniel Lombraña González[2],
Francisco Fernández de Vega[2], Maribel García Arenas[1], and
Juan Julián Merelo Guervós[1]

[1] University of Granada, ATC-ETSIIT
Periodista Daniel Saucedo Aranda s/n 18071, Granada, Spain
{juanlu,mgarenas,jmerelo}@geneura.ugr.es
[2] University of Extremadura
Avda. Sta. Teresa de Jornet n38 06800, Mérida, Spain
{daniellg,fcofdez}@unex.es

Abstract. This paper proposes a fine-grained parallelization of the Genetic Programming paradigm (GP) using the Evolvable Agent model (EvAg). The algorithm is decentralized in order to take full-advantage of a massively parallel Peer-to-Peer infrastructure. In this context, GP is particularly demanding due to its high requirements of computational power. To assess the viability of the approach, the EvAg model has been empirically analyzed in a simulated Peer-to-Peer environment where experiments were conducted on two well-known GP problems. Results show that the spatially structured nature of the algorithm is able to yield a good quality in the solutions. Additionally, parallelization improves times to solution by several orders of magnitude.

1 Introduction

Within the Evolutionary Computing area (EC), Genetic Programming (GP) applications are specially demanding as a consequence of the high cost associated to evaluating computer programs and the large population sizes required to guarantee a reliable convergence. In fact, the computational power required to tackle many real-life problems in GP may become so high as to prevent finding optimal solutions in reasonable time. For instance, Trujillo and Olague describe in [16] a computer vision problem using GP that needs more than twenty four hours in order to obtain a solution in a single computer. Times to solution can be even worse, lasting days, weeks or even months.

Fortunately, the nature of evolutionary algorithms is inherently suited to be parallelized offering a straightforward way to improve speed by just adding more computers. The main idea is to speed-up the execution times by sharing the workload of the evaluation of individuals among a pool of processors [5].

In that context, we proposed in [11] the Evolvable Agent model (EvAg) for tackling the challenge of massive parallelization of EAs via a fine-grained parallelization of the population in a Peer-to-Peer infrastructure (P2P) [15]. The

S. Silva et al. (Eds.): EuroGP 2011, LNCS 6621, pp. 108–117, 2011.

motivation behind the algorithm is tackling demanding instances of hard optimization problems in an efficient and accurate way via massive scalability of P2P systems. However, such a computing platform is devoid of any central server which challenges the central management of the evolutionary cycle (parent selection, breeding and rewarding). To address this issue, the EvAg model designates each individual as a peer and adopt a decentralized population structure defined by the P2P protocol newscast [9]. Then, any given individual has a limited number of neighbors and the mating choice is restricted within the local P2P neighborhood.

The massive use of such distributed environments is justified from the perspective of demanding EC applications and has received much attention from the scientific/technical community within the last decade. Under the term of P2P optimization, many optimization heuristics have been re-designed in order to take advantage of such computing platforms.

In this sense, Wickramasinghe et al. in [19] and Laredo et al. in [11] present respective fine-grained approaches to tackle Genetic Algorithms. Bánhelyi et al. propose in [1] a P2P parallelization of a Particle Swarm Optimization (PSO) and Branch and Bound algorithms. In the same line of P2P PSO, Scriven et al. present in [14] dynamic initialization strategies for multi-objective problems. Biazzini and Montresor propose in [2] a gossiping Differential Evolution algorithm able to adapt to network scale and to achieve good results even in presence of a high churn. Finally, and to the extent of our knowledge, only Folino and Spezzano have tried to parallelize Genetic Programming using a P2P philosophy in [6]. The proposal consists of a hybrid model combining islands with cellular EAs. Nevertheless, the scalability analysis is limited to ten peers and the algorithm yields the best performance with five peers which points to poor scalability.

Therefore, this paper aims to extend the study of viability of P2P GP but using thousands of peers for computation. To that end, the EvAg model has been empirically analyzed in a simulated P2P environment where experiments are conducted in two well known GP problems (tackled by Koza in [10]), the 11 bits multiplexer and the Even Parity 5. Results show that the algorithmic performance of the EvAg model is competitive with respect to a canonical approach. Additionally, times to solution are dramatically reduced at the fine-grained parallelization of individuals in the population.

The rest of the paper is structured as follows. A description of the EvAg model is presented in Section 2, in addition, Section 3 describes the main design components of the newscast protocol which acts as underlying population structure of the approach. Section 4 includes the experimental analysis of the model in two GP problems. Finally, we reach some conclusions and propose some future lines of work in Section 5.

2   Description of the Model

The EvAg model (proposed by the authors in [11]) is a fine-grained spatially structured EA in which every agent schedules the evolution process of a single individual and self-organizes its neighborhood via the newscast protocol.

As explained by Jelasity and van Steen in [9], newscast runs on every node and defines the self-organizing graph that dynamically maintains constant some graphs properties at a virtual level such as a low average path length or a high clustering coefficient from which emerges a small-world behavior [18]. This makes the algorithm inherently suited for parallel execution in a P2P system which, in turn, offers great advantages when dealing with computationally expensive problems at the expected speedup of the algorithm.

Every agent acts at two different levels; the evolutionary level for carrying out the main steps of evolutionary computation (selection, variation and evaluation of individuals [4]) and the neighbor level that has to adopt a neighbor policy for the population structure. In principle, the method places no restrictions on the choice of a population structure, however, such a choice will have an impact on the dynamics of the algorithm. In this paper, the newscast protocol will be considered as neighborhood policy (explained in Section 3).

Algorithm 1 shows the pseudo-code of an $EvAg_i \in [EvAg_1 \ldots EvAg_n]$ where $i \in [1 \ldots n]$ and n is the population size. Despite the model not having a population in the canonical sense, neighbors EvAgs provide each other with the genetic material that individuals require to evolve.

Algorithm 1. Pseudo-code of an Evolvable Agent $(EvAg_i)$

Evolutionary level

$Ind_{current_i} \Leftarrow$ Initialize Agent
while not *termination condition* **do**
    $Pool_i \Leftarrow$ Local Selection$(Neighbors_{EvAg_i})$
    $Ind_{new_i} \Leftarrow$ Recombination$(Pool_i, P_c)$
    Evaluate(Ind_{new_i})
    if Ind_{new_i} better than $Ind_{current_i}$ **then**
        $Ind_{current_i} \Leftarrow Ind_{new_i}$
    end if
end while

Local Selection$(Neighbors_{EvAg_i})$
$[Ind_{current_h} \in EvAg_h, Ind_{current_k} \in EvAg_k] \Leftarrow$ Random selected nodes from the newscast neighborhood

---

The key element at this level is the locally executable selection. Crossover and mutation never involve many individuals, but selection in EAs usually requires a comparison among all individuals in the population. In the EvAg model, the mate selection takes place locally within a given neighborhood where each agent selects the current individuals from other agents (e.g. $Ind_{current_h}$ and $Ind_{current_k}$ in Algorithm 1).

Selected individuals are stored in $Pool_i$ ready to be used by the recombination (and eventually mutation) operator. Within this process a new individual Ind_{new_i} is generated.

In the current implementation, the replacement policy adopts a replace if worst scheme, that is, if the newly generated individual Ind_{new_i} is better than the current one $Ind_{current_i}$, $Ind_{current_i}$ becomes Ind_{new_i}, otherwise, $Ind_{current_i}$ remain the same for the next generation.

Finally, every EvAg iterates until a termination condition is met.

3   Neighbor Selection Policy: Newscast

As mentioned in the previous section, newscast is the canonical underlying P2P protocol in the EvAg model. It represents a good example of purely decentralized protocol that has been shown to succeed in the main issues related to P2P computing such as massive scalability or fault tolerance [17]. This section describes the design of its main components.

Newscast is a self-organized gossiping protocol for the maintenance of dynamic unstructured P2P overlay networks proposed by Jelasity and van Steen in [9]. Without any central services or servers, newscast differs from other similar approaches (e.g. [8,13]) by its simplicity and performance:

1. The membership management follows a extremely simple protocol: In order to join a group, a given node just has to contact any node within the system from which gets a list of its neighbors members. Additionally, to leave the group, it will just stop communicating for a predefined time.
2. The dynamics of the system follow a probabilistic scheme able to keep a self-organized equilibrium at a macroscopic level. Such an equilibrium emerges from the loosely-coupled and decentralized run of the protocol within the different and independent nodes. The emerging macro-structure behaves as a small-world [18] allowing a scalable way for disseminating information and, therefore, making the system suitable for distributed computing.
3. Despite the simplicity of the scheme, newscast is fault-tolerant and exhibits a graceful degradation without requiring an extra mechanism other than its own emergent macro-behavior.

The small-world features emerging from the collective dynamics of the protocol has been shown in [12] to induce similar environmental selection pressures on the algorithm than panmictic populations, however, showing a much better scalability at the smaller node degree of small-world population structures.

Algorithm 2 shows the newscast protocol in an agent $EvAg_i$. There are two different tasks that the algorithm carries out within each node. The active thread which pro-actively initiates a cache exchange once every cycle and the passive thread that waits for data-exchange requests (the cache consists in a routing table pointing to neighbor nodes of $EvAg_i$).

Every cycle each $EvAg_i$ initiates a cache exchange. It selects uniformly at random a neighbor $EvAg_j$ from its $Cache_i$. Then $EvAg_i$ and $EvAg_j$ exchange their caches and merge them following an aggregation function. In this case, the aggregation consists of picking the freshest c items (i.e. c is the maximum degree of a node. In this paper $c = 40$) from $Cache_i \cup Cache_j$ and merging them into a single cache that will be replicated in $EvAg_i$ and $EvAg_j$.

Within this process, every EvAg behaves as a virtual node whose neighborhood is self-organized at a virtual level with independence of the physical network (i.e. overlay network in Figure 1). In this paper, we will assume the ideal case in which every peer host a single EvAg.

Algorithm 2. Newscast protocol in $EvAg_i$

Active Thread
loop
   wait one cycle
   $EvAg_j \Leftarrow$ Random selected node from $Cache_i$
   send $Cache_i$ to $EvAg_j$
   receive $Cache_j$ from $EvAg_j$
   $Cache_i \Leftarrow$ Aggregate $(Cache_i, Cache_j)$
end loop

Passive Thread
loop
   wait $Cache_j$ from $EvAg_j$
   send $Cache_i$ to $EvAg_j$
   $Cache_i \Leftarrow$ Aggregate $(Cache_i, Cache_j)$
end loop

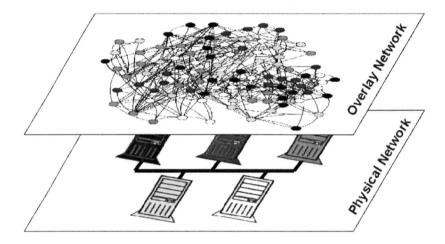

Fig. 1. Layered view of a P2P environment showing that every physical node can host more than a single EvAg. Same colors in virtual and physical nodes represent the mapping between both networks.

4   Experimental Setup and Results

In order to assess the viability of the P2P GP approach, we propose the experimental analysis of the model in a simulated P2P environment[1]. On one hand, simulations allow to reproduce dynamics of a massively scalable infrastructure at run-time of up to thousands peers; on the other hand, they allow the trace of experiments, otherwise hard to monitor (or even impossible) in fully decentralized environments as they are P2P systems.

That way, experiments were conducted for two well-known GP problems, the 11 bits multiplexer (11M) and the even parity 5 (EP5). The 11M problem consists

---

[1] Source code for the experiments is available at http://forja.rediris.es/svn/geneura/peerevogen, published under GPL v3 public license.

Table 1. Parameters of the algorithms

Population Size (P)	4000 individuals
Generations	51
Elitism	Yes
Crossover Probability	0.9
Reproduction Probability	0.1
Selection	Tournament (7)
Max Depth of Tree	17
Problem instances	11 Bits Multiplexer
	Even Parity 5
Algorithms	canonical generational GP
	EvAg GP
Node degree in newscast	40

in learning the boolean 11-multiplexer function, while the EP5 tries to build a program capable of calculating the parity of a set of 5 bits. Table 1 shows the parameter setup for the experiments which are Koza-I/II standard [10] (including optimal population sizes for both problems: choosing smaller or larger sizes will lead to worse results).

As a baseline for comparison, we have used a canonical generational GP with 1-elitism which runs sequentially in a single node of the simulator, while the EvAg GP will run on P nodes (being P the population size at Table 1). This allows not only the comparison of the best solution found at the maximum number of generations (equally parametrized in both approaches), but also, a forecast on the time that every approach would take. Obviously, results are for guidance only since they assume that the time required for communications is negligible with respect to the fitness computational time. We are of the opinion that such an assumption might be unrealistic for small easy problems, but turning feasible for real-like problems with an increasing ratio computation communication.

Figure 2 shows the best fitness convergence of the sequential and parallel approaches in terms of simulator cycles where every cycle takes the time of an individual evaluation. Therefore, differences in cycles represent the parallel execution of the EvAg GP in contrast with the sequential run of the canonical GP in a single node. Given that we have assumed no cost in communications, the algorithm speeds up in proportion with the population size (e.g. $speed - up = 4000$ in the problems under study). In this sense, the purpose is highlighting the possible benefits in time of a P2P parallelization rather than providing accurate execution times of the algorithm on a real platform (i.e. execution times will depend on the specific features of both, the problem evaluation cost and the underlying platform itself).

With respect to the fitness convergence, the EvAg GP shows a better progress in fitness and is able to outperform the best solution found by the canonical GP at the maximum number of generations. Such an improvement is rough in the EP5 but outstanding in the case of the 11M problem. Given that both algorithms have been equally parametrized, the most remarkable difference relies

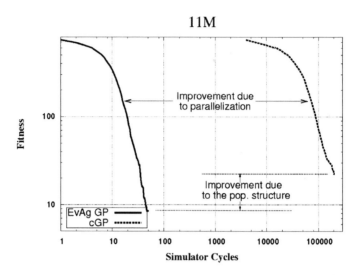

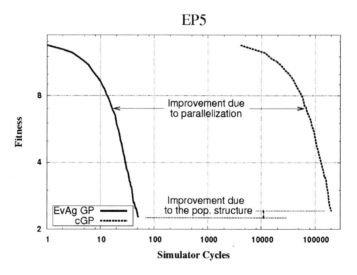

Fig. 2. Best fitness convergence of the Peer-to-Peer (EvAg GP) and canonical (cGP) approaches when tackling the 11 bits multiplexer (*11M*) and even parity 5 (*EP5*) problems. Representation, in a *log-log* scale, depicts ideal differences in simulator cycles between tackling the problems sequentially and in parallel. In addition, the EvAg approach is able to improve the solution at the complex network structure of the population. Results are averaged from 100 independent runs.

Table 2. Kolmogorov-Smirnov test on the normality of the distributions. We have considered $p - values > 0.1$ for a distribution to be normal.

Problem	Algorithm	Kolmogorov-Smirnov test	Normal distribution?
11M	cGP	D = 0.5, p-value = 6.118e-07	no
	EvAg GP	D = 0.89, p-value < 2.2e-16	no
EP5	cGP	D = 0.808, p-value < 2.2e-16	no
	EvAg GP	D = 0.7413, p-value = 9.548e-15	no

on the different population structures. In the case of the canonical approach is panmictic, meanwhile in the EvAg GP adopts the shape of a complex network defined by the newscast protocol.

In order to analyze the results with confidence, data has been statistically analyzed (each experiment has been run 100 times). Firstly, we have analyzed the normality of the data in Table 2 using the Kolmogorov-Smirnov test [3]. The small $p - values$ at the best fitness distributions show that results do not follow a normal distribution. Thus, to compare both approaches, we have used the non-parametric Wilcoxon test [7].

Table 3. Wilcoxon test comparing the best fitness distributions of equally parametrized canonical GP (cGP) and EvAg GP in 11M and EP5 problems. Results are obtained on 100 independent runs. We have considered that $p - values < 0.1$ refute the null hypothesis in the equivalence of the distributions.

Problem	Algorithm	Fitness	Wilcoxon test	Significantly different?
11M	cGP	22.4±43.84		
	EvAg GP	**8.53 ± 27.32**	W = 5790.5, p-value = 0.008	yes
EP5	cGP	2.43±1.57		
	EvAg GP	2.27±1.56	W = 1278.5, p-value = 0.3082	no

Table 3 shows the Wilcoxon analysis of the data showing significant differences in the 11M problem with the EvAg model outperforming the canonical approach, while they are equivalent in the EP5 problem. Therefore, it can be concluded that the P2P approach is at least algorithmically competitive against the canonical approach.

5   Conclusions and Future Works

This paper studies the viability of a Peer-to-Peer Genetic Programming approach in the context of fine-grained spatially structured Evolutionary Algorithms. To that end, a GP implementation of the Evolvable Agent model has been presented and assessed empirically using two well-known GP problems as a benchmark. Every EvAg acts at two levels; evolving a single individual with the mate selection locally restricted within a neighborhood and dynamically structuring such a neighborhood by means of the protocol newscast. This makes the approach suitable for a P2P execution in which every **EvAg** can be potentially placed in a different peer.

Experiments were conducted in a simulated P2P environment in which up to 4000 **EvAgs** evolve in parallel. In such a loosely-coupled scenario, results show that the parallel approach is able to outperform the search of a canonical approach (taken as a baseline for comparison). Given that both approaches have been equally parametrized following Koza standards, it must be the complex network structure of the population the cause why the **EvAg** model is able to get a better progress in fitness. Additionally, the parallel execution implies a speed-up on times to solution, outperforming the sequential approach in several orders of magnitude.

As a future line of work, and after the present analysis of viability, we plan to analyze the model dynamics under host churn in such a way that we can measure the impact of nodes failures on the algorithm performance.

Acknowledgements. This work has been supported by the Junta de Andalucia projects P08-TIC-03903 and P08-TIC-03928, Education and Science Project TIN2008-06681-C06-01, Consorcio Identic, project Extremadura@home, and Junta de Extremadura projects PDT-08A09, GRU-09105, GRU10029 and FEDER.

References

1. Bánhelyi, B., Biazzini, M., Montresor, A., Jelasity, M.: Peer-to-peer optimization in large unreliable networks with branch-and-bound and particle swarms. In: Giacobini, M., Brabazon, A., Cagnoni, S., Di Caro, G.A., Ekárt, A., Esparcia-Alcázar, A.I., Farooq, M., Fink, A., Machado, P. (eds.) EvoWorkshops 2009. LNCS, vol. 5484, pp. 87–92. Springer, Heidelberg (2009)
2. Biazzini, M., Montresor, A.: Gossiping differential evolution: a decentralized heuristic for function optimization in p2p networks. In: Proceedings of the 16th International Conference on Parallel and Distributed Systems (ICPADS 2010) (December 2010)
3. Crawley, M.J.: Statistics, An Introduction using R. Wiley, Chichester (2007)
4. Eiben, A.E., Smith, J.E.: Introduction to Evolutionary Computing. Springer, Heidelberg (2003)
5. Fernandez, F., Spezzano, G., Tomassini, M., Vanneschi, L.: Parallel genetic programming. In: Alba, E. (ed.) Parallel Metaheuristics, Parallel and Distributed Computing, ch. 6, pp. 127–153. Wiley-Interscience, Hoboken (2005)
6. Folino, G., Spezzano, G.: P-cage: An environment for evolutionary computation in peer-to-peer systems. In: Collet, P., Tomassini, M., Ebner, M., Gustafson, S., Ekárt, A. (eds.) EuroGP 2006. LNCS, vol. 3905, pp. 341–350. Springer, Heidelberg (2006)
7. Garcia, S., Molina, D., Lozano, M., Herrera, F.: A study on the use of non-parametric tests for analyzing the evolutionary algorithms behaviour: a case study on the CEC 2005 special session on real parameter optimization. Journal of Heuristics 15(6), 617–644 (2009)

8. The Gnutella Developer Forum GDF. The annotated gnutella protocol specification v0.4 (2001)
9. Jelasity, M., van Steen, M.: Large-scale newscast computing on the Internet. Technical Report IR-503, Vrije Universiteit Amsterdam, Department of Computer Science, Amsterdam, The Netherlands (October 2002)
10. Koza, J.R.: Genetic Programming: On the Programming of Computers by Means of Natural Selection. MIT Press, Cambridge (1992)
11. Laredo, J.L.J., Castillo, P.A., Mora, A.M., Merelo, J.J.: Exploring population structures for locally concurrent and massively parallel evolutionary algorithms. In: IEEE Congress on Evolutionary Computation (CEC 2008), WCCI 2008 Proceedings, pp. 2610–2617. IEEE Press, Hong Kong (2008)
12. Laredo, J.L.J., Eiben, A.E., van Steen, M., Guervós, J.J.M.: Evag: a scalable peer-to-peer evolutionary algorithm. Genetic Programming and Evolvable Machines 11(2), 227–246 (2010)
13. Ratnasamy, S., Francis, P., Handley, M., Karp, R., Shenker, S.: A scalable content addressable network. In: ACM SIGCOMM, pp. 161–172 (2001)
14. Scriven, I., Lewis, A., Mostaghim, S.: Dynamic search initialisation strategies for multi-objective optimisation in peer-to-peer networks. In: CEC 2009: Proceedings of the Eleventh conference on Congress on Evolutionary Computation, Piscataway, NJ, USA, pp. 1515–1522. IEEE Press, Los Alamitos (2009)
15. Steinmetz, R., Wehrle, K.: What is this peer-to-peer about. In: Steinmetz, R., Wehrle, K. (eds.) Peer-to-Peer Systems and Applications. LNCS, vol. 3485, pp. 9–16. Springer, Heidelberg (2005)
16. Trujillo, L., Olague, G.: Synthesis of interest point detectors through genetic programming. In: Cattolico, M. (ed.) Proceedings of the Genetic and Evolutionary Computation Conference, GECCO 2006, Seattle, Washington, USA, July 8-12, vol. 1, pp. 887–894. ACM, New York (2006)
17. Voulgaris, S., Jelasity, M., van Steen, M.: Agents and Peer-to-Peer Computing. In: Moro, G., Sartori, C., Singh, M.P. (eds.) AP2PC 2003. LNCS (LNAI), vol. 2872, pp. 47–58. Springer, Heidelberg (2004)
18. Watts, D.J., Strogatz, S.H.: Collective dynamics of "small-world" networks. Nature 393, 440–442 (1998)
19. Wickramasinghe, W.R.M.U.K., van Steen, M., Eiben, A.E.: Peer-to-peer evolutionary algorithms with adaptive autonomous selection. In: GECCO 2007, pp. 1460–1467. ACM Press, New York (2007)

Performance Models for Evolutionary Program Induction Algorithms Based on Problem Difficulty Indicators

Mario Graff[1] and Riccardo Poli[2]

[1] Division de Estudios de Posgrado
Facultad de Ingenieria Electrica
Universidad Michoacana de San Nicolas de Hidalgo, Mexico
mgraffg@lsc.fie.umich.mx
[2] School of Computer Science and Electronic Engineering
University of Essex, UK
rpoli@essex.ac.uk

Abstract. Most theoretical models of evolutionary algorithms are difficult to apply to realistic situations. In this paper, two models of evolutionary program-induction algorithms (EPAs) are proposed which overcome this limitation. We test our approach with two important classes of problems — symbolic regression and Boolean function induction — and a variety of EPAs including: different versions of genetic programming, gene expression programing, stochastic iterated hill climbing in program space and one version of cartesian genetic programming. We compare the proposed models against a practical model of EPAs we previously developed and find that in most cases the new models are simpler and produce better predictions. A great deal can also be learnt about an EPA via a simple inspection of our new models. E.g., it is possible to infer which characteristics make a problem difficult or easy for the EPA.

Keywords: Evolutionary Program-induction Algorithms, Genetic Programming, Performance Prediction, Hardness Measures.

1 Introduction

Evolutionary Program-induction Algorithms (EPAs) are search techniques for the automatic evolution of computer programs. Genetic Programming (GP) [10,16], Cartesian GP (CGP) [12], Grammatical Evolution [14] and Gene Expression Programming (GEP) [3], among others, are members of this family.

There has been an enormous increase in interest in EPAs over the last two decades, resulting in the appearance of new generic operations, representations, and so on. Also, researchers and practitioners from a number of disciplines have used EPAs on a huge variety of new problems. There are, however, also many unanswered questions regarding EPAs.

In principle one would expect theory to be able to shed light on old and new EPA techniques and problems. However, this is not so. The key reason is that

S. Silva et al. (Eds.): EuroGP 2011, LNCS 6621, pp. 118–129, 2011.

producing EPA theory is objectively a very hard and slow process where the precise details of an algorithm matter a great deal [11,17]. In practice, exact theoretical EPA models are almost impossible to apply to realistic situations.

This has important consequences from a practitioner's point of view. For instance, one is unable to estimate the performance of an algorithm on a particular problem without running the algorithms on that problem. Thus, we cannot *a priori* discern which is the most suitable algorithm to solve a particular problem out of a set of different algorithms and/or algorithms with different parameters. Also, the current lack of theoretically-sound guidelines forces practitioners to manually hand-tune algorithms, parameters and operators.

This paper proposes a solution to these problems. We introduce two practical models for the performance of EPAs based on *difficulty indicators*. With these we are able to precisely forecast the performance of a variety of EPAs for symbolic regression and Boolean induction problems and two performance measures — the expected end-of-run fitness and the success rate. These models also allow us to define procedures that can solve the algorithm selection problem [18] (i.e., the problem of deciding which algorithm to use from a set of available algorithms to solve a particular problem) via the creation of algorithm portfolios. Also, as we will see, our models are simple to analyse.

The rest of the paper is organised as follows. In Section 2, we review related work. Section 3 presents our new modelling technique and the process used to apply it. The problems and algorithms used to validate our approach are described in Section 4. Section 5 provides our experimental results. Section 6 shows how our models can be used to analyse algorithms. Some conclusions and possible directions for future work are given in Section 7.

2   Related Work

Our work is related to the problem of understanding what makes a problem easy or hard for EAs. One of the earliest attempts in this direction was made by Jones [9] who introduced a heuristic called *fitness distance correlation (fdc)*, as an algebraic indicator of problem difficulty for GAs. The study of *fdc* has been extended to GP (e.g., see [19]). This has shown that *fdc* is often a reliable indicator of problem hardness, but that it has also some flaws, the most severe one being that the computation of *fdc* requires the optimal solution(s) to be known beforehand. This is obviously unrealistic and prevents one from using the *fdc* to estimate how hard a problem is in practical applications.

A measure that does not require knowledge of optimal solution(s) is the *negative slope coefficient (nsc)* [20] which has been shown to be fairly reliable in a number of different benchmark problems in GP.

While these (and other) measures have shown some success at providing insights on how hard or easy a problem is, they do not really provide a estimation of the performance of an algorithm. For instance, they are unable to predict the success rate of a particular algorithm or the fitness of the best solution found during the evolution process.

There are two other research areas where models capable of making such predictions have been proposed: (a) the algorithm selection problem and the related algorithm portfolios (e.g., see [8]) and (b) the models of performance of evolutionary algorithms proposed in [2,6,7]. These are described in more detail in the next sub-sections.

2.1   Algorithm Selection and Portfolios

The algorithm selection problem and algorithm portfolios (which are a collection of algorithms that are run in parallel or in sequence to solve a particular problem) require the creation of models to forecast the performance of every algorithm in a set before deciding which one to use to solve a problem [8].

Such models take a variety of forms including: linear equations, Markov decision processes, Bayesian models, and so on. However, generally, all models are based on a set of features that are related to the hardness of the problems being solved. For instance, in SAT problems the feature set might include the number of variables, the number of clauses, the ratio between variables and clauses, and so on. This characteristic was a source of inspiration for the work in this paper.

2.2   Models of Performance for Evolutionary Algorithms

[2] proposed to model the performance of GAs based on a re-representation of the fitness function and a liner combination of the degrees of freedom of such a representation. The idea was that any algorithm whose selection is based on comparing the fitness of different solutions (such as GAs with rank selection, truncation selection or tournament selection) can be re-represented using the outcome of all possible comparisons between pairs of solutions. This approach presents some interesting features, but it does not scale well with the size of the search space. As a result, it cannot be applied directly to large search spaces (such as the spaces explored by EPAs).

In [6,7] we proposed an approach that has shown success in modelling the performance of EPAs. The key idea was to predict the performance, $P(\mathbf{t})$, of an EPA based on the similarity between the problem, $\mathbf{t}$, to be faced by the algorithm and a set of reference problems, $\mathcal{S}$, previously selected. In formulae:

$$P(\mathbf{t}) \approx a_0 + \sum_{\mathbf{p} \in \mathcal{S}} a_{\mathbf{p}} \cdot d(\mathbf{p}, \mathbf{t}), \tag{1}$$

where d is a similarity measure and a_i are coefficients. The reference problems and the coefficients were determined using a training set of pairs $(f, P(f))$, where f represents a problem and $P(f)$ is the performance of the EPA on f.

3   Modelling EPA's Performance

In this paper we want to combine the strengths of the approaches reviewed above. We want to use approximate linear models as in [2] and [6,7], together with sets of features that are related to the difficulty of problems, as is done in algorithm portfolios and the algorithm selection problem.

We start by asking if *fdc* and *nsc*, which have been proposed as difficulty indicators for EPAs' problems, can be used as features in our new performance models. Unfortunately, as we will see in the next subsection, the characteristics of these difficulty indicators prevent their use in this particular task. However, in Section 3.2, we propose a new set of difficulty indicators for Boolean functions that is suitable for creating performance models. We generalised these difficulty indicators to the domain of continuous functions in Section 3.3. Then in Section 3.4 we present our new model and the procedure used to apply it.

3.1   Can NSC and FDC Be Used in Performance Models?

As mentioned previously, the *fdc* requires the optimal solution(s) or the distance of each point from a global optimum to be known in advance. As a consequence, this indicator cannot be used for performance modelling with unseen/unexplored problems, which is really the main objective of performance modelling.

This limitation is not present in the *nsc*. However, the procedure used to compute the *nsc* is stochastic and involves a considerable computational load. This makes *nsc* less than ideal for the construction of performance models of EPAs.

3.2   Difficulty Indicators for Boolean Functions

Franco [4] has recently proposed an indicator for the difficulty of learning Boolean functions (in terms of the number of training epochs required) using feed-forward artificial neural networks. The indicator is based on evaluating a Boolean function with different inputs and counting in how many cases the function produces different outputs. This process is applied to every pair of input patterns that are at a specific distance. The Hamming distance was used to measure the distance between pairs of inputs and Franco limited his indicator to use only those inputs that are at a Hamming distance of 1 or 2.

In formulae, this indicator is defined as:

$$\tilde{\vartheta}(f) = \vartheta_1(f) + \vartheta_2(f),$$

where $\vartheta_i(f)$ counts the number of pair of outputs that are different when the Hamming distance of their inputs is i. That is,

$$\vartheta_i(f) = \frac{1}{2^N \times \binom{N}{i}} \sum_{\mathbf{j} \in I} \sum_{\{\mathbf{k} \in I: \ H(\mathbf{j}, \mathbf{k}) = i\}} |f(\mathbf{j}) - f(\mathbf{k})|, \qquad (2)$$

where $f(\mathbf{j})$ is the output of Boolean function f for input $\mathbf{j}$, N is the number of variables, I contains all the possible input patters (e.g., for 3 variables I contains 8 different input patterns), and $H(\mathbf{j}, \mathbf{k})$ is the Hamming distance between input $\mathbf{j}$ and $\mathbf{k}$. The term $\frac{1}{2^N \times \binom{N}{i}}$ is a normalisation factor that ensures the value of $\vartheta_i(f)$ is in the interval $[0, 1]$. The second summation is over all inputs $\mathbf{k}$ that are at a Hamming distance i from input $\mathbf{j}$.

The indicator $\tilde{\vartheta}(f)$ was found to correlate well with the time needed by a learning algorithm to train a feed-forward neural network. Based on this success, it is reasonable to wonder whether this difficulty indicator could help in assessing the hardness of Boolean induction problems also in relation to EPAs.

Clearly, the active ingredients in Franco's difficulty indicator are the terms $\vartheta_i(f)$. Therefore, we will use these terms as our features in the creation of difficulty-indicator-based performance models of EPAs.

3.3   Difficulty Indicators for Continuous Functions

We should note that the Boolean difficulty indicator described above is related to the concept of differences of Boolean functions [1]. For example, in the case of a Hamming distance of 1, Eq. (2) can be rewritten as:

$$\frac{1}{2^N \times \binom{N}{i}} \sum_{\mathbf{x} \in I} \sum_{i}^{N} \triangle f_{x_i}(\mathbf{x}),$$

where $\triangle f_{x_i}(\mathbf{x})$ is defined as $f(\mathbf{x}) \oplus f(\mathbf{x} \oplus \mathbf{e}_i)$, $\mathbf{e}_i$ is a unit vector whose i-th component is 1 and $\oplus$ is the exclusive disjunction.

Naturally, the counterpart of the Boolean difference for continuous functions is the discrete derivative. Based on this equivalence, it is reasonable to ask whether the discrete derivative can be used to produce difficulty indicators for continuous functions. Let us see what this entails.

The discrete derivative of function f w.r.t. a variable x_i can be written as

$$\triangle_h f_{x_i} = \frac{f(\mathbf{x} + h\mathbf{e}_i) - f(\mathbf{x})}{h}, \tag{3}$$

where $\mathbf{x}$ is the independent variable, h is the step and $\mathbf{e}_i$ is a unit vector.

Note: the term $f(\mathbf{x} + h\mathbf{e}_i) - f(\mathbf{x})$ in Eq. (3) matches the term $f(\mathbf{j}) - f(\mathbf{k})$ of Eq. (2). This suggests that we might be able to generalise Eq. (2) to continuous spaces if we grouped inputs according to the distance h. To achieve this we create groups of input-pairs having distances $h_1, h_2, \ldots$, where $h_i < h_{i+1}$. This process continues until the maximum distance between pairs of input patterns is reached. Then we define our difficulty indicator for continuous functions as:

$$\varrho_i(f) = \sum_{\mathbf{x} \in I} \sum_{j}^{N} |\triangle_i f_{x_j}(\mathbf{x})|, \tag{4}$$

where I contains all the input patters and N is the number of independent variables.

To sum up, in this and the previous section we have introduced two sets of difficulty indicators: the set $\{\vartheta_i\}$ for the case of Boolean functions and the set $\{\varrho_i\}$ for the case of continuous functions. In the next section, we will employ them to create difficulty-indicator-based performance models.

3.4   Difficulty-Indicators Models

Having introduced suitable difficulty indicators, we are now in a position to define our new models of performance. We model the performance of EPAs on Boolean functions using the equation:

$$P(f) \approx a_0 + \sum_{i=1}^{N} a_i \cdot \vartheta_i(f) \tag{5}$$

and the performance of EPAs on continuous functions using the equation:

$$P(f) \approx a_0 + \sum_{i}^{N} a_i \cdot \varrho_i(f), \tag{6}$$

respectively. In the Boolean case N is the number of inputs, while in the continuous case it is the number of different distance classes. In both cases a_i are coefficients that need to be identified and f is a problem.

3.5   Model Identification

To instantiate the equations introduced above, one needs a training set of problems, T, and a validation set, V. T is used to identify the coefficients a_i so as to obtain a good fit between predicted and actual performance. The set V is used to test the generality of the model. T and V are composed by pairs $(f, P(f))$ where $P(f)$ is obtained by running the algorithm being modelled on f and assessing its performance. This was done by performing 100 independent runs of the algorithms being modelled and averaging their performance (either the expected end-of-run best fitness or the success rate) obtained in such runs.

Given a training set T, one can apply an ordinary least square method to determine the coefficients a_i [13]. For example, in order to identify a_i of Eq. (5), one needs to solve the linear system $\mathbf{W}\mathbf{a} = \mathbf{p}$, where $\mathbf{a}$ is a vector representing the a_i coefficients (i.e., $\mathbf{a} = (a_0, a_1, \ldots, a_N)'$), $\mathbf{p}$ is a vector that contains the measure performance of the algorithm for every problem in the training set T (i.e., $\mathbf{p} = (P(f_1), \ldots, P(f_{|T|}))'$), and $\mathbf{W}$ is a matrix whose first column is all ones and the remaining columns are the result of computing ϑ_i (or ϱ_i) from 1 to N for every problem in T.

Although the procedure described above identifies $\mathbf{a}$, it does not identify and discard those difficulty indicators (i.e., ϑ_i or ϱ_i) that are not positively contributing to the accuracy of the model. Since their inclusion decreases the quality of the model, an alternative procedure is needed. In order to identify and use only the difficulty indicators that contribute to the model, we decided to use least-angle regression (LARS) and a cross-validation technique. LARS sorts the difficulty indicators based on their correlation with the performance of the algorithm while cross validation tells us how many of them to include in the model.

Finally, we measure the overall generalisation error using the Relative Square Error (RSE) which is defined as

$$\text{RSE} = \sum_{i}(P_i - \tilde{P}_i)^2 / \sum_{i}(P_i - \bar{P})^2,$$

where i ranges over T or V depending on whether we are measuring the training set or validation set error, respectively, P_i is the average performance recorded for problem i, $\tilde{P}_i$ is the performance predicted by the model, and $\bar{P}$ is the average performance over all problems.

4   Test Problems and Systems

We consider two radically different classes of problems, namely continuous symbolic regression of rational functions and Boolean function induction problems, and two typical performance measures: a normalised version of best of run fitness (more on this below) and the success rate.

A benchmark set was created for continuous symbolic regression by randomly generating and then sampling rational functions. We created 1,100 different rational functions using the following procedure. Two polynomials, $W(x)$ and $Q(x)$, were built by randomly choosing the degree of each in the range 2 to 8, and then choosing at random real coefficients in the interval $[-10, 10]$ for the powers of x up to the chosen degree. A rational function in our training set is then given by $f(x) = \frac{W(x)}{Q(x)}$. Each of the rational functions in the set was finally sampled at 21 points uniformly distributed in the interval $[-1, 1]$.

For each rational function f, we performed 100 independent runs recording the normalised best of run fitness (NBRF) which is computed by normalising the problem and the behaviour of the best individual and then summing the absolute differences between the components of these normalised functions.

The second benchmark used is the class of 4 input Boolean induction problems. We randomly selected 1,100 different Boolean functions from this set and for each we counted the number of times the algorithm found a program that encoded the target functionality in 100 independent runs. We took as our performance measure the success rate, i.e., the fraction of successful runs out of the total number of runs.

Each benchmark set was divided into two sets: a training set T composed of 500 elements, and a validation set V comprising the remaining 600 elements.

The algorithms modelled are different versions of GP, GEP, CGP, and a Stochastic Iterated Hill Climber (SIHC). Tab. 1 shows the parameters that are common to all the systems. Below, we briefly review the systems used:

GP. We used two different implementations of GP. One system is essentially identical to the one used by Koza [10]. The other is TinyGP with the modifications presented in [16] to allow the evolution of constants. For the Koza-style GP system, besides the traditional roulette-wheel selection, we also used tournament selection with a tournament size of 2.

GEP. We used three different configurations of GEP [3]. The first is almost equivalent to the one described in [3] with the difference that we replaced GEP's initialisation method with the ramped-half-and-half method used in GP [10]. We used this modification in all configurations. The second GEP uses tournament selection (tournament size 2). The third is a steady state system with tournament selection.

Table 1. Parameters of the EPAs used in our tests

Function set (rational problems)	$\{+, \ -, \ *, \ / \ \text{(protected)}\}$
Function set (Boolean problems)	$\{AND, \ OR, \ NAND, \ NOR\}$
Terminal set (rational problems)	$\{x, \ \mathbb{R}\}$
Random constants (i.e., $\mathbb{R}$)	100 constants drawn from the inteval $[-10, 10]$
Terminal set (Boolean problems)	$\{x_1, \ x_2, \ x_3, \ x_4\}$
Number of individuals evaluated	50000
Number of independent runs	100

CGP. We used the implementation of CGP available at `https://sites.google.com/site/julianfrancismiller/`.

SIHC. Our Stochastic Iterated Hill Climber is similar to the one in [15], but we modified the mutation operators. We used sub-tree mutation and a mutation operator similar to the one in [15], which we call *uniform mutation*. We also tested this system with different values for the maximum number of allowed mutations before the search restarts from a new random individual.

5 Results

We start this section by comparing the quality of the models proposed in terms of RSE values. As we will see, the new models do quite well in terms of making accurate predictions. Furthermore, in most cases they outperform the performance models we proposed in our earlier work [6,7] (see Eq. (1)).

Tab. 2 shows an accuracy comparison (in terms of RSE) between difficulty-indicator models (i.e., Eqs. (5) and (6)) and performance models (i.e., Eq. (1)) in the validation set for rational symbolic regression and Boolean induction. All difficulty-indicator models have an RSE value well below 1 indicating that they are predicting better than the mean. For the case of rational functions, we see that difficulty-indicator models obtained lower RSE values in all the cases except for the GEP system with tournament selection. In Boolean induction, difficulty-indicator models obtain lower RSE values in 18 out of 21 cases.

Difficulty-indicator models clearly make more accurate predictions than our earlier performance models. In addition, they are simpler to understand because they have fewer degrees of freedom. For instance, in the case of rational functions difficulty-indicator models use on average 14 coefficients whilst performance models use more than 100 coefficients. The situation is even more extreme in the case of Boolean induction problems where difficulty-indicator models have at most 4 coefficients whilst performance models require more than 100 coefficients.

While RSE values provide an objective assessment of the quality of the models, it might be difficult for the reader to appreciate the accuracy of our models from such values alone. To provide a more visual indication of the quality of the models, Fig. 1 shows scatter plots of the actual performance *vs* the performance estimated with our new models on the validation set for both classes of problems.

Table 2. Quality (RSE) of the difficulty-indicators models (Eqs.. (6) and (5)) and the performance models (Eq. (1)) in the validation set, for different EPAs and parameter settings (crossover rate p_{xo} and mutation rate p_m)

Configuration				Rational Functions		Boolean Functions	
Type	Selection	p_{xo}	p_m	Eq. (6)	Eq. (1)	Eq. (5)	Eq. (1)
Generational	Roulette	1.00	0.00	**0.4511**	0.5246	**0.2306**	0.2877
		0.90	0.00	**0.4552**	0.5375	**0.2289**	0.2962
		0.50	0.50	**0.4206**	0.4999	**0.2274**	0.2833
		0.00	1.00	**0.4074**	0.4907	**0.2493**	0.3058
			GEP	**0.4926**	0.5212	0.5265	**0.3745**
Generational	Tournament	1.00	0.00	**0.3506**	0.4082	**0.3166**	0.4065
		0.90	0.00	**0.3525**	0.4257	**0.3182**	0.3941
		0.50	0.50	**0.3448**	0.4130	**0.3267**	0.4010
		0.00	1.00	**0.3545**	0.4291	**0.3643**	0.4686
			GEP	0.4794	**0.4477**	0.5570	**0.4501**
Steady State	Tournament	1.00	0.00	**0.4436**	0.5778	**0.4433**	0.5401
		0.90	0.00	**0.4399**	0.5634	**0.4628**	0.5820
		0.50	0.50	**0.4794**	0.5967	**0.4729**	0.6379
		0.00	1.00	**0.5137**	0.6367	**0.5208**	0.6336
			GEP	0.4347	**0.4243**	0.7216	**0.5512**
Sys.	Mut.	Max. Mut.		Eq. (6)	Eq. (1)	Eq. (5)	Eq. (1)
SIHC	Sub-Tree	50		**0.2859**	0.4349	**0.3226**	0.4045
		500		**0.2885**	0.4540	**0.3025**	0.3989
		1000		**0.2676**	0.4378	**0.2855**	0.3787
		25000		**0.2935**	0.4587	**0.2416**	0.3120
	Uni.	25000		**0.4002**	0.4890	**0.2792**	0.3641
System				Eq. (6)	Eq. (1)	Eq. (5)	Eq. (1)
CGP				**0.4348**	0.5271	**0.7400**	0.8355

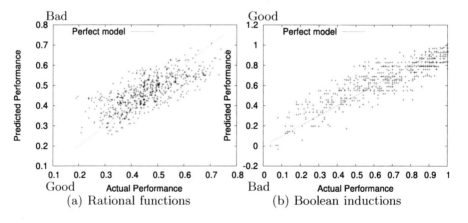

(a) Rational functions   (b) Boolean inductions

Fig. 1. Scatter plots of the predicted *vs.* actual performance in continuous regression problems (a) and Boolean induction problems (b) on the validation set for a GP system with 90% crossover, no mutation and roulette-wheel selection

The data in Fig. 1 refer to the GP system with 90% crossover rate, no mutation and roulette-wheel selection, but other parameter settings and systems provide qualitatively similar results. The solid diagonal line in each plot represents the behaviour of a perfect model. As can be seen, the points form tight clouds around the perfect models which is a clear qualitative indication of the accuracy of the predictions of our models of performance.

6   Eliciting Knowledge from Our Models

Our difficulty-indicator models can be used to analyse the behaviour of the systems being modelled. For instance, they could, for example, be used to create meaningful taxonomies of EPAs following the procedure presented in [5], by simply replacing the modelling technique used there with the difficulty-indicator models. Such taxonomies could reveal relationships, in terms of performance, of the different algorithms involved in the taxonomy.

However, here we want to focus on a simpler technique. By looking at the coefficients of the models, one can infer which characteristics of the problem make it difficult to solve for the algorithm under study. That is, the sign and the value of a coefficient indicate whether the corresponding characteristic makes a problem easier or harder for a particular EPA. For example, if $a_i > 0$ and the performance measure is the success rate, then a problem having a high value in factor ϑ_i (or ϱ_i) will be easier than a problem with a lower ϑ_i (or ϱ_i), and *vice versa* for $a_i < 0$.

Tab. 3 shows the a_i coefficients for different systems in the Boolean induction case. From the table one can see that the coefficients associated to ϑ_1 and ϑ_4 are negative indicating that a problem with a high value on those factors will be difficult. For example, the even-parity Boolean function, which is considered to be a hard problem, has indeed a high value of ϑ_1 (while $\vartheta_2 = 0$).

From the table, it may be also observed that the coefficients of the systems that only differ in the crossover and mutation rates are more similar than the ones that differ in the selection mechanism and type of evolution. Furthermore, one can see that the coefficients of the generational GP system with tournament selection are closer to the steady state GP system than to the generational system

Table 3. Coefficients of the difficulty-indicators models, in the Boolean induction problems, corresponding to the GP systems with 100% crossover and 100% mutation

Configuration			Coefficients			
Type of Evolution	Selection	p_{xo}	ϑ_1	ϑ_2	ϑ_3	ϑ_4
Generational	Roulette	1.00	-1.9744	1.2584	0.2382	-0.1529
Generational	Roulette	0.00	-1.8071	1.2427	0.2099	-0.2139
Generational	Tournament	1.00	-1.1809	1.1316	0.1316	-0.1258
Generational	Tournament	0.00	-0.9720	0.9342	0.1281	-0.1472
Steady State	Tournament	1.00	-0.7842	0.7720	0.2144	-0.1059
Steady State	Tournament	0.00	-0.6959	0.7386	0.1763	-0.1503

with roulette selection. This is an indication that the selection mechanism is playing a more important role than the type of evolution and the rates of the genetic operators in the EPAs under test. These conclusions are similar to those drawn in [5] from the analysis of algorithm taxonomies. However, here such conclusions were reached by the much simpler direct inspection of the coefficients of a model, further illustrating the practicality of our new models.

7   Conclusions

We have presented a new technique for building efficient and accurate models of the performance of EPAs. We modelled three versions of GP with multiple parameter settings, three versions of GEP, two versions of SIHC (one with multiple parameters settings) and one version of CGP. These algorithms were applied to two problem classes: symbolic regression of rational functions and Boolean induction. The new models are a hybrid between our previous models and the models based on difficulty measures used to solve the algorithm selection problem.

We compared the new models against our previous approach to model the performance of EPAs. In most of the cases the new difficulty-indicator models have better accuracy than the corresponding models obtained in our previous work. Furthermore, the new difficulty-indicator models require to identify fewer coefficients. This has a positive effect from the point of view of the generalisation ability of the models. Furthermore, it makes it easy to analyse our models and draw conclusions about what makes a problem hard or easy for a particular system.

Finally, we would like to briefly discuss possible future research avenues. Our approach is able to perform accurate predictions on the classes of problems tested here. However, our difficulty indicators are at present restricted to these two classes of problems. In future work, we will explore whether these indicators can be further generalised for use in other classes of problems.

Acknowledgements. The first author acknowledges support from CONACyT through the scheme of "Repatriación 2010".

References

1. Akers Jr., S.B.: On a theory of boolean functions. Journal of the Society for Industrial and Applied Mathematics 7(4), 487–498 (1959)
2. Borenstein, Y., Poli, R.: Information landscapes. In: Beyer, H.-G., O'Reilly, U.-M. (eds.) Proceedings of the Genetic and Evolutionary Computation Conference (GECCO-2005), Washington DC, USA, pp. 1515–1522. ACM, New York (2005)
3. Ferreira, C.: Gene expression programming: A new adaptive algorithm for solving problems. Complex Systems 13(2), 87–129 (2001)
4. Franco, L.: Generalization ability of boolean functions implemented in feedforward neural networks. Neurocomputing 70, 351–361 (2006)

5. Graff, M., Poli, R.: Automatic creation of taxonomies of genetic programming systems. In: Vanneschi, L., Gustafson, S., Moraglio, A., De Falco, I., Ebner, M. (eds.) EuroGP 2009. LNCS, vol. 5481, pp. 145–158. Springer, Heidelberg (2009)
6. Graff, M., Poli, R.: Practical performance models of algorithms in evolutionary program induction and other domains. Artif. Intell. 174(15), 1254–1276 (2010)
7. Graff, M., Poli, R.: Practical model of genetic programming's performance on rational symbolic regression problems. In: O'Neill, M., Vanneschi, L., Gustafson, S., Esparcia Alcázar, A.I., De Falco, I., Della Cioppa, A., Tarantino, E. (eds.) EuroGP 2008. LNCS, vol. 4971, pp. 122–132. Springer, Heidelberg (2008)
8. Hutter, F., Hamadi, Y., Hoos, H.H., Leyton-Brown, K.: Performance prediction and automated tuning of randomized and parametric algorithms. In: Benhamou, F. (ed.) CP 2006. LNCS, vol. 4204, pp. 213–228. Springer, Heidelberg (2006)
9. Jones, T., Forrest, S.: Fitness distance correlation as a measure of problem difficulty for genetic algorithms. In: Eshelman, L.J. (ed.) ICGA, pp. 184–192. Morgan Kaufmann, San Francisco (1995)
10. Koza, J.R.: Genetic Programming: On the Programming of Computers by Natural Selection. MIT Press, Cambridge (1992)
11. Langdon, W.B., Poli, R.: Foundations of Genetic Programming. Springer, Heidelberg (2002)
12. Miller, J.F., Thomson, P.: Cartesian genetic programming. In: Poli, R., Banzhaf, W., Langdon, W.B., Miller, J., Nordin, P., Fogarty, T.C. (eds.) EuroGP 2000. LNCS, vol. 1802, pp. 121–132. Springer, Heidelberg (2000)
13. Olver, P.J., Shakiban, C.: Applied Linear Algebra. Prentice-Hall, Englewood Cliffs (2006)
14. O'Neill, M., Ryan, C.: Grammatical evolution. IEEE Transactions on Evolutionary Computation 5(4), 349–358 (2001)
15. O'Reilly, U.-M., Oppacher, F.: Program search with a hierarchical variable length representation: Genetic programming, simulated annealing and hill climbing. In: Davidor, Y., Männer, R., Schwefel, H.-P. (eds.) PPSN 1994. LNCS, vol. 866, pp. 397–406. Springer, Heidelberg (1994)
16. Poli, R., Langdon, W.B., McPhee, N.F.: A field guide to genetic programming (2008), Published via http://lulu.com and freely available at http://www.gp-field-guide.org.uk (With contributions by J. R. Koza)
17. Poli, R., McPhee, N.F.: General schema theory for genetic programming with subtree-swapping crossover: II. Evolutionary Computation 11(2), 169–206 (2003)
18. Rice, J.R.: The algorithm selection problem. Advances in Computers 15, 65–118 (1976)
19. Tomassini, M., Vanneschi, L., Collard, P., Clergue, M.: A study of fitness distance correlation as a difficulty measure in genetic programming. Evolutionary Computation 13(2), 213–239 (Summer 2005)
20. Vanneschi, L., Clergue, M., Collard, P., Tomassini, M., Vérel, S.: Fitness clouds and problem hardness in genetic programming. In: Deb, K., et al. (eds.) GECCO 2004. LNCS, vol. 3103, pp. 690–701. Springer, Heidelberg (2004)

Examining Mutation Landscapes in Grammar Based Genetic Programming

Eoin Murphy, Michael O'Neill, and Anthony Brabazon

Natural Computing Research and Applications Group,
Univeristy College Dublin, Ireland
{eoin.murphy,m.oneill,anthony.brabazon}@ucd.ie

Abstract. Representation is a very important component of any evolutionary algorithm. Changing the representation can cause an algorithm to perform very differently. Such a change can have an effect that is difficult to understand. This paper examines what happens to the grammatical evolution algorithm when replacing the commonly used context-free grammar representation with a tree-adjunct grammar representation. We model the landscapes produced when using integer flip mutation with both representations and compare these landscapes using visualisation methods little used in the field of genetic programming.

1 Introduction

Three very important components of any evolutionary algorithm are the representation, the variation operations and the fitness function. The interaction of these components within the algorithm forms a complex process and the modification of any one of them can have a major effect on how the algorithm performs. Such an effect may not be immediately obvious and is difficult to understand. Koza and Poli [12] said that visualising the program search space would be useful and help us understand how the algorithm operates.

Grammatical Evolution (GE) [8, 21] has recently been extended to make use of tree-adjunct grammars (TAG) [10, 11] in place of the usual grammar type, context-free grammars (CFG) [16]. TAGs have shown promise in the field of Genetic Programming (GP) [5, 6, 7, 17] as well as other fields in natural computing [1]. This promise carried over when TAGs were incorporated into GE, i.e., Tree-Adjunct Grammatical Evolution (TAGE), in the form of an increased ability to find fitter solutions in fewer generations and an increased success rate [16]. Previous work has examined how the TAG representation overcomes some of the structural difficulties present in GP [8], but the full extent of how TAGs affect GE is unclear.

Landscapes are a tool to help understand complex processes [9]. They have been employed here in an attempt to further understand how the use of TAGs in GE affects performance. Using a single variation operation, Integer Flip Mutation (IFM), the landscapes of a number of different problems are examined for both TAGE and GE. The IFM operation is where the integer value of a

S. Silva et al. (Eds.): EuroGP 2011, LNCS 6621, pp. 130–141, 2011.
© Springer-Verlag Berlin Heidelberg 2011

codon is replaced with a new random integer value. Viewing the entire search space/landscape is difficult due to its large size and high complexity. To alleviate this problem, this study employs a method of visualisation little used in the field of GP, heat maps.

This paper compares the single IFM landscapes of GE and TAGE for a series of problems in an attempt to further understand how the change in representation affects each algorithm's ability to search.

This section is followed by a brief introduction to the landscape model used in this study in section 2 along with a description of GE and TAGE in sections 3 and 4; The experiments performed are outlined in section 5 along with the results; These are followed by a discussion in section 6 and some conclusions and future work in the final section.

2   Landscapes

The landscape model used in this paper is as defined by Jones [9], where he quotes Nilsson [18], *"In its broadest sense, problem solving encompasses all of computer science because any computational task can be regarded as a problem to be solved."*, Pearl [22], *"Every problem-solving activity can be regarded as the task of finding or constructing an object with given characteristics"*, and Rich [23], *"Every search process can be viewed as a traversal of a directed graph in which each node represents a problem state and each arc represents a relationship between the states represented by the nodes it connects"*, stating that from the above statements one can conclude that search is ubiquitous and that it can be described as a process on a graph structure [9]. It is for this reason that he adopts a graph as a view of his landscape model.

The full description of the landscape model is outlined at length in [9]. It is sufficient to say for this study that the landscape model can be written as

$$L = (R, \phi, f, F, >_F) \tag{1}$$

where R is the representation space, ϕ is the operator (in this case a genetic operator), the function f which maps a multi-set of R, $M(R): \mapsto F$ for some set F, the fitness space, and a partial ordering $>_F$ over F. The landscape, L can be viewed as a directed labeled graph where the set of vertices, V, is a subset of $M(R)$ and an edge exists between the vertex v and the vertex w if $p(\phi(v, w)) > 0$.

In this study the landscapes are defined using the space of chromosomes paired with either a CFG or TAG as R. The object space, O, is the solution/phenotypic space. ϕ is the IFM operator, and f is the fitness function. The landscapes can be viewed as graph structures both where $V \subseteq M(R)$, and where $V \subseteq O$ (each vertex is a phenotype, but edges are dependent on ϕ and R).

3   Grammatical Evolution

GE is a grammar-based approach to GP, combining aspects of Darwinian natural selection, genetics and molecular biology with the representational power of grammar formalisms [2, 15, 21]. The grammar, written in Backus-Naur form, enables GE to define and modify the legal expressions of an arbitrary computer language. Moreover, the grammar also enables GE to modify the structure of the expressions generated, something that is not trivial for other forms of GP. In addition, the separation of the genotype from the phenotype in GE allows genetic operations to be applied not only to the phenotype, as in GP, but also to the genotype, extending the search capabilities of GP. GE is considered to be one of the most widely applied GP methods today [15].

3.1   Grammatical Evolution by Example

Representation in GE consists of a grammar and a chromosome, see Fig. 1. A genotype-phenotype mapping uses values from the chromosome to select production rules from the grammar, building up a derivation tree. The phenotype can be extracted from this tree's frontier.

The mapping begins with the start symbol, <e>. The value of the first codon, 12, is read from the chromosome. The number of production rules for the start symbol are counted, 2, <e><o><e> and <v>. The rule to be chosen is decided according to the mapping function i mod c, where i is the current codon value and c is the number of choices available, e.g, 12 mod 2 = 0, therefore the zeroth rule is chosen. <e> is expanded to <e><o><e>. This expansion forms a partial derivation tree with the start symbol as the root, attaching each of the new symbols as children. The next symbol to expand is the first non-terminal leaf node discovered while traversing the tree in a depth first manner. However, it should be noted that there is on-going study into variations on the method used to choose which node to expand next [19, 20]. In this case the left-most <e> is chosen. The next codon, 3, is read, expanding this <e> to <v> and growing the tree further. The next symbol, <v> is expanded using the next codon, 7. 7 mod 2 = 1, so the rule at index 1, Y, is chosen.

Derivation continues until there are no more non-terminal leaf nodes to expand, or until the end of the chromosome has been reached. If there are non-terminal leaf nodes left when the end of the chromosome has been reached, derivation can proceed in one of a few different manners. For example, a bad fitness can be assigned to the individual, so it is highly unlikely that this individual will survive the selection process. Alternatively the chromosome can be

Fig. 1. Example GE grammar, chromosome and resulting derivation tree

wrapped, reusing it a predefined number of times. If after the wrapping limit has been reached and the individual is still invalid, it could then be assigned a bad fitness. The complete derivation tree for this example is shown in Fig. 1.

4 Tree-Adjunct Grammatical Evolution

TAGE, like GE, uses a representation consisting of a grammar and a chromosome. The type of grammar used in this case is a TAG rather than a CFG. A TAG is defined by a quintuple (T, N, S, I, A) where:

1. T is a finite set of terminal symbols;
2. N is a finite set of non-terminal symbols: $T \cap N = \emptyset$;
3. S is the start symbol: $S \in N$;
4. I is a finite set of finite trees. The trees in I are called *initial trees* (or α trees). An initial tree has the following properties:
   - the root node of the tree is labeled with S;
   - the interior nodes are labeled with non-terminal symbols;
   - the leaf nodes are labeled with terminal symbols;
5. A is a finite set of finite trees. The trees in A are called *auxiliary trees* (or β trees). An auxiliary tree has the following properties:
   - the interior nodes are labeled with non-terminal symbols;
   - the leaf nodes are labeled with terminal symbols apart from the foot node which is labeled with the same non-terminal symbol as the root; the convention in [10] is followed and foot nodes are marked with *.

An initial tree represents a minimal non-recursive structure produced by the grammar, i.e., it contains no recursive non-terminal symbols. Inversely, an auxiliary tree of type X represents a minimal recursive structure, which allows recursion upon the non-terminal X [14]. The union of initial trees and auxiliary trees forms the set of *elementary trees*, E; where $I \cap A = \emptyset$ and $I \cup A = E$.

During derivation, composition operations join elementary trees together. The adjunction operation takes an initial or derived tree a, creating a new derived tree d, by combining a with an auxiliary tree, b. A sub-tree, c is selected from a. The type of the sub-tree (the symbol at its root) is used to select an auxiliary tree, b, of the same type. c is removed temporarily from a. b is then attached to a as a sub-tree in place of c and c is attached to b by replacing c's root node with b's foot node. An example of TAG derivation is provided in Section 4.1.

4.1 Tree-Adjunct Grammatical Evolution by Example

TAGE generates TAGs from the CFGs used by GE. Joshi and Schabes [10] state that for a *"finitely ambiguous CFG*[1] *which does not generate the empty string, there is a lexicalised tree-adjunct grammar generating the same language and tree set as that CFG"*. An algorithm was provided by Joshi and Schabes [10] for generating such a TAG. The TAG produced from Fig. 1 is shown in Fig. 2.

---

[1] A grammar is said to be *finitely ambiguous* if all finite length sentences produced by that grammar cannot be analysed in an infinite number of ways.

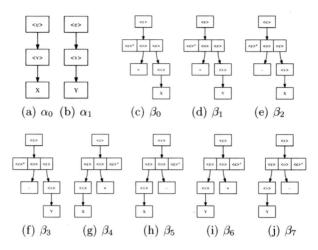

(a) α_0 (b) α_1 (c) β_0 (d) β_1 (e) β_2

(f) β_3 (g) β_4 (h) β_5 (i) β_6 (j) β_7

Fig. 2. Initial and auxiliary tree sets of the TAG produced from the CFG in Fig. 1

Derivation in TAGE is different to GE. A TAGE derivation tree is a tree of trees. That is to say a node in a TAGE derivation tree contains an elementary tree. The edges between those nodes are labeled with a node address of the tree in the parent derivation node. It is at this address that the beta tree in the child node is to be adjuncted. The derived tree in TAGE is a tree of symbols, similar to GE's derivation tree, resulting from the application of the adjunction operations defined in the TAGE derivation tree.

Given the TAG G, where $T = \{x, y, +, -\}$, $N = \{< e >, < o >, < v >\}$, $S =< e >$ and I and A are shown Fig. 2, derivation, using the chromosome from Fig. 1, operates as follows. An initial tree is chosen to start derivation. The first codon value, 12, is read and is used to choose an initial tree based on the number of trees in I. Using the same mapping function as GE, 12 mod 2 = 0, the zero-th tree, α_0, is chosen from I. This tree is set as the root node of, t, the derivation tree, see Fig. 3(a).

Next we enter the main stage of the algorithm. A location to perform adjunction must be chosen. The set N is created of the adjunct-able addresses available within all nodes(trees) contained within t. An adjunct-able address in a tree is the breadth first traversal index of a node labeled with a non-terminal symbol of which there is an auxiliary tree of that type, and there is currently no auxiliary tree already adjoined to the tree at that index. In this case N = $\{\alpha_0 0\}$, so a codon is read and an address is selected from N, 3 mod 1 = 0 indicating which address to choose, N[0]. Adjunction will be performed at $\alpha_0 0$, or index 0 of tree α_0, <e>. An auxiliary tree is now chosen from A that is of the type 1, i.e., the label of it's root node is 1, where 1 is the label of the node adjunction is being performed at. In this case 1 = <e>. Since there are 8 such trees in A, 7 mod 8 = 7, β_7 is chosen. This is added to t as a child of the tree being adjoining to, labeling the edge with the address 0, see Fig. 3(b). The adjunct-able addresses in β_7 will be added to N on the next pass of the algorithm. This process is repeated

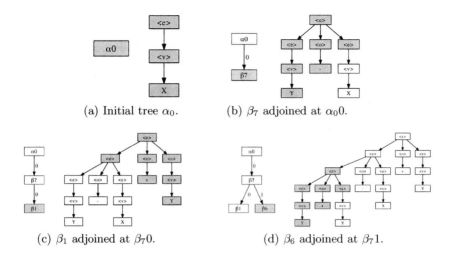

(a) Initial tree α_0.   (b) β_7 adjoined at $\alpha_0 0$.

(c) β_1 adjoined at $\beta_7 0$.   (d) β_6 adjoined at $\beta_7 1$.

Fig. 3. The derivation tree and corresponding derived tree at each stage of derivation in TAGE. The shaded areas indicate the new content added to the tree at each step.

until all remaining codons have been read. The resulting derivation and derived trees at each stage of this process can be seen in Fig. 3.

5 Experiments and Results

The aim of this study is to compare GE and TAGE IFM landscapes in order to ascertain some insight into how TAGE improves the algorithm's performance. In order to compare landscapes bounds must be set on the size of the landscapes. Since the size and form of solutions are rarely known a priori, the grammars used in GE tend to be recursive. As a result the structural space of possible solutions is infinite, and hence the landscape is infinite, restricted only by the number of codons available to the mapping procedure. This applies to both TAGE and CFG since they generate the same language.

5.1 Experimental Setup

In order to restrict the landscapes a specific number of codons, N, is selected as the maximum length of a TAGE chromosome. A value for N is chosen for each problem examined in order to sufficiently restrict the size of the landscapes. At each chromosome length, from one to N, an enumeration of all possible chromosomes is performed, building up the representation space, R. It is required for TAGE to process each increasing length of chromosome since with mutation alone, the number of codons used when mapping cannot change and hence TAGE would not be able to represent the same set of phenotypes as GE.

The enumeration is performed by initially selecting a chromosome of all zeros. At each position along the chromosome, every possible IFM is independently performed. That is to say, the mapping procedure is stopped at the each codon and the total number of possible choices at that codon is counted. This indicates how many different IFMs can be applied at each codon, creating the set of all chromosomes one IFM away from the original. Each of these neighbouring chromosomes are mapped, if both the original and the neighbour is valid, i.e., if the chromosome maps to an executable solution (for TAGE this is not an issue, since all chromosomes are valid), an edge/connection is recorded between them. If the neighbour has never been observed, it is added to a set of chromosomes from which new initial chromosomes are drawn to repeat this process.

Once this set of chromosomes is depleted, the chromosome length is incremented and the process repeated with a new initial chromosome. The process halts when the all chromosomes of length N have been processed.

The resulting phenotypes are used to repeat the above process for GE. Rather than setting a chromosome length limit, the length is incremented until the set of phenotypes generated contains the set of phenotypes generated by TAGE.

5.2   Problems

Standard GE was compared to TAGE using four classic benchmark problems taken from the GP literature. The CFGs used by GE and to generate TAGs for each problem are shown in Fig. 4.

Even-5-parity: The five input even-parity boolean function, in which the best fitness is obtained when the correct output is returned for each of the 2^5 test cases. A value of 3 was used for N.

Symbolic Regression: The classic quartic function, $x + x^2 + x^3 + x^4$. Fitness is the sum of the error across 20 test cases drawn from the range $[-1, 1]$. Successful solutions have an error less than 0.01, as described in [13]. A value of 5 was used for N.

Six Multiplexer: The classic GP two input and four output line boolean function. Fitness is measured by how many of the 64 test cases generate correct outputs. A value of 3 was used for N.

Max: This problem, as described in [3], aims to evolve a tree whose growth is constrained by a depth limit, that when the tree's phenotype is executed, returns the largest value possible. A function set of $\{+, *\}$ and a terminal set of $\{0.5\}$ are used. A max tree depth of 8 was used for the purposes of these experiments. A value of 9 was used for N.

5.3   Visualisations

Viewing the landscapes as 2D graphs is not feasible due to their large size and high complexity. 2D heat maps are used instead to map the connections in the landscape. Heat maps are little used in GP literature and are an effective way of graphically representing data in a 2 dimensional map, where the values of the variable being visualised are represented as colours.

```
Even-5 parity grammar:                      Max grammar:
<prog>   ::= <expr>                          <prog> ::= <expr>
<expr>   ::= <expr> <op> <expr>              <expr> ::= <op><expr><expr>
         | ( <expr> <op> <expr> )                   | <var>
         | <var>                            <op>   ::= + | *
         | <pre-op> ( <var> )               <var>  ::= 0.5
<pre-op> ::= not
<op>     ::= "|" | & | ^
<var>    ::= d0 | d1 | d2 | d3 | d4

Symbolic Regression grammar:                Six Multiplexer grammar:
<expr> ::= (<op><expr><expr>)               <B> ::= (<B>)&&(<B>)
       | <var>                                    | (<B>)"||"(<B>)
<op>   ::= + | - | *                               | !(<B>)
<var>  ::= x0 | 1.0                                | (<B>) ? (<B>) : (<B>)
                                                   | a0 | a1 | d0| d1 | d2 | d3
```

Fig. 4. CFG grammars in Backus-Naur form used for all the benchmark problems

Rather than using the genotypic landscape, i.e., where each vertex represents a single genotype from the representation, the phenotypic landscape is used, since comparing the genotypes of two different representations may not be useful and both representations in question generate the same phenotypic space.

Connection Maps (CM) are heat maps where the set of commonly generated phenotypes label each axis. If the genotypes of two phenotypes are one IFM away, the shared cell is marked. CMs give insight into how well connected each phenotype is within the landscape. The denser the CM, the greater the representation's ability to move from one phenotype to another.

The CMs for both setups for each of the problems can be seen in Figs. 5 and 6. The axes of these figures are labeled with the phenotypes in ascending order of length, from the top left to the bottom right.

Frequency Maps aim to address one of problems with the CMs described above. CMs do not take into account that there may be more than one connection between two phenotypes. This can occur due to GE and TAGE having redundant mappings. However, neutral mutation was not allowed in this study. The frequency of connections between phenotypes is important since if one connection from a phenotype has a high frequency and all of the other connections

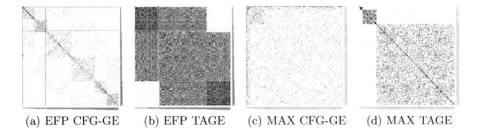

(a) EFP CFG-GE   (b) EFP TAGE   (c) MAX CFG-GE   (d) MAX TAGE

Fig. 5. Connection Maps: Even Five Parity (a) (b) for a max TAGE chromosome length of 3; Max (c) (d) for a max TAGE chromosome length of 9

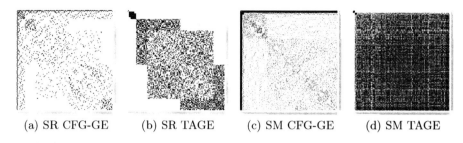

(a) SR CFG-GE   (b) SR TAGE   (c) SM CFG-GE   (d) SM TAGE

Fig. 6. Connection Maps: Symbolic Regression (a) (b) for a max TAGE chromosome length of 5; Six Multiplexer (c) (d) for a max TAGE chromosome length of 3

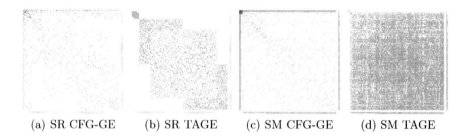

(a) SR CFG-GE   (b) SR TAGE   (c) SM CFG-GE   (d) SM TAGE

Fig. 7. Frequency Maps: Symbolic Regression (a) (b); Six Multiplexer (c) (d)

from that phenotype have a relatively low frequency of connections then there is a much higher probability that a mutation will follow the connections of high frequency. Frequency maps colour each cell from 25% grey (0) to red (200+) depending on the cell's degree of connectivity. The upper bound of 200 connections was to ensure a feasible colour delta when colour coding the maps due to the large number of relatively low frequency cells and a small number of much higher frequency cells. Frequency maps for only the symbolic regression and six multiplexer problems can be seen in Fig. 7 due to space restrictions.

6   Discussion

The CMs in figures 5 and 6 show that phenotypes in the TAGE landscapes, across the problems/grammars examined, are much more connected than the same phenotypes in the GE landscapes. This might not necessarily improve TAGE's ability to move from one phenotype to another of better fitness since the concept of fitness is not present in the CM plots. It was however noted by Murphy et al. [16] that TAGE maintains a much larger fitness variance within the population than GE. It was suggested that this variance, as a result of a more diverse population, could help TAGE avoid getting stuck at local optima [16]. The high degree of connectivity visible here could be attributed with helping to increase diversity within the population.

Interestingly, it can also be seen that mutation alone is not sufficient for TAGE to explore the entire search space. Unlike GE where an IFM can reduce the effective size of the chromosome, TAGE makes use of the entire chromosome and as a result IFM cannot change the size of a TAGE derivation tree. In order to enable a full exploration of the search space additional operators capable of changing the length of the chromosome would be needed. This would be as simple as a codon insertion/deletion operator or more complex such as one point crossover [16]. The clusters of connections in the top left corner of each of the TAGE sub-figures are the connections between the shorter chromosomes generated during setup, the remainder of the cells are white due to the lack of connections with the phenotypes of the larger chromosome.

Furthermore, the frequency maps in Fig. 7 show that in GE, the phenotypes produced from a smaller amount of the chromosome have a disproportionately high frequency of connections amongst themselves (see the red cells in the top left corner of (a) and (c)), and to a lesser extent with the rest of the phenotypes (left and top borders of (a) and (c)). In some cases the frequency of connections of these cells are orders of magnitude greater than the frequency of connections of the larger phenotypes. This indicates that the CFGs used in this study have a bias towards shorter phenotypes. A bias that doesn't appear in the frequency maps of TAGE's landscapes. This feature of TAGE may help avoid some of the initialisation problems experienced by GE outlined by Harper [4]. For example, when the grammar is dermined to be *explosive* randomised individuals tend to be short having a lasting effect on the algorithms performance.

7   Conclusions

IFM landscapes were generated for a number of problems using both CFG-based GE and TAGE. Viewing an entire landscape directly is very difficult [12]. As such, the landscapes were restricted in size, and a number of different plots were employed to enable indirect analysis and comparison of the landscapes.

For the problems and grammars used in this study, it was found that phenotypes in the TAGE landscapes have a much higher degree of connectivity to the rest of the phenotypes than their counterparts in the GE landscapes. This may help explain the increased diversity within TAGE populations observed previously. Moreover, it was discovered that the connectivity in the TAGE landscapes is much more evenly distributed between the other phenotypes in the landscape. Whereas in the GE landscape, shorter phenotypes are much more densely connected not only between themselves, but also, to a lesser extent, to the rest of the landscape.

This study presented a method for comparing large and highly complex landscapes using specific visualisation methods. This method of comparison can not only be further applied to the field of GE, but also to broader fields such as GP and genetic algorithms. Such an extension might enable better comparisons of each of the fields for a given problem, e.g., GP versus GE.

Future work arising from this study includes extending the method to other operators, allowing a better comparison of GE and TAGE; incorporating fitness into the CM method; and as mentioned above, comparing other representations with both GE and TAGE.

Acknowledgments

This research is based upon works supported by the Science Foundation Ireland under Grant No. 08/IN.1/I1868.

References

[1] Abbass, H., Hoai, N.X., McKay, R.I.: AntTAG: A new method to compose computer programs using colonies of ants. In: Proceedings of 2002 World Congress on Computational Intelligence, vol. 2, pp. 1654–1666. IEEE Press, Los Alamitos (2002)

[2] Dempsey, I., O'Neill, M., Brabazon, A.: Foundations in Grammatical Evolution for Dynamic Environments. SCI. Springer, Heidelberg (2009)

[3] Gathercole, C., Ross, P.: An adverse interaction between crossover and restricted tree depth in genetic programming. In: GECCO 1996: Proceedings of the First Annual Conference on Genetic Programming, pp. 291–296. MIT Press, Cambridge (1996)

[4] Harper, R.: GE, explosive grammars and the lasting legacy of bad initialisation. In: IEEE Congress on Evolutionary Computation (CEC 2010), Barcelona, Spain, July 18-23, IEEE Press, Los Alamitos (2010)

[5] Hoai, N.X.: Solving the symbolic regression with tree-adjunct grammar guided genetic programming: The preliminary results. In: Australasia-Japan Workshop on Intelligent and Evolutionary Systems, University of Otago, Dunedin, New Zealand, November 19-21 (2001)

[6] Hoai, N.X., McKay, R.I., Essam, D., Chau, R.: Solving the symbolic regression problem with tree-adjunct grammar guided genetic programming: The comparative results. In: Proceedings of the 2002 Congress on Evolutionary Computation CEC 2002, May 12-17, pp. 1326–1331. IEEE Press, Los Alamitos (2002)

[7] Hoai, N.X., McKay, R.I., Abbass, H.A.: Tree adjoining grammars, language bias, and genetic programming. In: Ryan, C., Soule, T., Keijzer, M., Tsang, E.P.K., Poli, R., Costa, E. (eds.) EuroGP 2003. LNCS, vol. 2610, pp. 335–344. Springer, Heidelberg (2003)

[8] Hoai, N.X. (Bob) McKay, R.I., Essam, D.: Representation and structural difficulty in genetic programming. IEEE Transactions on Evolutionary Computation 10(2), 157–166 (2006)

[9] Jones, T.: Evolutionary Algorithms, Fitness Landscapes, and Search. PhD thesis, University of New Mexico (1995)

[10] Joshi, A.K., Schabes, Y.: Tree-Adjoining Grammars. Handbook of Formal Languages, Beyond Words 3, 69–123 (1997)

[11] Joshi, A.K., Levy, L.S., Takahashi, M.: Tree adjunct grammars. Journal of Computer and System Sciences 10(1), 136–163 (1975)

[12] Koza, J.R., Poli, R.: Genetic programming. In: Search Methodologies: Introductory Tutorials in Optimization and Decision Support Techniques, ch. 5, pp. 127–164. Springer, Heidelberg (2005)

[13] Koza, J.R.: Genetic Programming. MIT Press, Cambridge (1992)

[14] Kroch, A., Joshi, A.K.: The Linguistic Relevance of Tree Adjoining Grammar, Technical Report, University of Pennsylvania (1985)

[15] McKay, R., Hoai, N., Whigham, P., Shan, Y., O'Neill, M.: Grammar-based genetic programming: a survey. Genetic Programming and Evolvable Machines 11, 365–396 (2010)

[16] Murphy, E., O'Neill, M., Galvan-Lopez, E., Brabazon, A.: Tree-adjunct grammatical evolution. In: 2010 IEEE World Congress on Computational Intelligence, IEEE Computational Intelligence Society, Barcelona, Spain, July 18-23, pp. 4449–4456. IEEE Press, Los Alamitos (2010)

[17] Nguyen, X.H. (Bob) McKay, R.I.: A framework for tree-adjunct grammar guided genetic programming. In: Post-graduate ADFA Conference on Computer Science, Canberra, Australia, pp. 93–100 (2001)

[18] Nilsson, N.J.: Problem-Solving Methods in Artificial Intelligence. McGraw-Hill Pub. Co., New York (1971)

[19] O'Neill, M., Brabazon, A., Nicolau, M., Garraghy, S.M., Keenan, P.: πGrammatical Evolution. In: Deb, K., et al. (eds.) GECCO 2004. LNCS, vol. 3103, pp. 617–629. Springer, Heidelberg (2004)

[20] O'Neill, M., Fagan, D., Galvan, E., Brabazon, A., McGarraghy, S.: An analysis of Genotype-Phenotype Maps in Grammatical Evolution. In: Esparcia-Alcázar, A.I., Ekárt, A., Silva, S., Dignum, S., Uyar, A.Ş. (eds.) EuroGP 2010. LNCS, vol. 6021, Springer, Heidelberg (2010)

[21] O'Neill, M., Ryan, C.: Grammatical Evolution: Evolutionary Automatic Programming in a Arbitrary Language. Genetic programming, vol. 4. Kluwer Academic Publishers, Dordrecht (2003)

[22] Pearl, J.: Heuristics: intelligent search strategies for computer problem solving. Addison-Wesley Longman Publishing Co., Inc., Boston (1984)

[23] Rich, E.: Artificial intelligence. McGraw-Hill, Inc., New York (1983)

ReNCoDe: A Regulatory Network Computational Device

Rui L. Lopes and Ernesto Costa

Centro de Informática e Sistemas da Universidade de Coimbra,
Polo II - Pinhal de Marrocos, 3030-290 Coimbra, Portugal
{rmlopes,ernesto}@dei.uc.pt

Abstract. In recent years, our biologic understanding was increased with the comprehension of the multitude of regulatory mechanisms that are fundamental in both processes of inheritance and of development, and some researchers advocate the need to explore computationally this new understanding. One of the outcomes was the Artificial Gene Regulatory (ARN) model, first proposed by Wolfgang Banzhaf. In this paper, we use this model as representation for a computational device and introduce new variation operators, showing experimentally that it is effective in solving a set of benchmark problems.

1   Introduction

Long time ago researchers saw the potential of using nature-inspired algorithms to solve learning, design and optimization problems, giving rise to a new research area called Evolutionary Computation ([3]). There are a plethora of different algorithms each of them tuned for a specific problem or class of problems, yet they all rely on a similar approach: (1) randomly define an initial population of solution candidates; (2) select, according to fitness, some individuals for reproduction with variation; (3) define the survivors for the next generation; (4) repeat steps (2) and (3) until some condition is fulfilled. Typically, the objects manipulated by the algorithms are represented at two different levels. At a low level, the genotype, the representations are manipulated by the variation operators; at a high level, the phenotype, the objects are evaluated to determine their fitness and are selected accordingly. Because of that, we need a mapping between these two levels.

One of the criticisms addressed to standard evolutionary algorithms, is related to the fact that they are a simplification of nature's principles, for they implement a simple and direct, one-to-one, mapping between the genotype and the phenotype. There is nothing in between, and there are no regulation links. Today we are aware of the crucial role of regulation in the evolution and development of complex and well adapted creatures ([2]). Regulation is a consequence of external (i.e., epigenetic) or internal (i.e., genetic) influences. Some computational explorations have been proposed to deal with the inclusion of regulatory mechanisms into the standard evolutionary algorithm. Wolfgang Banzhaf ([1]) proposed an

S. Silva et al. (Eds.): EuroGP 2011, LNCS 6621, pp. 142–153, 2011.

artificial gene regulatory (ARN) model as a novel representation for genetic programming, and showed how it could be used computationally in different settings ([8], [11]). In this paper, we present a new computational model which uses the ARN as representation, aimed at solving some of the previous models' limitations, namely the automatic definition of inputs and outputs, and we propose different variation operators, showing experimentally that it is effective and efficient in solving a set of benchmark problems.

The paper is organized as follows. Section 2 describes the ARN model as proposed by W. Banzhaf. Then, Section 3 describes our contributions, elucidating how we extract a program from a network and the new type of operators. In Section 4 we briefly refer to the problems used and the experimental setup. The results are presented and analyzed in Section 5. Finally, in Section 6 we draw some conclusions and present some ideas for future work.

2   Banzhaf's ARN

Genome. The Artificial Regulatory Network (ARN) ([1]) is composed of a binary genome and proteins. The genome can be generated randomly or by a process of duplication with variation (i.e., mutation), also called *DM*. In the latter case we start with a random 32-bit binary sequence, that is followed by several duplication episodes. The mutation rate is typically of 1%. So, if we have 10 duplication events then the final length of the genome is $2^5 \times 2^{10} = 32768$. The genome itself is divided in several regions. Each of those regions comprehend a regulatory site, the promoter and the gene itself. The first 32 bits of the regulation zone are the enhancer site, while the following 32 bits are the inhibitory site. The promoter comes downstream and has the form $XYZ01010101$. This means that only the last 8 bits are fixed. The gene is a five 32-bit long sequence, i.e., a 160-bit string.

Gene expression. The phenotype - genotype mapping is defined in the following way: each gene produces a 32-bit protein using a majority rule. If we consider the gene divided into 5 parts, at position i, say, the protein's bit will have a value corresponding to the most frequent in each of these 5 parts, at the same position. Figure 1 gives an idea of the representation.

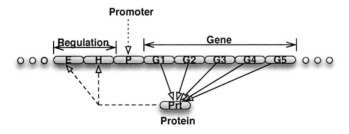

Fig. 1. Artificial Regulatory Network, after W. Banzhaf

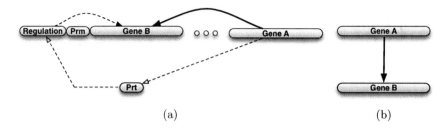

(a)   (b)

Fig. 2. Gene - Protein - Gene interaction

Regulation. The proteins can bind to the regulatory region. The strength of the binding is computed by calculating the degree of complementarity between the protein and each of the regulatory regions, according to formula 1:

$$e_i, h_i = \frac{1}{N} \sum_{j=1}^{N} c_j e^{\beta(\mu_j - \mu_{max})} \tag{1}$$

where N is the number of proteins, c_j the concentration of protein j, μ_j is the number of bits that are different in each position of the protein and of the regulation site, μ_{max} is the maximum match achievable, and β is a scaling factor. The production of a protein along time depends on its concentration, which in turn is a function of the way it binds to the regulatory regions. It is defined by the differential equation $\frac{dc_i}{dt} = \delta(e_i - h_i)c_i$.

Genes interact mediated by proteins. If, say, gene **A** expresses protein p_A and that protein contributes to the activation of gene **B**, we say that gene **A** regulates **B** (see figure 2).

Notice that in order for a link to exist between any two genes, the concentration of the corresponding protein must attain a certain level, and that depends on the strength of the binding.

Computational Device. Using this process we can build for each genome the corresponding gene regulatory network. These networks can be studied in different ways. We can be concerned by topological aspects (i.e., to study the degrees distribution, the clustering coefficient, small world or scale free, and the like) or the dynamics of the ARNs (i.e., attractors, influence of the protein-gene binding threshold) ([10], [8]). This is interesting, but from a problem-solving perspective what we want is to see how the model can be used as a computational device. In order to transform an ARN into a computational problem-solver we need to clarify what we put in the system (including the establishment of what is the relationship with the environment) and what we extract from the system. At the same time we need to define the semantics, that is, the meaning of the computation in which the network is engaged. Finally, and as a consequence of the points just identified, it is also fundamental to determine if we are interested in the input/output relationship or if what we want is the output. A solution for the latter situation was proposed in [7] in the context of optimization problems.

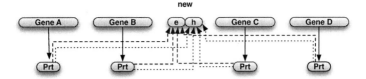

Fig. 3. Fabricating an output

The idea is to define (randomly) two new contiguous 32-bit sequences in the genome. The first one being a new inhibitory site, and the second one a new activation site. All generated proteins can bind to these sites. The levels of activation and of inhibition can be computed as before (see equation 1) , but there is no gene (thus no protein) attached (see figure 3).

The state of this site is just the sum of all bindings (see equation 2) and is defined as the output. This additional binding is thus a method to extract a meaning from the variation of the proteins' concentrations along time.

$$s(t) = \sum_i (e_i - h_i) \qquad (2)$$

To use the model as a representation formalism for genetic programming one needs to define what are the inputs and what are the outputs. For that purpose Banzhaf's model was extended in two directions ([11]). First, some extra proteins, not produced by genes but contributing to regulation, were introduced and act as inputs. Second, the genes were divided into two sets, one producing proteins that are used in regulation (i.e., transcriptional factors), and a second one with proteins without regulatory function which are used as outputs. These two types of genes are distinguished by having different promoters (see figure 4).

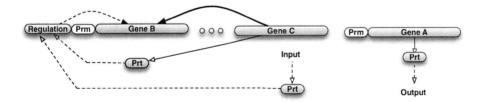

Fig. 4. The modified ARN

This model was tested with the classic control problem known as the pole balancing problem.

3   Architecture

The computational model of the ARN just described attained good results, yet it has some limitations, for e.g., the need for the artificial definition of new proteins, which can be overcome. Our work will explore two directions: on one hand, we will use the regulatory network, after a simplification, as an executable representation like in genetic programming; on the other hand, we will introduce special variation operators, besides mutation, that will increase the efficiency of the algorithm.

Extracting Circuits from ARNs. The networks resultant from ARN genomes are very complex, composed of multiple links (inhibition and excitation) between different nodes (genes). In order to extract a circuit from these networks they must first be reduced, input and output nodes must de identified, and we must ascribe a semantic to its nodes. The final product will be an executable feed-forward circuit.

We start by transforming every pair of connections, excitation (e) and inhibition (h), into one single connection with strength equal to the difference of the originals (e-h). Every connection with negative or null strength will be discarded. Then a directed graph is built adding only the nodes with active edges and the strongest edge between each pair of nodes. Next, the output node is chosen: the one with the highest connectivity. After that the circuit is built backwards from the output function until the terminals (nodes without input edges) are reached. If, at any point of this process, there is a deadlock (every function is input to some other), again the gene with highest connectivity is chosen, discarding its role as input to others, and thus resulting in a feed-forward circuit of *influences* between genes.

To complete the process, a mapping is needed linking nodes (i.e., genes) to functions and terminals. To that end we use the gene-protein correspondence using the protein's signature to obtain the function/terminal by a majority vote process. As an example, to code the function set { +, -, *, / } only two bits are needed. These are obtained by splitting the 32-bit protein into sixteen 2-bit chunks and applying the majority rule in each position of the chunks. This will provide us with an index to the function/terminal set.

Evolvability and Genetic Operators. Using a standard Evolutionary Strategy (ES), where mutation is the only operator changing the genomes, a tendency to increase the number of genes and the binding strength of the proteins was observed. This resulted in very complex circuits, suggesting not only that most of the mutations were perceived in the resulting phenotype (therefore loosing the potential of the neutral walks in the search space), but also that some of the small punctual mutations were translated into rather big modifications in the phenotype (and thus not exhibiting causality). In order to overcome these issues three biologically inspired operators were designed with different goals in mind.

First, there is the possibility of not using an operator, in which case only mutation is applied, maintaining the genome's length fixed. This is referred later as *idle*. Second, in order to increase the noncoding regions of the genome, a *junk* operator was added. This introduces a stream of zeros (noncoding DNA), at a random location in the genome. Due to the way the promoters are defined this implies that the number of genes will be lower, and increases the potential for neutral mutations. Third, inspired in nature, a *transposon-like* operator was introduced, that copies and pastes a random portion of the genome at a random location. Finally, a *delete* operator removes a random genome part, giving the system enough flexibility to manipulate the genome size and remove sections that may be decreasing fitness.

4   Problems and Experimental Setup

In order to evaluate the algorithm's efficacy and efficiency with the different operators it was applied to three benchmark problems: symbolic regression[6], artificial ant[9], and the cart-centering pole-balancing problems[5]. These will be described in the next subsections, followed by the experimental setup.

Symbolic Regression. The symbolic regression problem consists of evolving a mathematical formula to fit a polynomial expression to a set of points over some range.

The function set is composed by the arithmetic operators $\{+, -, *, /\}$, while the terminal set has only the input $\{x\}$. Protected division is used, returning 0 whenever the denominator is 0. Usually the arithmetic operators have a fixed arity of 2. In this work, however, the arity is defined by the regulatory network, in order to take more advantage out of function reusability. This will be clarified in Sect. 5.

As an example, the expression $x^6 - 2x^4 + x^2$ over the range $[-1, 1]$, with a 0.1 step was used as the target. The fitness function used was the mean absolute error over the set of points.

Artificial Ant. The goal in the artificial ant problem is to evolve a strategy that an agent will follow to collect food along a trail in a toroidal 2D grid. The artificial ant will be placed at the upper left corner facing east, in this grid of 32x32 cells where there is a specific trail of food pellets with a few gaps in between, in this case the Santa Fe Ant trail.

The functions set is composed by $\{\text{if-food-ahead, progn}\}$ and the terminal set consists of $\{\text{left, right, move}\}$. The function if-food-ahead allows the ant to check wether there is food in the cell directly in front of it and perform some action or group of actions according to this result. With progn the ant will perform a set of consecutive actions, depending on the arity of the function. This is the replacement for the usual prog2 and prog3, with variable arity, defined automatically by the network. The terminals allow the agent to turn 90 degrees left or right without moving, and to move to the adjacent cell it is facing. If a cell contains a food pellet it is immediately eaten when the ant stands on it.

Using the described functions and terminals a foraging strategy is build and each ant is given a determined number of steps, 400 in this case, to apply it repeatedly searching for the 89 food pellets available in the trail. The fitness function used for this problem is the remaining number of food pellets, after the foraging strategy is applied.

Cart Centering - Pole Balancing. In this problem there is a cart with an inverted pendulum, the pole, in a finite length 1D track. The goal is to keep to pole balanced and the cart inside the track limits by successively applying a constant force in either direction (left or right) [11].

The function set used is the same as in the symbolic regression problem. The terminal set is composed by the position and velocity of the cart, as well as the angle and angular velocity of the pole. Using these building-blocks a controller is constructed which output is translated into a push to the left if negative, or to the right if positive.

Each individual was tested over 50 randomized trials, returning the mean fitness of the set. The individual fitness of each trial is defined as in [11]: $F(x) = 120000/number\_of\_successful\_steps$.

After the evolution process the controller is tested for generalization over $4 \times 5^4 = 625$ trials, spread over the variable spectrum, during 1000 steps, as described in [12,11].

Experimental Setup. A standard evolutionary strategy was used for every experiment that will be described: (50+100)-ES. The process selects the best 50 solutions generating two new individuals from each using the bit-flip mutation and the genetic operators described earlier.

Table 1. Parameter values for the benchmark problems

Parameter	Regression	Artificial Ant	Cart-Pole
Number of Runs	30	30	50
Initial Population	100	100	100
Max. Evaluations	10^6	10^6	10^5
Number of DMs	6	5	6
Mutation Rate	0.01	0.01	0.01
Protein Bind Threshold	16	16	16
Genome Length	Fix/Var	Fix/Var	Var
Operator Type	Fix/Var	Fix/Var	Fix
Operator Length	-	-	40

Table 1 summarizes the parameterizations used to solve the different problems. For the symbolic regression and the artificial ant all experiments were run 30 times, initialized from 30 sets of 100 offline generated genomes. These genomes were generated with 6 and 5 DM events respectively, and a mutation rate of 1%. The genomes for the cart-pole problem were randomly generated for every run, with 6 DM events. Evolution terminates when a solution for the problem is found or the maximum number of evaluations is reached.

The two operators *junk* and *transposon* were tested independently and compared for different sizes of copying/deleting, including variable size (from 1 bit to the length of the genome). The base probabilities to select which operation will be applied (*idle, junk/transp, del*) are defined as follows: 50% of the cases only mutation will be applied (*idle*); 30% of the times the operator (*junk* or *transposon*) will be applied before mutation; the remaining 20% will apply the delete operator before mutation.

Statistical Validation. The different experiments with operator length that were run with both the Ant and Regression problems, aim at understanding wether the operator introduction increased efficiency or not. To compare the average evaluation number of the experiments the results' statistical significance must be verified with non-parametric tests. First, the Kruskal-Wallis test was applied for each problem and each operator, at a significance level of 5%. Then, for the problems with statistical significance in any of the versions ($p < 0.05$), post-hoc tests were run. These consisted in 5 pair-wise comparisons of the *Idle* version as control group against the *Fixed* versions, using the Mann-Whitney test with the Bonferroni correction for multiple comparisons (therefore the significance level is now $0.05/5 = 0.01$) [4].

5   Results and Analysis

In this section the results obtained for the three benchmark problems will be presented and briefly discussed. First, efficacy and efficiency of the operators will be analyzed with the symbolic regression and the artificial ant problems, presenting also examples of evolved circuits. Finally, the generalization results for the cart-pole problem controller will be shown.

Symbolic Regression and Artificial Ant. For both the problems, as described in the previous section, different operator sizes were tried. Table 2 shows that the algorithm is able to solve the problems in the given time (maximum number of evaluations), using both operators, for most of the experimented sizes. In Table 3 we can see the average effort (number of evaluations) for each version of each problem.

Although the application of one of the operators appears to perform better than the version without operators (*Idle*), the difference may not be significant. This is clarified in Table 4, which presents the statistical validation of the results, using the non-parametric Kruskal-Wallis test.

In the Artificial Ant problem it is clear that the operators introduced were successful at improving the evolvability of the binary genome, showing that it is important to have a flexible length genotype. As we can see from Table 3 both operators perform better than the version without them (where the genomes have fixed length). The *transposon-like* operator is more conservative than the *junk* and for the biggest duplication sizes experimented is not significantly better compared to not using it (see Table 5).

In the regression problem, the *junk* operator did not perform as well as in the previous problem. Moreover, although the averages are lower when an operator is used with moderated size (Table 3), the difference is not statistically significant for any of the experimented sizes (Table 4).

Table 2. Summary of the efficacy results for the Artificial Ant (ANT) and Regression (SYMB) problems. For each different operator/operator size is presented the amount of experiments that reached a solution (with N = 30).

Artificial Ant				Regression		
Op.	Size	Success Rate (%)		Op.	Size	Success Rate (%)
Junk	Var.	100		Junk	Var.	100
	Idle	86.(6)			Idle	100
	20	100			20	100
	40	100			40	100
	80	100			80	93.(3)
	120	100			120	83.(3)
	200	93.(3)			200	83.(3)
Transp	Var.	96.(6)		Transp	Var.	100
	Idle	93.(3)			Idle	100
	20	100			20	100
	40	100			40	100
	80	96.(6)			80	100
	120	100			120	100
	200	100			200	100

Table 3. Summary of the results for the Artificial Ant (ANT) and Regression (SYMB) problems. For each different operator size the minimum, mean and standard deviation are presented.

Artificial Ant					Regression				
Op.	Size	Min.	Mean	Std.Dev.	Op.	Size	Min.	Mean	Std.Dev.
Junk	Idle	8500	227155	347911.017	Junk	Idle	700	47253.33	67099.179
	Var.	2100	50578.33	92459.687		Var.	1700	76350	100278.501
	20	3500	30441.67	56273.927		20	3200	52306.67	92426.138
	40	2500	16485	14738.591		40	1400	94916.67	180569.030
	80	1500	54545	136523.971		80	4600	184720	316263.296
	120	2300	48195	104943.276		120	3700	230073.33	366158.183
	200	2500	81691.67	250079.688		200	4400	230056.67	384972.093
Transp	Idle	3500	130321.67	246709.449	Transp	Idle	4400	76190	154310.136
	Var.	7500	120105	194630.643		Var.	4000	44023.33	61152.228
	20	4000	46961.67	58305.923		20	2800	43956.57	93006.925
	40	1700	31918.33	60022.873		40	2000	40436.67	67218.811
	80	6000	72948.33	177657.373		80	2100	39350	72626.022
	120	4500	57545	57298.356		120	500	27840	41376.468
	200	2600	44851.67	44945.428		200	3100	50213.33	105132.031

Table 4. Kruskal-Wallis test grouping cases by operator size

Problem	Variant	p
Ant	Junk	0.000
	Transp	0.007
Regression	Junk	0.173
	Transp	0.489

Table 5. Mann-Whitney tests, using the *Idle* version as control group against the *Fixed* versions for the Artificial Ant

Size	Junk	Transposon
	p	
20	0.000	0.044
40	0.000	0.001
80	0.000	0.367
120	0.000	0.610
200	0.000	0.156

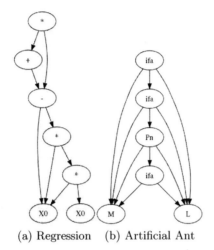

(a) Regression   (b) Artificial Ant

Fig. 5. Example of solution circuits for both problems. In a) the functions are as described in Sect. 4. The terminal *X0* represents the input x. In b) the terminals are represented by {M, L, R}, for move, left, and right, respectively. The function set is represented by {ifa, Pn}, for the corresponding if-food-ahead and progn.

Finally, Figure 5 shows a typical solution for each of these problems. In those, we can see that some nodes (and their sub-trees if they exist) are reused in different layers of the circuit, providing a functionality similar to GP automatically defined functions (ADF).

Cart Centering - Pole Balancing. In the case of the cart centering and pole balancing problem the results for one set of parameters will be shown. As described in Sect. 4 when the maximum number of evaluations was reached the controller's evolution was stopped and it was tested for generalization.

Table 6 summarizes the generalization results and the plot of the generalization tests for the best run is presented in Fig. 6. As we can see there is a big improvement in the generalization results of the controller, compared to the results described in [11]. Moreover, the average value of generalization failures is very close to the best found.

Table 6. Summary of the generalization tests for the cart-pole problem, by number of failures

Op.	N	Best	Mean	Std. Dev.
Transp	50	128	146.32	12.629

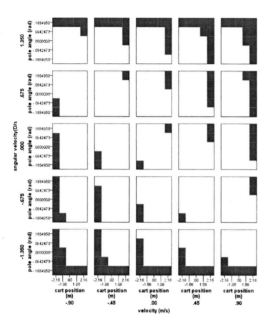

Fig. 6. Best generalization result for the cart-pole problem. Each plot has the position and angle as, respectively, x and y. In the plot matrix, from left to right, we have increasing velocity; from top to bottom we have decreasing angular velocity. The filled positions represent the initial position from which the controller failed to balance the pole.

6   Conclusions and Future Work

In this work we described a method to extract a circuit from an ARN, which was named Regulatory Network Computational Device. This method uses an ARN as the genome representation, and the protein signatures and their binding strengths are mapped to functions and connections between them. The efficacy of this technique was assessed by successfully solving three different benchmark problems. The second contribution was the introduction of biologically inspired operators aimed at increasing the method's efficiency. Both operators proved to increase performance. However, the results were not statistically significant for the regression problem.

Although the first results were promising, further research is needed to address some issues of the proposed model. The operators should be investigated aiming at reducing the results dispersion. Since the differences in the length of the operators were not conclusive, much more values should be experimented for this parameter. Apparently the number of DM events needed to initialize the solutions is important but appears to be problem dependent. This should be carefully analyzed with different values for this parameter, as well as its influence in the operators' performance.

Finally, to address different classes of problems, it is important to introduce feedback. This is already being developed, using one level of loops that are otherwise removed during the circuit extraction process. The results of some preliminary experiments have been encouraging.

References

1. Banzhaf, W.: Artificial Regulatory Networks and Genetic Programming. In: Riolo, R.L., Worzel, B. (eds.) Genetic Programming Theory and Practice, ch. 4, pp. 43–62. Kluwer, Dordrecht (2003)
2. Davidson, E.H.: The regulatory genome: gene regulatory networks in development and evolution. Academic Press, London (2006)
3. Eiben, A.E., Smith, J.E.: Introduction to Evolutionary Computing. Springer, Heidelberg (2003)
4. Field, A.P., Hole, G.: How to design and report experiments. Sage Publications Ltd., Thousand Oaks (2003)
5. Koza, J., Keane, M.: Genetic breeding of non-linear optimal control strategies for broom balancing. Analysis and Optimization of Systes 144, 47–56 (1990)
6. Koza, J.R.: Genetic Programming II: Automatic Discovery of Reusable Programs (Complex Adaptive Systems). MIT Press, Cambridge (1994)
7. Kuo, P., et al.: Evolving dynamics in an artificial regulatory network model. In: Yao, X., et al. (eds.) PPSN 2004. LNCS, vol. 3242, pp. 571–580. Springer, Heidelberg (2004)
8. Kuo, P.D., et al.: Network topology and the evolution of dynamics in an artificial genetic regulatory network model created by whole genome duplication and divergence. Biosystems 85(3), 177–200 (2006)
9. Langdon, W.: Why ants are hard. Cognitive Science Research Papers, 193–201 (1998)
10. Nicolau, M., Schoenauer, M.: Evolving specific network statistical properties using a gene regulatory network model. In: Raidl, G., et al. (eds.) GECCO 2009: Proceedings of the 11th Annual Conference on Genetic and Evolutionary Computation, pp. 723–730. ACM, Montreal (2009)
11. Nicolau, M., et al.: Evolving Genes to Balance a Pole. In: Esparcia-Alcázar, A.I., Ekárt, A., Silva, S., Dignum, S., Uyar, A.Ş. (eds.) EuroGP 2010. LNCS, vol. 6021, pp. 196–207. Springer, Heidelberg (2010)
12. Whitley, D., et al.: Alternative evolutionary algorithms for evolving programs: evolution strategies and steady state GP. In: Proceedings of the 8th Annual Conference on Genetic and Evolutionary Computation, GECCO 2006, pp. 919–926. ACM, New York (2006)

Statistical Distribution of Generation-to-Success in GP: Application to Model Accumulated Success Probability

David F. Barrero, Bonifacio Castaño*, María D. R-Moreno,
and David Camacho**

Universidad de Alcalá, Departamento de Automática
Crta. Madrid-Barcelona, Alcalá de Henares, Madrid, Spain
david@aut.uah.es, bonifacio.castano@uah.es,
mdolores@uah.es, david.camacho@uam.es

Abstract. Many different metrics have been defined in Genetic Programming. Depending on the experiment requirements and objectives, a collection of measures are selected in order to achieve an understanding of the algorithm behaviour. One of the most common metrics is the accumulated success probability, which evaluates the probability of an algorithm to achieve a solution in a certain generation. We propose a model of accumulated success probability composed by two parts, a binomial distribution that models the total number of success, and a lognormal approximation to the generation-to-success, that models the variation of the success probability with the generation.

Keywords: Generation-to-success, success probability, measures, models, performance metrics.

1 Introduction

Understanding the behaviour of Evolutionary Algorithms (EAs) is far from being trivial. One consequence of this fact is the difficulty of selecting a set of properties able to explain what is happening in the algorithm. It is not easy to understand why the algorithm fails (or success), or which is the effect of changing a parameter of the algorithm. Actually, there is no a general consensus about what should be measured, and thus different studies use different measures.

Some authors, however, have tried to provide some clues about which metrics should be used in which cases [1], even a classification of measures has been proposed by Bartz-Beielstein [2]. He differentiates effectivity and efficiency measures. The former informs about whether or not the algorithm is able to find a solution while the latter measures how many resources are required to achieve it. Despite the lack of a general consensus about which measures should be used

---

* Bonifacio Castaño is with the Departamento de Matemáticas, Universidad de Alcalá.
** David Camacho is with the Departamento de Informática, Universidad Autónoma de Madrid.

S. Silva et al. (Eds.): EuroGP 2011, LNCS 6621, pp. 154–165, 2011.

in the Genetic Programming (GP) experimentation, there are some widely used metrics, such as the success rate, the mean best fitness and the mean average fitness [1]. Metrics should be used with care since they represent different views of the object of study, and, as Luke observed with fitness and success rate [3], there may not be correlation between these measures.

Several metrics have been proposed to measure efficiency. In GP, Koza's computational effort [4] is a particularly popular measure. It estimates the minimum number of individuals that have to be processed to achieve, at least, one success with a certain fixed probability. Computational effort has been widely used in GP research to measure, fixing an algorithm, the difficulty of a problem or, on the contrary, fixing the problem, the performance of an algorithm. Other measures such as the success effort [5], tries to merge the measurement of efficiency and effectivity into one integrated statistic. All these composed measures share the use of the success probability.

Due to the stochastic nature of GP, there is no guarantee that the algorithm would achieve a success: the algorithm might find, or not, a solution. The success probability provides information about this property, and its calculation is straightforward; it is just the ratio between the number of success and the number of trials, or runs, in the experiment.

The main contribution of this paper is the proposal of a model of success probability in generational GP. This model considers the existence of two different -although related- problems: whether the algorithm is able to find a solution, and, given that it has been found, when that happens. The model is based on the observation that the run time, measured as generations required to find a solution, follows a lognormal distribution. If the generation is fixed, a classical binomial distribution is derived from our model [6]. Following it we discuss some practical applications of the model. For instance, given that the generation where the algorithm finds a solution (i.e. the *generation-to-success*) could be described with a known probability distribution, it would be determined when the algorithm is more likely to find a solution, and therefore, use this information to set the maximum number of generations in a well grounded way.

The paper is distributed as follows. Firstly some definitions are introduced to aid identify the semantics of the terms, then, in section 3 a general model of accumulated success probability is proposed. This general model assumes that the run time to find a solution is described by a distribution function which is empirically identified in section 4. After that, the complete model is presented in section 5, followed by a validation of the results using some experiments. Section 7 discusses some applications of the proposed model. The paper finishes with a description of the related work, an outline of the conclusions and the future work.

2 Initial Definitions

In order to clarify the terminology used in this work, we first define some terms that will be widely used. We define the *generation-to-success* as the generation in which the execution of the algorithm achieves its first success. We do not think that other similar terms, such as run time, convergence generation or generation-to-termination [5] are suitable for our purposes because they include all the executions, regardless whether they found a solution or not, and we need to exclude from the definition all runs that were unable to find a solution.

A *run* is a single execution of an EA, while an *experiment* is a collection of n independent runs. Due to the random nature of EAs, many of their properties are stochastical, and thus they cannot be characterized using a run, but with an experiment. One of these properties is the *accumulative success probability*, or, using Koza's notation [4], $P(M, i)$, where M is the population size, and i the generation number. $P(M, i)$ is calculated as the ratio between the number of successful runs in generation i, $k(M, i)$, and the number of runs n in the experiment.

$$P(M, i) = \frac{k(M, i)}{n} \tag{1}$$

This estimator of the success probability is also its maximum likelihood estimator [7]. We define *Success Rate* (SR) as the accumulated success probability in an infinite number of generations, so $SR = \lim_{i \to \infty} P(M, i)$. The reader would agree with us if we state that running the algorithm for an infinite number of generations is not a general practice. Usually an experiment is run for a fixed finite number of generations, G, then the SR is given by $SR = \lim_{i \to \infty} P(M, i) \approx P(M, G)$. Since the true SR can hardly be measured in experiments, $P(M, G)$ is just an approximation to SR, and thus it can be seen from a statistical perspective as an estimator $\hat{SR} = P(M, G)$.

There is a relationship among SR, accumulated success probability and generation-to-success that is described in the next section.

3   A General Model of Accumulated Success Probability

Modeling the behaviour of success probability during execution of the generational algorithm is more difficult than modeling the number of success in a fixed generation. The latter case involves two variables: the number of success and the generation where they were found, while the former only involves the number of success.

Some of the properties of the dynamic behaviour in GP are shown by the generation-to-success. Given that a run k that has been a success, its generation-to-success g_k is a discrete random variable that can take any non-negative integer number. Let us suppose that its probability mass function is known, as well as its cumulative distribution function (CDF).

Accumulated success probability can be decomposed into two different, but related, components. One reflects how likely is the algorithm to find a solution; the other one is a probability that gives the proportion of runs that get a success before generation i. With these considerations we can state that, given an algorithm that is run for G generations, the accumulative success probability can be expressed as

$$P(M, i) = SR\ F(i) \tag{2}$$

where $F(i)$ is the CDF of the generation-to-success, and represents the probability $P(g_k < i)$ and $SR \approx \frac{k(M,G)}{n}$ is the estimation of the SR calculated in generation G. This equation provides an alternative representation of the accumulated success probability. Equation (2) has a underlying binomial nature. If we fix the generation number to, for instance, the last generation, G, and assuming that G is large enough, $F(G) \approx 1$ and thus $P(M, G) \approx k(M, G)/n$. By definition, $k(M, G)$ is a binomial random variable.

Unfortunately, the model given by (2) does not provide a close form of $P(M, i)$, but rather changes the problem, we need the CDF of generation-to-success to make (2) useful. Nevertheless, this problem is easier to address because it is a classical problem of model fitting [7]. An empirical study with some statistical tools could provide the knowledge we need to complete the model.

4   Empirical Model of Generation-to-Success

In order to fit the generation-to-success with a known distribution, we study this random variable in four classical GP problems introduced by Koza in his first book [4]: Artificial ant with the Santa Fe trail, 6-multiplexer, even-4-parity and a polynomial regression with no Ephemeral Random Constants (ERC)[1]. We have used the implementation of these problems found in ECJ v18 with its default parameter settings.

One problem that is found is the lack of an exact knowledge about the true accumulated probability of the four problems under study. This information is necessary to have a reference to compare the model with. Since the true value of $P(M, i)$ is extremely difficult (if not impossible) to achieve, we have used another approach. Each problem has been run a large number of times, in this way the value of accumulated success probability is rather accurate, and therefore we can suppose this is the exact one without a great loss of accuracy. Three problems were run $100,000$ times whereas the fourth problem (4-parity) was run $5,000$ times because the population was larger and required more computational resources.

Measurements obtained using all the runs are the best estimations we have available for this study, so the best estimation of the SR is named $\hat{p}_{best}$. Table 1 presents the number of runs (n), number of success found (k) and best estimation available of the SR $(\hat{p}_{best})$. Looking at the SR shown in Table 1 we deduce that the difficulty of the problems ranges from very hard (4-parity, with SR around 0.06) to very easy (6-multiplexer, $SR \approx 0.96$). There are two problems whose hardness is intermediate, the artificial ant ($SR \approx 0.13$) and the regression ($SR \approx 0.29$).

To find the distribution that better describes the generation-to-success, we have first performed an exploratory experiment representing the generation-to-success of the four problems under study as histograms overlapped with some

---

[1] All the code, configuration files, scripts and datasets needed to repeat the experiments can be obtained in http://atc1.aut.uah.es/~david/eurogp2011/

Table 1. Best estimations of the four problem domains under study. It reports number of runs (n), number of successful runs (k) and best estimation of SR $\hat{p}_{best}$.

	Artificial ant	6-Multiplexer	4-Parity	Regression
n	100,000	100,000	5,000	100,000
k	13,168	95,629	305	29,462
$\hat{p}_{best}$	0.13168	0.95629	0.061	0.29462

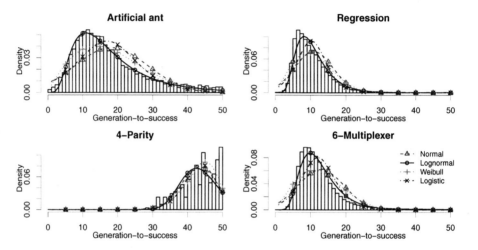

Fig. 1. Histograms of the generation-to-success of the four study cases compared with different probability density functions. The parameters of the functions have been calculated using maximum likelihood estimation. All the available successful runs have been used in the histogram and model fit.

common statistical distributions. The parameters of the distributions have been calculated with all the successful runs and using a maximum-likelihood method implemented in R's function `fitdistr()`. Histograms for the four study cases are depicted in Fig. 1. We tried to fit data against several different distributions (Poisson, Students' t, ...), but for clarity, only the four fittest distributions are shown in Fig. 1. All the runs of the study cases were used to plot the histogram and fit the distributions.

Fig. 1 suggests that the distribution that better fits our experimental data is the lognormal. Other distributions such as the logistic or Weibull might fit data as well. However, the lognormal describes the data surprisingly well. Looking at the other study cases, the lognormal distribution fits data quite well in all the cases, specially the artificial ant. The 4-parity problems presents one particularity: the number of generations used to run this study case were not enough, and thus the histogram appears truncated, but even in this case the

Table 2. Pearson's χ^2 test for fit of generation-to-success against a lognormal distribution. Parameters of the distribution were calculated using maximum likelihood estimation. χ^2 has been calculated with a confidence level $\alpha = 0.05$. Values of χ^2 that provide evidence to reject the null hypothesis have been marked in bold.

Problem	$\hat{\mu}$	$\hat{\sigma}$	df	χ^2	$\chi^2_{\alpha,df}$
Artificial ant	2.74	0.59	4	6.56	9.49
6-Multiplexer	2.46	0.43	4	0.73	9.49
4-Parity	3.76	0.12	4	**18.89**	9.49
Regression	2.29	0.44	4	0.88	9.49

generation-to-success is well fitted by a lognormal distribution, however in this case the other distribution functions also fit well. Another particularity of the 4-parity histogram is the rugosity of its shape, the other histograms are notably smoother. It is because the histograms only show successful runs, and in case of the 4-parity they are much lower (305) than, let say, the artificial ant (13, 168).

In order to give a more rigorous validation of the model, a Pearson's χ^2 test for fit against the lognormal distribution for the four study cases was done. The test used 300 random runs sampled from the datasets, and the parameters were estimated using maximum-likelihood with the whole datasets. Table 2 shows the results, including the estimated $\hat{\mu}$ and $\hat{\sigma}$ of the lognormal distribution. The degrees of freedom (df) are given by $df = k - p - 1$, where k is the number of bins and p the number of estimated parameters. The size and limits of the bins were manually set for each problem, they were selected to contain at least 7 observations to assure a fair test. Table 2 shows that the only problem where $\chi^2 > \chi^2_{\alpha,df}$, and thus we have to reject the null hypothesis, is the artificial ant, which is also the problem that fits worse to the lognormal due to the low number of generations used and the low number of successes found (see Fig. 1).

An explanation to the lognormality of the generation-to-success might be found in the Reliability Theory. The failure rate of a physical component usually follows the shape of an inverted tub. When the component is new, manufacturing defects produce high failure rates, but once the component has reached its maturity, its failure rate decreases and remains stable for a period of time. At the end of the component life cycle the degradation of its materials and other factors increases again its failure rate. A similar behaviour is observed in GP. In its early stages, evolution has not had enough time to find a solution, the search is almost a random search. Then, after some point, the algorithm begins to find more solutions, but if that solution is not achieved, the algorithm might have converged to a local maxima and it is unlikely to move out of there, reducing the instant success probability.

This relationship between GP and Reliability Theory provides another clue for the lognormality of the convergence generation. The modeling of the probability density of lifetime is a critical issue in the Reliability Theory, so this topic has been subject of intense research. There are three models that are widely used: exponential, when the component is memoryless (which is clearly not the case here); Weibull and lognormal. Experimentation reported in this section showed that the generation-to-success of the four study cases are well fitted with a lognormal distribution. This is not, of course, a solid theoretical explanation of the lognormality of generation-to-success, but shows a certain coherence in the model and is an interesting topic for further research.

Experimentation done in this section shows that it is reasonable to model the generation-to-success using a lognormal distribution. With this result we can complete the general model previously shown.

5   A Specific Model of Accumulated Success Probability

Experimentation reported above showed that the generation-to-success in the four study cases was well described by a lognormal distribution, so it seem reasonable to assume a lognormal distribution from this point. If we make this assumption, then it is straightforward to then deduce a model of $P(M, i)$ from (2) that could be used in practice. It is well known that the lognormal CDF [8] is given by

$$F(i; \hat{\mu}, \hat{\sigma}) = \Phi\left(\frac{\ln i - \hat{\mu}}{\hat{\sigma}}\right) \tag{3}$$

where $\Phi(...)$ is the standard normal CDF. If there are m runs that have converged in the experiment, and $g_k, k = 1, ..., m$ is the generation-to-success of run k, then

$$\hat{\mu} = \frac{\sum_{k=1}^{m} \ln g_k}{m} \tag{4}$$

and

$$\hat{\sigma} = \sqrt{\frac{\sum_{k=1}^{m} (\ln g_k - \hat{\mu})^2}{m}} \tag{5}$$

Using (2) and (3) yields that the accumulated success probability can be expressed as

$$P(M, i) = \frac{k(M, i)}{n} \Phi\left(\frac{\ln i - \hat{\mu}}{\hat{\sigma}}\right) \tag{6}$$

All the parameters involved in this equation are known by the experimenter.

6   Experimental Validation of the Model of Accumulated Success Probability

Although data has been fitted a lognormal distribution, there is no experimental support to claim that the model of accumulated success probability given by (6) is a correct model. So, additional experimental evidence is collected in this section.

Fig. 2 shows a comparison between $P(M, i)$ calculated using the standard maximum-likelihood method and the lognormal approximation. All the samples available in the datasets were used to calculate $P(M, i)$ with both methods.

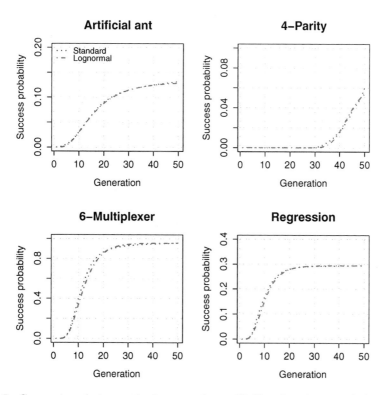

Fig. 2. Comparison between the best maximum-likelihood estimator of the accumulated success probability and the model approached using a lognormal distribution

It can be seen in Fig. 2 that both methods achieve very similar results, and thus, in the study cases under consideration, when using a large number of runs, our proposal achieves estimations of $P(M, i)$ pretty close to the standard method. Nevertheless, this experiment shows an unrealistic scenario since the computational cost of running an experiment with real-world problems imposes a maximum number of runs much lower than the used in this experiment.

A collection of experiments were simulated using different values of n. Given the whole set of runs stored in the previous experiments, 25, 50, 100 and 200 runs were resampled with replacement, $P(M, i)$ calculated using both methods and finally they were depicted in Fig. 3. To give more elements to compare with, the best estimation of $P(M, i)$ (calculated with all the runs) was also included in Fig. 3.

As can be seen in Fig. 3, when there are a high number of runs, the differences between the three curves tend to disappear, and the estimation with both methods tend to be closer to the best estimation available. More interesting is the relationship between the two methods, they yield similar estimations of the accumulated success probability, which is logical because they use the same data; if one method makes a bad estimation of the accumulated success probability,

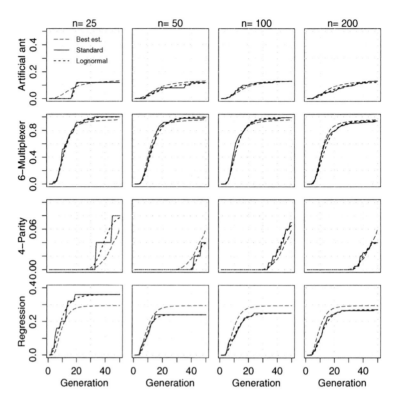

Fig. 3. Comparison between the best maximum-likelihood estimator of the accumulated success probability and the model approached using a lognormal distribution

the other one also makes a bad estimation. It leads us to an almost tautological conclusion: there is no magic. With a small number of runs there are not much information available, and without information, it is simply not possible to reach good conclusions.

Despite the lack of magic of the proposed method, Fig. 3 shows an interesting property: the proposed method is able to interpolate values using the experimental data, yielding much smoother curves than the standard method. And apparently, it can be done without sacrificing the accuracy of the measure. This fact is rather clear, for instance, in the 4-parity problem with $n = 25$. A similar property is the ability of the lognormal approximation to extrapolate values of the accumulated success probability. This is interesting in early generations, where there are no success due to a low, but not null, success probability. In these cases the standard method yields a null estimation of the success probability while the lognormal yields a non zero value.

Another interesting fact that can be found in Fig. 3 is the excellent estimation made in the 4-parity problem. Despite the fact that the experiment was run for too few generations and it was the domain with the poorest model fit, it generates a nice approximation to the maximum-likelihood estimator of the accumulated success probability. This fact could be quite interesting to reduce the number of generations needed to study the performance of GP using less computational resources, however we feel that this issue requires more study.

7   Discussion

Experiments that were carried out showed evidence about the good properties of the model and the lognormal nature of generation-to-success. In particular, how the generation-to-success of the four problems studied fits a lognormal distribution opens some interesting applications. For instance, it might be used to set the maximum number of generations that the algorithm should be run using a well grounded criteria.

The experimenter would carry out an exploratory experiment with a relatively low number of runs, then the estimators $\hat{\mu}$ and $\hat{\sigma}$ could be calculated. The determination of the number of runs needed to estimate the parameters is a classical problem in statistical inference, and the transformation between normal and lognormal is straightforward, $X \sim N(\mu, \sigma) \Rightarrow e^X \sim LN(\mu, \sigma)$ and $X \sim LN(\mu, \sigma) \Rightarrow \ln(X) \sim N(\mu, \sigma)$ [8], using this transformation the number of runs in the exploratory experiment can be determined using

$$n = \frac{z_{\alpha/2}^2 \, s^2}{e^2} \tag{7}$$

where e is the desired level of precision for the estimation of the mean, given in the same unit than s, s the standard error of the samples and $z_{\alpha/2}$ is the upper-$\alpha/2$ critical point from $N(0, 1)$ [9]. The probability of getting at least one success in generation i is given by $P(M, G) \, F(i; \hat{\mu}, \hat{\sigma})$, while the probability of not getting any success in generation G is

$$\varepsilon = 1 - P(M, G) \, F(G; \hat{\mu}, \hat{\sigma}) \tag{8}$$

This equation provides an estimation of the error ε that is introduced by limiting the maximum number of generations to G. Moreover, if we set a maximum error that is tolerable, we could calculate G from (8), yielding the maximum number of generations that the algorithm should be executed to achieve that error. This is a grounded method able to determine the maximum number of generations.

8   Related Work

Despite the importance of understanding the statistical properties of the several measures used in the experimentation of GP, there are not too many studies available. One of the metrics that has motivated some papers is Koza's computational effort. Angeline first observed that computational effort [10] is actually a random variable, and concluded that the stochastic nature of the computational effort should be handled with statistical tools. Some time after, Keijzer [11] calculated computational effort using confidence intervals (CIs), achieving a re-markable conclusion: when success probability is low, CIs of the computational effort are almost as large as the computational effort. In order words, the vari-ability of the computational effort is similar to its magnitude, and thus, in that case, the high dispersion of computational effort degrades its reliability.

To the author's knowledge, the only systematic attempt made to understand why the computational effort presents the variability observed by Keijzer was presented by Christensen [12]. He identified three sources of variability and pro-vided empirical data that gave some light to the circumstances that reduces the reliability of the computational effort. More research in this area was done by Walker [13,14], who studied how to apply CIs to the calculus of computa-tional effort, and Niehaus [15], who investigated the statistical properties of the computational effort in steady-state algorithms.

9   Conclusions and Future Work

One of the most important metrics in GP is the accumulated success probability. This measure reflects how likely is the algorithm to find a solution in a given generation, and has been widely used in research and practice to understand the performance, or compare different algorithms.

When we fix the generation in GP, the number of runs that finds a solution can be modeled using a binomial distribution, but its variation with the generation is no longer described with a binomial, and a more complex model is required. The dynamic behaviour of the algorithm can be modeled using the generation-to-success, or generation in which the run achieves its first success. Experiments suggested that generation-to-success can be fitted with a lognormal distribution.

The combination of SR, which is a binomial random variable, and the CDF of generation-to-success can model the accumulative success probability. Experi-ments carried out show that the accumulated success probability calculated with this model yields curves quite close to the maximum-likelihood method, with a smoother shape and it is able to extrapolate values in zones where no success run has been found. The lognormality of generation-to-success opens some in-teresting applications, such as a well grounded stopping condition in GP.

The experimentation that has been done used some well known problem do-mains in GP, and despite we do not find any reason that may limit the generality of the results. The accumulated success of probability is used in other Evolution-ary Computing disciplines, such as Genetic Algorithms, and it seems reasonable that we could expect the same behaviour of the generation-to-success. It could also be interesting to answer why the generation-to-success follows a lognormal distribution.

Acknowledgments. This work was partially supported by the MICYT project ABANT (TIN2010-19872) and Castilla-La Mancha project PEII09- 0266-6640.

References

1. Eiben, A.E., Smith, J.E.: Working with Evolutionary Algorithms. In: Introduction to Evolutionary Computing, pp. 241–258. Springer, Heidelberg (2009)
2. Bartz-Beielstein, T.: Tuning evolutionary algorithms: overview and comprenhensive introduction. Technical Report 148/03, University of Dortmund (2003)
3. Luke, S., Panait, L.: Is the perfect the enemy of the good? In: Genetic and Evolutionary Computation Conference, pp. 820–828. Morgan Kaufmann, San Francisco (2002)
4. Koza, J.: Genetic Programming: On the programming of Computers by Means of Natural Selection. MIT Press, Cambrige (1992)
5. Walker, M., Edwards, H., Messom, C.H.: Success effort and other statistics for performance comparisons in genetic programming. In: IEEE Congress on Evolutionary Computation, pp. 4631–4638. IEEE, Singapore (2007)
6. Barrero, D.F., Camacho, D., R-Moreno, M.D.: Confidence Intervals of Success Rates in Evolutionary Computation. In: GECCO 2010: Proceedings of the 12th Annual Conference on Genetic and Evolutionary Computation, Portland, Oregon, USA, pp. 975–976. ACM, New York (2010)
7. Montgomery, D.C., Runger, G.C.: Applied Statistics and Probability for Engineers, 4th edn. John Wiley & Sons, Chichester (2006)
8. Limpert, E., Stahel, W.A., Abbt, M.: Log-normal distributions across the sciences: Keys and clues. BioScience 51, 341–352 (2001)
9. Smith, M.F.: Sampling Consideration. In: Evaluating Cooperative Extension Programs. In: Florida Cooperative Extension Service Bulletin PE-1, Institute of Food and Agricultural Sciences. University of Florida (1983)
10. Angeline, P.J.: An investigation into the sensitivity of genetic programming to the frequency of leaf selection during subtree crossover. In: GECCO 1996: Proceedings of the First Annual Conference on Genetic Programming, pp. 21–29. MIT Press, Cambridge (1996)
11. Keijzer, M., Babovic, V., Ryan, C., O'Neill, M., Cattolico, M.: Adaptive logic programming. In: Proceedings of the Genetic and Evolutionary Computation Conference (GECCO 2001), pp. 42–49. Morgan Kaufmann, San Francisco (2001)
12. Christensen, S., Oppacher, F.: An analysis of koza's computational effort statistic for genetic programming. In: Foster, J.A., Lutton, E., Miller, J., Ryan, C., Tettamanzi, A.G.B. (eds.) EuroGP 2002. LNCS, vol. 2278, pp. 182–191. Springer, Heidelberg (2002)
13. Walker, M., Edwards, H., Messom, C.H.: Confidence intervals for computational effort comparisons. In: Ebner, M., O'Neill, M., Ekárt, A., Vanneschi, L., Esparcia-Alcázar, A.I. (eds.) EuroGP 2007. LNCS, vol. 4445, pp. 23–32. Springer, Heidelberg (2007)
14. Walker, M., Edwards, H., Messom, C.: The reliability of confidence intervals for computational effort comparisons. In: GECCO20'07: Proceedings of the 9th Annual Conference on Genetic and Evolutionary Computation, pp. 1716–1723. ACM, New York (2007)
15. Niehaus, J., Banzhaf, W.: More on computational effort statistics for genetic programming. In: Ryan, C., Soule, T., Keijzer, M., Tsang, E.P.K., Poli, R., Costa, E. (eds.) EuroGP 2003. LNCS, vol. 2610, pp. 164–172. Springer, Heidelberg (2003)

Learnable Embeddings of Program Spaces

Krzysztof Krawiec

Institute of Computing Science, Poznan University of Technology, Poznań, Poland
krawiec@cs.put.poznan.pl

Abstract. We consider a class of adaptive, globally-operating, semantic-based embeddings of programs into discrete multidimensional spaces termed *prespaces*. In the proposed formulation, the original space of programs and its prespace are bound with a learnable mapping, where the process of learning is aimed at improving the overall locality of the new representation with respect to program semantics. To learn the mapping, which is formally a permutation of program locations in the prespace, we propose two algorithms: simple greedy heuristics and an evolutionary algorithm. To guide the learning process, we use a new definition of semantic locality. In an experimental illustration concerning four symbolic regression domains, we demonstrate that an evolutionary algorithm is able to improve the embedding designed by means of greedy search, and that the learned prespaces usually offer better search performance than the original program space.

1 Introduction

Genetic Programming (GP) is an evolution-inspired search in space of programs guided by their performance in some environment. Its inherent feature is that a program (code, syntax) and its effect (semantics, behavior) are usually separated by multiple levels of abstraction, as opposed to conventional optimization tasks where the relationship between genotype and phenotype is typically much more direct. As a result, even minute changes of code can radically alter program's semantics and fitness, which makes GP fitness landscapes quite chaotic, often hampers scalability of search algorithms, and limits applicability of this methodology to difficult real-world problems.

This problem has been noticed in different contexts and its aftermaths identified and labeled by various terms. In the global perspective, it has been observed that fitness landscapes in GP typically have low fitness-distance correlation [13]. As a result, most crossovers are destructive and only a small fraction of them lead to increase in fitness [4]. In the local perspective, GP's genotype-phenotype mapping has low locality [6], which makes particularly hard designing mutation operators that behave in a way they should, i.e., have only minor effect in fitness.

In the eyes of many, this is an inherent property of programming and programming languages, and the attempts aimed at circumventing it are futile. This stance is particularly common among such adversaries of operational semantics like E. Dijkstra, who famously stated that "In the discrete world of computing, there is no meaningful metric in which 'small' changes and 'small' effects go hand in hand, and there never will be" [3]. Nevertheless, a growing body of research evidence in GP, mostly concerning design of

S. Silva et al. (Eds.): EuroGP 2011, LNCS 6621, pp. 166–177, 2011.

search operators, seems to suggest otherwise. There is a repertoire of search operators designed to be more aware of the abovementioned problems, like context-preserving crossover [1], semantically-aware crossover [9], or geometrical crossover [5], to name only a few. On the other end, there is a long list of alternative program representations for GP, like Linear Genetic Programming [2] and Cartesian Genetic Programming [7], some of which have been designed to, among others, circumvent the above problem.

The above approaches are *symbolic* in the sense that they manipulate symbols given *a priori* with problem specification (typically instructions), and usually they do that in abstraction from the actual meaning (semantics) of those symbols. While there are many reasons to believe that representing programs by structures of symbols is convenient for human programmers, it is not necessarily the best choice for an automated search process. Symbolic representations lack a natural and semantically coherent similarity relation. They are redundant, which itself does not have to be problematic, but their redundancy is difficult to capture – for instance, reversals of instruction order can be neutral for program outcome. Finally, they are highly epistatic – distant code fragments often strongly interact with each other, with complex effects on program semantics.

The primary objective of this study is to investigate alternative program representation spaces, in which programs are aligned in a way that is less related to their syntax and more to their semantics. Our main contribution is a concept of program representation space (*prespace*) that to some extent abstracts from the symbolic program code and delegates the process of automated programming to another domain. In short, rather than designing new search operators or devising new structures for program representations, we switch to prespace, embed the programs in it, and learn a mapping from that space to the original space, a mapping with some desirable properties.

2   Programs, Semantics, and Locality of Semantic Mapping

In abstraction from any specific search algorithm, we formalize in this section the notion of search space and its properties. Let P be a *program space,* and let $s(p)$ define the *semantics* of program $p \in P$, where $s : P \to S$ is the *semantic mapping* (a.k.a. genotype-phenotype mapping) and S is a metric *semantic space*. For the rest of this paper, we assume that s is surjective (i.e., $S \equiv image(P)$). Typically, s is also not invertible and $|P| \gg |S|$. By assuming that s is endowed with the entire knowledge required for program execution, we are not interested here in program syntax and the process of program interpretation.

We define *problem instance* as a program space augmented by a fitness function $f : P \to \mathbb{R}$. Multiple problem instances can be defined for a given program space P. In GP, the fitness f of a program p is often directly based on a distance between $s(p)$ and a predefined point in S.

Let $N(p)$ be the *neighbourhood* of p ($N : P \to 2^P \setminus \{\emptyset\}$, $p \notin N(p)$). Typically, $N(p)$ is the set of all programs that can be built form p by introducing small changes in it (like substituting a single instruction with another). From EC perspective, it is common to identify $N(p)$ with the set of all mutants of p. The pair (P, N) is also referred to as *configuration space*, and the triple (P, N, f) as *fitness landscape* (see, e.g., [8]).

We propose the following expression for measurement of locality in the neighbourhood of program p:

$$l(N,p,s) = \frac{1}{|N(p)|} \sum_{p' \in N(p)} \frac{1}{1 + \|s(p') - s(p)\|} \tag{1}$$

where $\|\|$ denotes a metric in S. The rationale for this particular formula is that, by mapping semantic distances to interval $(0,1]$, l becomes less sensitive to their absolute values, which in general can be arbitrarily high. Clearly, the maximal l is 1.0, which is the case when all neighbours of p have the same semantics as p. On the other hand, l close to zero implies that all neighbours of p are semantically very different from it.

To generalize this concept to the entire program space, we define *semantic locality* (or *locality* for short) of neighbourhood N as an average of $l(N,p,s)$ over all programs:

$$L(N,P,s) = \frac{1}{|P|} \sum_{p \in P} l(N,p,s) \tag{2}$$

Because the semantic mapping s (as well as its codomain S) is given as a part of problem specification and cannot be changed, we shortly write $L(N,P)$ where it is unambiguous. In general $L \in (0,1]$, but $L = 1$ should not be expected in practice, as this can happen only if all programs in P have the same semantics.

Locality is an important and desirable property of program space and the associated neigbourhood, as it correlates with smoothness of fitness landscape, which in turn influences performance of a search algorithm. This is particularly true when the fitness of program p is defined based on distance between $s(p)$ and some *target semantics*, like, e.g., a vector of desired program outputs in the tasks of symbolic regression and logical function synthesis. Thus, high locality suggests smooth or close-to-smooth, and for some classes of problems even convex, fitness landscape [5]. A search algorithm is more likely to converge to good solutions for problems with high locality. Though this is particularly true for local search algorithms that rely exclusively on neighbourhoods, exploitation ability of global search algorithms, including evolutionary algorithms, can be also sensitive to this aspect. On the other hand, it is worth emphasizing that high locality does not *guarantee* good search performance – easy problems typically have high locality, but the reverse is in general not true.

We conclude this section with remark that the notion of locality can be alternatively based directly on fitness function instead of semantics (see, e.g., [11]). In that perspective, locality is high if the neighbouring programs have similar fitness values. Such formulation is not concerned with how is the fitness derived from program code and, in particular, does not assume the existence of semantic space. However, by the same token it is very problem-specific: conclusions concerning locality of one problem instance cannot be generalized to other problem instances. The semantics-based formulation, on the other hand, requires semantic space, but allows us to draw more general conclusions that abstract from specific fitness function, because the measures defined above depend only on P, N, and s, which can be shared by many problem instances (fitness functions).

Table 1. Properties of the original program representation spaces for different instruction sets, averaged over 10 randomly generated sets of 10 fitness cases: the mean semantic distance between syntactical neighbours $\overline{\|s(p') - s(p)\|}$, the number of semantically unique programs (semantic variability) $|S|$, and locality $L(N,P)$

Instruction set	$\overline{\|s(p')-s(p)\|}$	$\|S\|$	$L(N,P)$	Instruction set	$\overline{\|s(p')-s(p)\|}$	$\|S\|$	$L(N,P)$
$\{+,-\}$	47.8	229	0.844	$\{*,/\}$	21.3×10^6	162	0.729
$\{+,*\}$	70.1×10^4	631	0.837	$\{-,/\}$	340.7	970	0.806

2.1   Example 1: Semantic Locality of Symbolic Regression Program Spaces

For illustration let us consider a constrained version of genetic programming applied to univariate symbolic regression task. For the sake of simplicity, we fix the program length and constrain our program space to full binary trees of depth 3, with 7 inner nodes and 8 leaves, which we unroll to linear sequences of 15 instructions. We use two terminal symbols (the constant 1.0 and the independent variable) and consider four sets of non-terminals, $\{+,-\}, \{+,*\}, \{*,/\}$, and $\{-,/\}$, which we refer to as instruction sets in following. The semantics of a program is the vector of outputs it produces for 10 values of the independent variable (fitness cases), drawn at random from the interval $[1, 10]$. The metric in S is the Euclidean distance.

There are $|P| = 2^7 \times 2^8 = 32768$ syntactically unique programs that can be built with this representation. We assume that p' is a neighbour of p if and only if they differ at exactly one instruction, i.e., the Hamming distance between p' and p is one. Note that this implies $|N(p)| = 15, \forall p \in P$.

Table 1 presents the properties of this program representation space for the considered instruction sets, averaged over 10 randomly generated sets of 10 fitness cases. Let us start from noting that the mean semantic distance between neighbours $(\overline{\|s(p') - s(p)\|})$ justifies our way of measuring locality in Eq. (1) – without mapping each distance individually to interval $(0, 1]$, small distances would be dominated by outliers (the variance of semantic distance, not reported here, has similar order of magnitude as the mean).

The second observation is that due to redundancy in the genotype-phenotype mapping, the actual number of semantically unique programs (semantic variability) is much smaller than $|P|$, ranging from 162 to 970 depending on the instruction set. As expected, $|S|$ is particularly small when instruction set contains mutually inverse operations that can cancel each other, and so reduce the semantic variability (the case of $\{+,-\}$ and $\{*,/\}$). On the other hand, the cause for $\{-,/\}$ having the highest semantic variability is probably the lack of symmetry of these instructions: for instance, $a - b$ and $b - a$ can produce different semantics, while $a + b$ and $b + a$ cannot.

The last column of Table 1 presents the locality of configuration space (N,P). Undoubtedly, by lacking clear correlation with semantic diversity or mean semantic distance, $L(N,P)$ does tell something extra about the structure of the space. In general, the observed values confirm our expectations: replacing a single occurrence of $*$ with $/$ in a program is more likely to profoundly change its semantics than, for instance, replacing $+$ with $-$.

3 Embedding Programs in Prespaces

As neighbourhood definition is the primary determinant of locality, most efforts devoted to overcoming the problem of low locality in GP focused on designing new search operators that, in a more or less direct way, redefine the notion of neighbourhood (see review in Section 1). Apart from a few exceptions [5,9], such design typically involves only the original program space.

Design of search operators can be characterized as a 'program-centric' approach, which focuses on building a new candidate solution from one or more given solutions. In this study, we propose a different, 'space-centric' direction. Rather than designing new operators for the existing space of programs, we switch to another space and try to build a mapping from that space onto P such that its composition with semantic mapping provides good locality. Figure 1 outlines the overall idea of this approach.

More formally, our objective is to design three formal objects: (i) a space X, which, by analogy to pre-image, we term *prespace*, (ii) a neighbourhood relation N_X in that space, and (iii) a mapping u from X to P, such that together they maximize $L(N_X, X)$. To clearly tell apart Ps from Xs, we rename P as the *original* program space.

Before we proceed, let us first note that because locality cannot distinguish semantically equivalent programs, it does not matter for it which of the semantically equivalent programs is returned by $u(x)$ for a given $x \in X$. More formally, let $[p] = \{p' \in P : s(p') = s(p)\}$ be the *semantic equivalence class* of program p. If $u(x) = p$, then as long as we redefine u in such a way that $u(x) \in [p]$, locality in X remains the same. Thus, it is enough to consider as codomain of u the space of the semantically unique representatives of programs in P, which we define as

$$[P] = \{\hat{p} : p \in P\}$$

where $\hat{p}$ denotes the representative of equivalence class of p (appointed in any deterministic way). This property is important from practical perspective, because $|[P]| = |S|$, while typically $|P| \gg |S|$, and thus considering only the representatives is computationally much less demanding. Thus, in following $u : X \to [P]$.

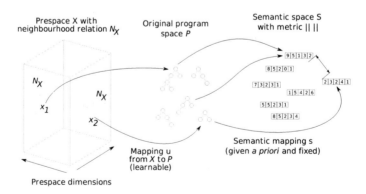

Fig. 1. The conceptual sketch of learnable embedding of program space

As there are many ways in which X, N_X, and u can be defined, our task is clearly underconstrained. To narrow the spectrum of possible designs, we limit our considerations to bijective mappings u, which implies that $|X| = ||[P]||$. In such case u is a permutation of X and can be considered as a form of *embedding* of $[P]$ in X. The obvious advantage of this choice is that u can be implemented as a lookup table and the cost of calculating $u(x)$ for a given x is negligible.

Concerning the topology of X, it seems convenient to endow it with a natural neighbourhood definition. For that purpose, we assume X to be a d-dimensional hypercube of size n, i.e., $X \equiv [1,n]^d \cap \mathbb{Z}^d$. Of course, for a given d, there does not have to exist n such that $n^d = ||[P]||$, but we will handle this technicality in the example that follows.

We define the neighbourhood $N_X(x)$ of location x as the set of locations in X such that their city-block distance from x is 1, where we assume that the distance is calculated modulo n on each dimension, so that the hypercube is effectively toroidal. We find this definition natural, as it introduces minimal modifications in x's location.

The above design choices fix X and N_X, so the search for an optimal triple (X, N_x, u) boils down to finding a permutation u that maximizes $L(N_x, X)$. In other words, we want to *learn the embedding* of $[P]$ in X, which can be formally written as:

$$u^* = \arg\max_u L(N_X, X, s \circ u) \tag{3}$$

where u^* is the optimal embedding. In other words, we look for such u^* that $s \circ u^*$ is the best locality-preserving mapping from X to S via P.

It is easy to notice that for $d = 1$ the above task becomes equivalent to traveling salesperson problem, where u defines a particular order of 'visiting' the elements of $[P]$. This implies that the problem of finding u^* is NP-complete, and it is quite easy to demonstrate that this is true also for $d > 1$. Finding u^* via exhaustive search is futile even for relatively small $||[P]||$, as the sheer number of possible permutations renders considering all mappings infeasible. This, however, should not be considered critical, because even a suboptimal mapping can potentially improve locality and bring benefits for search performance, which we verify further in this paper.

Finally, it is essential to point out that X and N_X implicitly define a new program representation space, where each program is represented by a d-tuple of coordinates that identifies its location in the prespace. In general, this representation is completely detached from program code, or more precisely, from the symbols (instructions) that the considered programs are composed of. Neighbours in X can correspond to syntactically very different programs in P. Program space P becomes transparent from the viewpoint of X, and serves as a proxy for reaching into the semantic space S. In particular, N becomes irrelevant (cf. Fig. 1).

3.1 Example 2: Learning Prespaces for Symbolic Regression

In this example we demonstrate the process of prespace design and learning on the symbolic regression domains analyzed in Example 1. We embed programs in a d-dimensional toroidal hypercube X for $d = 2, \ldots, 6$, using the same neighbourhood definition as in Example 1: $N_X(x)$ is the set of $2d$ 'immediate' neighbours of x, i.e., such that their city-block distance to x is 1.

Table 2. Locality for prespaces found using greedy heuristics, for each instruction set and prespace dimension d

$d =$	2	3	4	5	6
$\{+,-\}$	0.746	0.747	0.744	0.750	0.749
$\{+,*\}$	0.723	0.721	0.716	0.715	0.716
$\{*,/\}$	0.768	0.743	0.733	0.732	0.727
$\{-,/\}$	0.708	0.699	0.698	0.697	0.698

Because the cardinality of semantic space is quite accidental (cf. third column of Table 1), it typically cannot be expressed as n^d for integral d and n, so that $|X| = ||P|| = |S|$ holds. To solve this difficulty, for a given d we use the smallest n such that $n^d \geq ||P||$ and fill the extra $n^d - ||P||$ locations in X with the clones of randomly selected representatives from $||P||$. Although this could be alternatively circumvent by using structures that are less regular than hypercubes, that would impact the formal clarity.

We consider two algorithms to find an approximate solution to formula (3). The first of them is a simple greedy technique. The algorithm first assigns a randomly selected program representative to the origin of the coordinate system spanned by X. Then it iterates over consecutive locations in X, column by column, row by row, etc., and assigns to the current location the program that minimizes the semantic distance from the program assigned in the previous iteration. This approach is naive, as it is strictly local, cannot do any backtracking, and ignores the multidimensional topology of X by considering only one neighbour (previous location). Nevertheless, it is computationally cheap and it will serve us as a convenient reference point.

Table 2 presents locality of prespaces found using the above greedy algorithm for the four considered instruction sets and different prespace dimensionality. Locality turns out to vary across instruction sets, but remains almost insensitive to dimensionality, particularly for $d \geq 4$. This result is expected: for each embedded program, only two of its neighbours, its predecessor and its successor in the ordering determined by algorithm execution, are somehow optimized with respect to locality. As neighbourhood size is $2d$, algorithm's decisions can significantly contribute to locality for low dimensions, while for larger dimensions its effects become much less important.

Our second method of finding an approximate solution to (3) employs evolutionary algorithm (EA) with individuals directly encoding possible permutations of $[P]$, i.e., various mappings u. The algorithm evolves for 200 generations a population of 10000 individuals, each of them directly encoding some permutation of representatives from $[P]$ in the prespace X. Technically, the genome is a vector of integers of length n^d (thanks to the measures discussed above), representing the 'unrolled' hypercube X, with each gene pointing to a representative from $[P]$[1]. We use generational EA with tournament selection (tournament size 7), edge-replacing crossover applied with probability 0.4, mutation with probability 0.4, and soft elitism implemented by tournament and

---

[1] One could consider more sophisticated representations that take into account the toroidal topology in X and the fact that locality is insensitive to various transformations of X (e.g., mirroring along an axis), but the resulting reduction in genome length is probably not worth it.

Table 3. Locality for prespaces found using EA and the relative increase w.r.t. Table 2, for each instruction set and prespace dimension d

$d =$	2	3	4	5	6
$\{+,-\}$	0.867 +16%	0.841 +13%	0.827 +11%	0.819 +9%	0.786 +5%
$\{+,*\}$	0.762 +5%	0.750 +4%	0.740 +3%	0.736 +3%	0.741 +3%
$\{*,/\}$	0.853 +11%	0.806 +8%	0.783 +7%	0.777 +6%	0.756 +4%
$\{-,/\}$	0.735 +4%	0.723 +3%	0.715 +2%	0.718 +3%	0.705 +1%

reproduction with probability 0.2. The mutation swaps two randomly selected elements of permutation with probability 0.01 per element. The fitness of an individual that encodes a permutation u is $L(N_X, X, s \circ u)$. The starting population is initialized by clones of the mapping found by the greedy heuristics described earlier.

Table 3 presents the locality of best-of-run individuals (embeddings) found by EA, averaged over 10 evolutionary runs, together with relative increases with respect to the values reported in Table 2 for the greedy algorithm. For each instruction set and dimensionality, EA improved locality, which is encouraging given the difficulty of the optimization problem caused by the enormous number of permutations. The observed increases tend to be greater for instruction sets with lower semantic variability (cf. Table 1). This can be explained by the fact that genome length is directly determined by semantic variability (as $n^d \geq |S|$ has to hold), and, other things being equal, evolution is more effective for shorter genomes.

It is also evident that the higher the dimensionality, the more difficult it becomes to find an embedding that significantly improves locality. We observe this because neighbourhoods in high-dimensional spaces are larger and optimizing (3) is more constrained – it is more difficult to simultaneously optimize locality along all dimensions.

4   Learnable Embeddings and Prespaces as Tools for Search

In the previous sections, we relied on locality as the only gauge of quality of program representation space, both original and prespaces. From practical perspective however, we are usually more interested in how the choices concerning representation space and neighbourhood determine the actual *performance* of a search algorithm. In this context, the question we would like to answer is: are the prespaces better 'searcheable'?

To answer this question, we empirically compare the performance of a specific search algorithm in the original space and in prespaces on a sample of problem instances. Because our focus in this study is on locality, our choice is hill climbing (*HC*), a simple parameter-free local search procedure.

Given specific maximized fitness function $f \in [0,1]$, an initial search point p_0, and neighbourhood definition N, $HC(f, p_0, N)$ generates in each iteration all $p' \in N(p_j)$ and moves to $p_{j+1} = \arg\max_{p' \in N(p_j)} f(p')$ if $f(p_{j+1}) > f(p_j)$. Otherwise, it terminates with failure. We assume early detection of success, meaning that as soon as any of generated search points (whether p_j or one of its neighbours) yields $f = 1$, the algorithm terminates with success. Thanks to this feature, no unnecessary evaluations are carried out and runs that use different neighbourhoods can be fairly compared in terms of effort. Note also that, for simplicity, no attempts are made to escape plateaus.

Note that the above formulation refers to abstract search points. In practice, we invoke it as $HC(f, p_0, N)$ with some $p_0 \in P$ for the original program space and as $HC(f, x_0, N_X)$ with some $x_0 \in X$ for a prespace. For brevity, we do not redefine formally f for the prespaces – the fitness of each $x \in X$ is clearly determined by its location and the mapping u.

Using HC, we estimate the 'searchability' of the original representation space P using the following procedure:

For each target semantics $t \in S$:

1. Define fitness function as $f(p) \equiv 1/(1 + \|t - s(p)\|)$
2. Repeat 150 times:
   (a) Sample a starting search point p_0 from P (with uniform distribution)
   (b) Run $HC(f, p_0, N)$ and store its outcome (success/failure + best fitness)

For prespaces, we replace P with X, $s(p)$ with $s(u(x))$, and $HC(f, p_0, N)$ by $HC(f, x_0, N_X)$.

This procedure deserves two remarks. First, by choosing ts from the codomain of s, and not simply from $\mathbb{R}^{10}$ (recall we use 10 fitness cases), we consider only target semantics that are *realizable* using the available instruction set and program length limit. More formally, for each t considered, there exists at least one program p in P (and thus at least one point in X) such that $s(p) = t$. Secondly, the examination is quite thorough, as we consider *all* realizable semantics.

Based on the above procedure, we calculate the expected search effort required to find the optimum solution, which is simply the total number of evaluations (fitness function calls) summed over all runs, divided by the number of successful runs. The results are reported in Table 4 for the representation spaces analyzed in Examples 1 and 2, named for convenience as follows:

- *Original space*: the original 15-dimensional program representation space with neighbourhood N,
- *Greedy prespace*: d-dimensional prespace X with neighbourhood N_X and mapping found by the greedy heuristics,
- *Evolved prespace*: d-dimensional prespace X with neighbourhood N_X and mapping found by EA.

Table 4 clearly suggests that prespaces provide lower search effort for three of four considered instruction sets. As this applies also to greedy prespaces, this effect should be mainly attributed to the change of search space topology: switching from the 15-dimensional binary space of raw program encodings to d-dimensional prespace turns out to be beneficial. However, this alone cannot explain why an evolved prespace always provides lower search effort than the corresponding greedy prespace. This effect must be then due to the locality-driven evolutionary optimization of (3). This in turn implies that locality of a search space correlates with its 'searchability' meant as the expected effort required to find the optimum.

The reduction of search effort provided by the evolved prespaces is typically quite impressive, up to over 6-fold for $\{*, /\}$ and $d = 4$. The case of $\{+, *\}$ is an exception:

Table 4. Comparison of search effort, i.e., the expected number of evaluations required to find the optimal solution by *HC* (the lower, the better), for the original and learned representations spaces, averaged over 150 runs for each target semantics (problem instance)

$d =$	Original space —	Greedy prespace 2	3	4	5	6	Evolved prespace 2	3	4	5	6
$\{+,-\}$	469	221	209	205	208	<u>201</u>	150	143	<u>140</u>	144	175
$\{+,*\}$	468	707	675	649	<u>612</u>	663	631	565	549	<u>548</u>	557
$\{*,/\}$	788	165	152	<u>147</u>	154	161	154	130	<u>124</u>	128	149
$\{-,/\}$	1704	1141	1060	1014	978	<u>970</u>	964	877	867	<u>804</u>	905

here, neither greedy nor evolved prespace provided effort reduction. Our working hypothesis is that the culprit is here the greedy heuristics, which, we remind, is also used as population initializer for the EA. For some reasons that deserve further investigation, the particular initial permutation produced by the greedy algorithm seems to be extremely inconvenient for this instruction set. We come to this conclusion by analyzing Table 3, where EA only slightly improves the mapping found by the greedy algorithm. Although similarly low increases occur for $\{-,/\}$, in that case they may be due to the high semantic variability that requires longer genome and thus makes evolution slower.

In a broader perspective, the result for $\{+,*\}$ indicates that in general high locality of representation space cannot *guarantee* good search performance. This should not come as a surprise, as locality is a very low-level characteristics of search space and cannot capture all aspects of genotype-phenotype mapping (and, in consequence, of fitness landscape). It is easy to come up with deceptive problem instances that exhibit high locality but have global optima that are extremely difficult to discover.

5 Discussion

The above demonstration shows that learnable embeddings can improve locality of a search space and make the search process more effective. Because the outcome of the learning of embedding indirectly depends on the actual semantics of instructions, learnable embeddings can adapt to domain-specific knowledge, which is not the case for most of typical GP search operators.

There are more reasons for which semantics is an important component of learnable embeddings. Guiding the process of optimization of prespace-to-program-space mapping (Eq. (3)) by means of fitness function instead of semantic locality, though conceivable, would almost certainly lead to inferior locality. Fitness landscape alone provides much less detailed information on similarity of program behaviors. Also, such mapping cannot be generalized to other problem instances.

It is interesting to note that some far-fetched analogs of learnable embeddings can be found in evolutionary computation. For that instance, HyperNEAT [12] substitutes the direct search in space of neural network weights by a search in the space of pattern generators which serve as oracles providing weight values. An apparently closer example is grammatical evolution [10], where individuals encode production numbers of an *a priori* given grammar and use them to build programs. However, grammatical evolution still operates directly on program code, while learnable embeddings use programs only as a proxy for semantic space.

The appealing feature of learnable embeddings is that they operate globally. In this way, they can explicitly gather semantically similar programs in close proximity in the prespace. The resulting structure is absolute in the sense that similarly behaving programs occupy, in case of an ideal mapping, a topologically continuous fragment of prespace that has specific coordinates. In design of search operators, such structures are typically only relative to the currently considered (e.g., mutated) program(s).

The price we pay for globality is high complexity of mapping. Although several technical tricks have been used here to speed up computation, like caching the genotype-phenotype mapping and pre-computing of distances between all pairs of semantics required for Eq. (1), prespaces in their current formulation are far from effective in terms of computational effort and should not be one's choice for solving a single instance of a GP problem[2]. The overall cost of learning mapping and performing a search with its use is far greater than the computational effort of searching the program space P directly. Nevertheless, this can be potentially alleviated by replacing the analysis of all programs with some form of sampling.

On the other hand, as soon as a good embedding is found, its utilization for solving a particular problem instance is very effective. Because all unique semantics had to be calculated for the sake of learning of embedding, they are already known, and the search algorithm (like the local search algorithm above) does not have to run the programs anymore – all it does is measuring the distances of known semantics corresponding to the search points in X to the target semantics t. This can be very advantageous, given that our approach produces prespaces for an entire *class* of problems (meant here as an instruction set plus program length limit), rather than for a single problem instance, so the designed prespace can be re-used multiple times for different problem instances.

Finally, we hypothesize that high-locality mappings designed for code *fragments* (subprograms, modules) can potentially be used, via some form of hierarchical composition, for building larger programs that would to some extend inherit the locality of their constituent components.

Many interesting open questions can be addressed in the future research on this topic. What, for example, is the optimal prespace dimensionality for given program length and instruction set? (Note that the best evolved prespaces in Table 4 intriguingly group on dimensionality 4 and 5). Is it possible to attribute some meaningful interpretation to dimensions of the learned prespace? Which other heuristics, apart from evolutionary optimization of directly encoded permutations, can be useful for learning of mappings? And, can anything be gained by considering non-bijective mappings? The last question is particularly interesting, because non-bijective mappings imply redundancy, and many successful computational intelligence methods use inherently redundant representations. Question remains whether learnable embeddings are redundant in the 'right way', i.e., whether their redundancy is more useful than the redundancy present in conventional GP encodings.

---

[2] An ongoing work led to elaboration of a much faster algorithm for learning of embedding, which will be published elsewhere.

Acknowledgments. This work has been supported by grant no. N N519 441939.

References

1. Banzhaf, W., Nordin, P., Keller, R., Francone, F.: Genetic Programming: An Introduction. On the automatic Evolution of Computer Programs and its Application. Morgan Kaufmann, San Francisco (1998)
2. Brameier, M., Banzhaf, W.: Linear Genetic Programming. Genetic and Evolutionary Computation, vol. XVI. Springer, Heidelberg (2007)
3. Dijkstra, E.W.: On the cruelty of really teaching computing science. circulated privately (December 1988)
4. Johnson, C.: Genetic programming crossover: Does it cross over? In: Vanneschi, L., Gustafson, S., Moraglio, A., De Falco, I., Ebner, M. (eds.) EuroGP 2009. LNCS, vol. 5481, pp. 97–108. Springer, Heidelberg (2009)
5. Krawiec, K., Lichocki, P.: Approximating geometric crossover in semantic space. In: Raidl, G., et al. (eds.) GECCO 2009: Proceedings of the 11th Annual Conference on Genetic and Evolutionary Computation (2009)
6. Looks, M.: Competent Program Evolution. Doctor of science, Washington University, St. Louis, USA (December 11, 2006)
7. Miller, J.F.: An empirical study of the efficiency of learning boolean functions using a cartesian genetic programming approach. In: Banzhaf, W., Daida, J., Eiben, A.E., Garzon, M.H., Honavar, V., Jakiela, M., Smith, R.E. (eds.) Proceedings of the Genetic and Evolutionary Computation Conference, Orlando, Florida, USA, July 13-17, vol. 2, pp. 1135–1142. Morgan Kaufmann, San Francisco (1999)
8. Moraglio, A., Poli, R.: Topological interpretation of crossover. In: Deb, K., et al. (eds.) GECCO 2004. LNCS, vol. 3102, pp. 1377–1388. Springer, Heidelberg (2004)
9. Nguyen, Q.U., Nguyen, X.H., O'Neill, M.: Semantic aware crossover for genetic programming: The case for real-valued function regression. In: Vanneschi, L., Gustafson, S., Moraglio, A., De Falco, I., Ebner, M. (eds.) EuroGP 2009. LNCS, vol. 5481, pp. 292–302. Springer, Heidelberg (2009)
10. O'Neill, M., Ryan, C.: Grammatical Evolution: Evolutionary Automatic Programming in a Arbitrary Language. Genetic programming, vol. 4. Kluwer Academic Publishers, Dordrecht (2003)
11. Rothlauf, F.: On the locality of representations. Technical Report, University of Mannheim, Department of Information Systems 1 (2003)
12. Stanley, K.O., D'Ambrosio, D.B., Gauci, J.: A hypercube-based encoding for evolving large-scale neural networks. Artif. Life 15, 185–212 (2009)
13. Vanneschi, L., Tomassini, M.: Pros and cons of fitness distance correlation in genetic programming. In: Barry, A.M. (ed.) GECCO 2003: Proceedings of the Bird of a Feather Workshops, Genetic and Evolutionary Computation Conference, Chigaco, July 11, pp. 284–287. AAAI, Menlo Park (2003)

Parallel Linear Genetic Programming

Carlton Downey and Mengjie Zhang

Victoria University of Wellington, Wellington, NZ
{Carlton.Downey,Mengjie.Zhang}@ecs.vuw.ac.nz

Abstract. Motivated by biological inspiration and the issue of code disruption, we develop a new form of LGP called *Parallel LGP (PLGP)*. PLGP programs consist of n lists of instructions. These lists are executed in parallel, after which the resulting vectors are combined to produce program output. PGLP limits the disruptive effects of crossover and mutation, which allows PLGP to significantly outperform regular LGP.

1 Introduction

Derived from genetic algorithms [5], Genetic Programming (GP) [2,6] is a promising and nature inspired approach to constructing reliable solutions to a range of problems quickly and automatically, given only a set of human labeled instances on which an evolved program can be evaluated. GP uses ideas analogous to biological evolution to search the space of possible programs to evolve a good program for a particular task. Since the 1990s, GP has been successful for solving many machine learning problems [7,9,11,12,13].

In conventional GP, programs are trees of operators. This form of GP is known as Tree based GP (TGP). Linear Genetic Programming (LGP) is an alternative form of GP where individuals in the population are sequences of instructions. Structuring programs as sequences of instructions has many advantages over structuring them as trees. LGP has been shown to significantly outperform TGP on machine learning tasks such as multiclass classification [4,8].

LGP performs well on multiclass classification problems because of the power of its flexible program structure. LGP programs consist of a sequence of instructions which operate on a list of registers. This allows multiple outputs and permits results computed early in program execution to be reused later. These two properties make LGP a powerful problem solving technique.

Unfortunately LGP has a significant weakness inherent to its program structure, which severely limits its effectiveness. The execution of an instruction is heavily influenced by the execution of all previous instructions. This means that modifying an early instruction can be highly disruptive because it has the potential to upset the execution of all subsequent instructions.

LGP programs can be viewed as a tangled web of dependencies. Each instruction depends on the output of several other instructions in order to produce a result. GP operators can completely destroy this fragile balance by modifying one or more instructions critical to effective execution. Disrupting the program to

S. Silva et al. (Eds.): EuroGP 2011, LNCS 6621, pp. 178–189, 2011.

a large extent results in useless output, and wastes the time required to evaluate the program, greatly slowing down convergence.

Clearly the structure of LGP programs is fatally flawed, and a new program structure is required. In this paper we describe a new form of LGP which overcomes the problems of code disruption while still possessing all the power of conventional LGP.

1.1   Objectives

In this paper, we aim to develop a new form of LGP which removes, or at least limits the disruptive effects of mutation and crossover. This new form of LGP should posses all the power of conventional LGP, but should possess far fewer dependencies. Specifically, this paper has the following research objectives:

- To isolate and identify the features of conventional LGP program structure which significantly compromise performance.
- To develop a new program structure for LGP which limits these problems while preserving the power of conventional LGP.
- To compare the performance of our new program structure to a conventional program structure over a range of parameter settings on several classification problems.

1.2   Organization

The rest of the paper is organized as follows. Section 2 describes some necessary background on LGP and Classification, as well as covering GP related work to classification. Section 3 discusses the problem of code disruption, and describes our new program structure which prevents code disruption. The experiment design and configurations are provided in Section 4 and the results are presented in Section 5 with discussion. Section 6 concludes the paper and gives future work directions.

2   Background

2.1   LGP

In LGP the individuals in the population are programs in some imperative programming language. Each program consists of a number of lines of code, to be executed in sequence. The LGP used in this paper follows the ideas of register machine LGP [2]. In register machine LGP each individual program is represented by a sequence of register machine instructions, typically expressed in human-readable form as C-style code. Each instruction has three components: an operator, 2 arguments and a destination register. To execute the instruction, the operator is applied to the two arguments and the resulting value is stored in the destination register. The operators can be simple standard arithmetic

$$r[1] = 3.1 + f1;$$
$$r[3] = f2 \ / \ r[1];$$
$$r[2] = r[1] \ * \ r[1];$$
$$r[1] = f1 \ - \ f1;$$
$$r[1] = r[1] - 1.5;$$
$$r[2] = r[2] + r[1];$$

Fig. 1. An example LGP program

operators or complex specific functions predefined for a particular task. The arguments can be constants, registers, or features from the current instance. An example LGP program is shown in figure 1.

After an LGP program has been executed the registers will each hold a real valued number. For presentation convenience, the state of the registers after execution is represented by a floating point (`double`) vector **r**. These numbers are the outputs of the LGP program and can be interpreted appropriately depending on the problem at hand. A step by step example of LGP program execution can be found in figure 2.

Program Execution

Program Inputs

f1	f2	f3
0.1	3.0	1.0

(a)

	Program		Registers		
index	Instruction		$r[1]$	$r[2]$	$r[3]$
0	-		0	0	0
1	$r[1] = 3.1 + f1;$		3.2	0	0
2	$r[3] = f2 \ / \ r[1];$		3.2	0	0.94
3	$r[2] = r[1] \ * \ r[1];$		3.2	10.24	0.94
4	$r[1] = f1 \ - \ f1;$		0	10.24	0.94
5	$r[1] = r[1] - 1.5;$		-1.5	10.24	0.94
6	$r[2] = r[2] + r[1];$		-1.5	8.74	0.94

(b)

Program Outputs

$r[1]$	$r[2]$	$r[3]$
-1.5	8.74	0.94

(c)

Fig. 2. Example of LGP program execution on a specific training example

2.2 Classification

Classification problems involve determining the type or *class* of an object instance based on some limited information or *features*. Solving classification problems involves learning a classifier, a program which can automatically perform classification on an object with unknown class. Classification problems form the basis of empirical testing in this paper.

2.3 Classification Using LGP

LGP is particularly well suited to solving multiclass classification problems. The number of outputs from an LGP program is determined by the number of registers, and the number of registers can be arbitrarily large. Hence we can map each class to a particular output in the form of a single register. Classification then proceeds by selecting the register with the largest final value and classifying the instance as the associated class. For example if registers (r1, r2, r3) held the values (-1.5, 8.74, 0.94) then the object would be classified as class 2, since register 2 has the largest final value (8.74).

3   Parallel LGP

In this section we introduce a new form of LGP called parallel LGP (PLGP)[1] based on the concept of dividing programs into multiple parts which can be executed in parallel.

3.1   Motivation

While LGP is an effective technique for solving problems across many problem domains, it is hamstrung by several key weaknesses. We motivate our new form of LGP by discussing two of these weaknesses in some depth.

Disruption. One weakness is the disruption of program structure caused by the application of genetic operators. Genetic operators work by taking an existing LGP program and producing a new LGP program by modifying one or more instructions. This modification is what causes the formation of good solutions, but it can equally result in the formation of poor solutions. Poor solutions are usually the result of modifications disrupting a key part of the program. In standard LGP, disruption of program structure occurs when modification of one instruction has a negative impact on the execution of other instructions.

GP systems produce new solutions by improving on existing solutions. This improvement is a stepwise process where each generation we aim to improve on the previous generation by exploiting the good code present in individuals. Typically a good evolution is one which preserves the majority of existing program strengths while introducing new and beneficial code. Disruption retards this evolutionary process because it makes it difficult to evolve new code without disrupting existing code.

Code disruption is a particular problem with large PGLP programs with many instructions. Every single instruction after the modification point is vulnerable to code disruption. Longer programs generally have more instructions after the modification point. This means they have more instructions vulnerable to code disruption, and hence a higher chance of being disrupted. Unfortunately we often desire large LGP programs. Large programs have more power as they are able to explore a larger solution space. To solve complicated problems we often require complex solutions which can only be expressed by large programs.

---

[1] There is an existing technique called Parallel GP (PGP) based on executing distinct GP programs in parallel[1]. Namewise, PGP is similar to PLGP, however the two techniques are unrelated.

Structural Introns. A second weakness is that LGP programs have significant amounts of non-effective code, otherwise known as structural introns [3]. Structural introns are instructions which have no effect on program output due to their position in the instruction sequence. An instruction is *not* a structural intron if no subsequent instruction overwrites its value. Hence instructions near the start of a program are more likely to be structural introns, and larger LGP programs have a higher proportion of structural introns.

While it has been argued that such structural introns may be desirable on a small scale [3], they are certainly undesirable on a large scale. As LGP program size increases it becomes overwhelmingly likely that the changes caused by GP operators will be opaque to the fitness function. This in turn will greatly slow convergence resulting in solutions with poor fitness.

Solution. We desire a form of LGP where both the disruptive effects of the genetic operators, and the amount of non-effective code are independent of program size and program fitness. This is in contrast to conventional LGP where the negative influence of both *increases* with program size and fitness.

The genetic operators in LGP are most effective when applied to short sequences of instructions. By strictly limiting the maximum length of instruction sequences we can achieve the most benefit from the application of genetic operators, and hence achieve the best performance. However, many interesting problems require a large number of instructions in order to evolve an effective solution. Therefore the only way to increase the number of instructions *without* increasing the length of instruction sequences is to *allow each LGP program to consist of multiple short instruction sequences*. This idea forms the core of our new program representation.

3.2   Program Structure

Parallel LGP (PLGP) is an LGP system where each LGP program consists of n dissociated factors. A PLGP program consists of n LGP programs which are evaluated independently, to give n results vectors. These vectors are then summed to produce a single results vector. Formally let V_i be the ith results vector and let S be the summed results vectors. Then $S = \sum_{i=0}^{n} V_i$.

An example of PLGP program execution is shown in figure 3 and contrasted to an example of LGP program execution. In LGP all instructions are executed in sequence using a single set of registers, to produce a single results vector as output. In PLGP the instructions in *each program factor* are executed on *their own set of registers* to produce n (in this case 3) results vectors. In our example the program factors are separated by horizontal lines, so our PLGP program consists of 3 factors. These results vectors are then summed to produce the final program output. Notice how our LGP program and our PLGP program have *the same instructions* but produce *different output*. This is because in our LGP program the results of earlier computations are stored in the registers and can be reused by later computations, while in PLGP each factor begins execution with all registers initialized to zero.

LGP program execution

$$
\begin{bmatrix}
r[1] = 3.1 + f1; \\
r[3] = f2 \;/\; r[1]; \\
r[2] = r[1] \;*\; r[1]; \\
r[1] = f1 \;-\; f1; \\
r[1] = r[1] - 1.5; \\
r[2] = r[2] + r[1];
\end{bmatrix}
=
\begin{bmatrix}
-1.5 \\
8.74 \\
0.94
\end{bmatrix}
$$

PLGP program execution

$$
\begin{bmatrix}
r[1] = 3.1 + f1; \\
r[3] = f2 \;/\; r[1]; \\
r[2] = r[1] \;*\; r[1]; \\
r[1] = f1 \;-\; f1; \\
r[1] = r[1] - 1.5; \\
r[2] = r[2] + r[1];
\end{bmatrix}
=
\begin{bmatrix}
3.2 \\
0 \\
0.94
\end{bmatrix}
+
\begin{bmatrix}
0 \\
0 \\
0
\end{bmatrix}
+
\begin{bmatrix}
-1.5 \\
-1.5 \\
0
\end{bmatrix}
=
\begin{bmatrix}
1.7 \\
-1.5 \\
0.94
\end{bmatrix}
$$

Fig. 3. Contrasting PLGP program execution to LGP program execution

By separating the program code into several dissociated factors and summing the resulting vectors we obtain independence between instructions in different factors. Modifications to one part of the program have no influence on the output of other parts of the program. Hence programs may consist of a large number of instructions, but each instruction is executed as if it was part of a very short program.

3.3 Crossover for PLGP

An important issue to consider for our new LGP system is how crossover will be performed. More specifically, which components of the program we will allow code exchange to occur between.

One option is to allow free exchange of code between any two factors of any two programs. In other words the crossover operator would select one factor from each program at random and perform normal crossover between these two factors. In this view the factors which comprise the program have no ordering, and we can view a program as a *set* of factors.

The second option is to strictly limit code exchange to occur between equivalent factors. In other words we impose an ordering on the factors of each program, so that each program has a first factor, a second factor, etc. In this view the factors of the program are strictly ordered, and hence a program can be viewed as an *ordered list* of factors.

Which option we choose determines how we view the population as a whole. If we allow free crossover then we can view all genetic code as belonging to a single 'genetic population' where any instruction can exchange genetic material with any other instruction. If we restrict crossover based on factor number then we are effectively creating a number of genetic sub-populations where there is no interpopulation genetic flow. In other words genetic material can be exchanged within each sub population, but there is no transfer of genetic material *between* sub populations. We borrow some terminology from the area of Cooperative Coevolution and term this kind of PLGP as PLGP with *enforced sub populations (ESP)*.

Using ESP has been shown to give improved classification accuracy when applied in the area of cooperative coevolution. The theory is that by limiting the exchange of genetic material to within sub populations we encourage speciation. This in turn increases the likelihood of crossover between compatible segments, and hence improves the likelihood of a favorable crossover outcome. Unpublished work by the authors demonstrates that PLGP using ESP significantly outperforms constraint free crossover. Unfortunately this paper lacks adequate space to discuss these results in depth.

4   Experimental Setup

In order to compare the effectiveness of different GP methods as techniques for performing multiclass classification, a series of experiments was conducted in the important problem domain of object recognition.

4.1   Data Sets

We used three image data sets providing object classification problems of varying difficulty in the experiments. These data sets were chosen from the UCI machine learning repository [10].

The first data set is *Hand Written Digits*, which consists of 3750 hand written digits with added noise. It has 10 classes and 564 attributes. The second set is *Artificial Characters*. This data set consists of vector representations of 10 capital letters from the English language. It has a high number of classes (10), a high number of attributes/features (56), and 5000 instances in total. The third set is *Yeast*. In this data set the aim is to predict the protein localization sites based on the results of various tests. It has 10 classes, 8 attributes and 1484 instances. These tasks are highly challenging due to a high number of attributes, a high number of classes and noise in some sets.

4.2   Parameter Configurations

The parameters in table 1a are the *constant parameters*. These are the parameters which will remain constant throughout all experiments. These parameters are either experimentally determined optima, or common values whose reasonableness is well established in literature [6].

The parameters in table 1b are the *experiment specific parameters*. Each column of the table corresponds to the parameter settings for a specific experiment. Each experiment has two components, an LGP stage and a PLGP stage. In the LGP stage we determine the classification accuracy of an LGP system using

Table 1. Experimental Parameters

Parameter	Value
Population	500
Generations	400
Mutation	45%
Elitism	10%
Crossover	45%
Tournament Size	5

(a) Constant

	Exp 1	Exp 2	Exp 3	Exp 4	Exp 5	Exp 6
Total Instructions	10	20	35	50	100	400
# PLGP factors	2	4	5	5	10	20
PLGP factor Size	5	5	7	10	10	20

(b) Experiment Specific

programs of the specified length. In the PLGP stage we repeat our experiment but we use PLGP programs of equivalent length.

We allow terminal constants in the range [-1,1], and a function set containing Addition, Subtraction, Multiplication, Protected Division, and If. The data set is divided equally into a training set, validation set, and test set, and results are averaged over 30 runs. All reported results are for performance on the test set. The fitness function is simply the number of missclassified training examples. Finally all initial programs in the population consist of randomly chosen instructions.

PLGP Program Topologies. It is important to note that there are many different ways of arranging the instructions in a PLGP program. For instance a PLGP program with 10 instructions could consist of either 2 factors of 5 instructions, or 5 factors of 2 instructions. We refer to these different arrangements as *Program Topologies*. We lack the room to give an in depth analysis in this paper, however unpublished work by the authors shows that a large number of reasonable topologies give near optimal results. The program topologies used in our experiments have been previously determined to be optimal to within some small tolerance.

5   Results

The following graphs compare the performance of LGP with PLGP as classification techniques on the three data sets described earlier. Figure 4 compares performance on the Hand Written Digits data set, Figure 5 compares performance on the Artificial Characters data set and Figure 6 compares performance on the Yeast data set. Each line corresponds to an experiment with programs of a certain fixed length. Program lengths vary from very short (10 instructions) to very long (400 instructions).

5.1   Discussion

Program Size. *LGP performs at least as well as PLGP when small programs are used.* In the Hand Written Digits data set LGP significantly outperforms

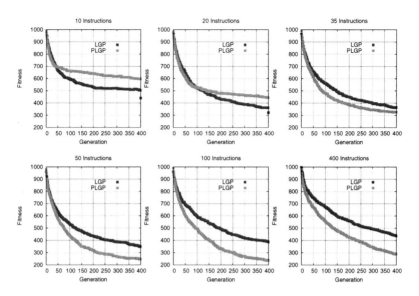

Fig. 4. LGP vs. PLGP on the *Hard Written Digits* data set

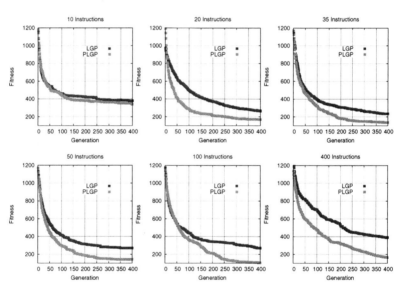

Fig. 5. LGP vs. PLGP on the *Artificial Characters* data set

PLGP for programs of length 10-20. In the Artificial Characters data set the performance of LGP and PLGP is comparable for programs of length 10. In the Yeast data set the performance of LGP and PLGP is comparable for programs of length 10-35.

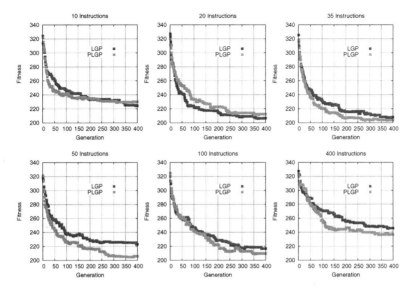

Fig. 6. LGP vs. PLGP on the *Yeast* data set

Short programs will not suffer from code disruption or non-effective code in the way larger programs do. This means our motivation for developing the PLGP technique does not hold for short programs. Also, dividing short programs into many factors strips them of much of their usefulness. The power of LGP programs lies in their ability to reuse results calculated earlier in the program and stored in the registers. Dividing very short programs into many factors means each factor will only be a handful of instructions long, rendering very short PLGP programs relatively powerless. However the flip side of this result is that short LGP programs are not what we are really interested in. Typically short programs are not powerful enough for any sort of interesting application. On difficult problems neither LGP or PLGP can achieve good performance using short programs.

PLGP significantly outperforms LGP for longer programs. In all three data sets PLGP significantly outperforms LGP whenever program length exceeds some minimum. For the Hand Written Digits this holds for all programs larger than 20 instructions. For the Artificial Characters data set this holds for all programs larger than 10 instructions. For the yeast data set this holds for all programs larger than 35 instructions.

Longer LGP programs are prone to problems such as code disruption and non-effective code. By dividing the program into parallel factors we can alleviate these problems. Hence it is expected that PLGP will outperform LGP when program size is large. In addition, the severity of these problems is proportional to the length of instruction sequence. Hence larger LGP programs will suffer these effects to a greater degree.

Optimal Size. There is an optimal size for programs. Short programs are insufficiently expressive: it is not possible to easily *express* good solutions using only a very small number of instructions. Long programs are overly expressive: while it is possible to express good solutions the search space is too large, making it difficult to *find* good solutions. Hence there is an optimal size for LGP programs, a balance between expressive power and search space complexity.

The optimal size for PLGP programs is significantly larger that the optimal size for LGP programs. In our three experiments the optimal size for LGP programs occurs between 20-50 instructions. However in these same three experiments the optimal size for PLGP programs occurs between 50-400 instructions.

The issues of code disruption and non effective code negatively impact the performance of large LGP programs. Conversely the PLGP program structure alleviates these problems, allowing large PLGP programs to give good performance. Hence the optimal size of LGP programs is smaller than the optimal size of PLGP programs.

Optimal Performance. It is clear that PLGP gives rise to better fitness solutions than LGP. This is intrinsically linked to the optimal program size for each of these two methods: PLGP has a higher optimal program size than LGP. Both program representations are equally able to *express* powerful solutions, however only PLGP is able to actually *find* these powerful solutions. By avoiding code disruption and non effective code PLGP allows us to exploit a range of powerful solutions not available to us with LGP.

6   Conclusions and Future Work

PLGP is a technique designed to minimize building block and program disruption by the addition of parallelism into the standard LGP technique. Longer LGP programs are easily disrupted since modifications to early instructions result in massive changes to program output. By executing many program parts in parallel, PLGP prevents code disruption from occurring. This allows PLGP to effectively exploit larger programs for significantly superior results. Our empirical tests support this: long PLGP programs significantly outperform long LGP programs on all data sets. In addition our results show that by exploiting the ability of PLGP to utilize large programs it is possible to obtain a significant overall improvement in performance. Both theory and results clearly demonstrate the benefits of this new parallel architecture.

PLGP offers many exciting opportunities for future work. PLGP programs are naturally suited to caching. They consist of several dissociated factors and at each generation only a single factor is modified. Therefore by caching the output of each factor it should be possible to cut execution time by an order of magnitude.

When summing the factors of a PLGP program it should be possible to introduce factor weights. By replacing the sum with a *weighted sum*, it is possible to deterministically improve performance by optimizing these weights. This weighted PLGP could either be used during evolution, or it could be applied post evolution to the best individual as a final optimization procedure.

References

1. Andre, D., Koza, J.R.: A parallel implementation of genetic programming that achieves super-linear performance. Information Sciences 106(3-4), 201–218 (1998)
2. Banzhaf, W., Nordin, P., Keller, R.E., Francone, F.D.: Genetic Programming: An Introduction on the Automatic Evolution of computer programs and its Applications. Morgan Kaufmann Publishers, Dpunkt-Verlag, San Francisco, Heidelburg (1998)
3. Brameier, M., Banzhaf, W.: Linear Genetic Programming. Genetic and Evolutionary Computation, vol. XVI. Springer, Heidelberg (2007)
4. Fogelberg, C., Zhang, M.: Linear genetic programming for multi-class object classification. In: Zhang, S., Jarvis, R.A. (eds.) AI 2005. LNCS (LNAI), vol. 3809, pp. 369–379. Springer, Heidelberg (2005)
5. Holland, J.H.: Adaptation in Natural and Artificial Systems: An Introductory Analysis with Applications to Biology, Control, and Artificial Intelligence. University of Michigan Press, MIT Press, Ann Arbor, Cambridge (1975)
6. Koza, J.R.: Genetic programming: on the programming of computers by means of natural selection. MIT Press, Cambridge (1992)
7. Krawiec, K., Bhanu, B.: Visual learning by evolutionary and coevolutionary feature synthesis. IEEE Transactions on Evolutionary Computation 11(5), 635–650 (2007)
8. Olague Caballero, G., Romero, E., Trujillo, L., Bhanu, B.: Multiclass object recognition based on texture linear genetic programming. In: Giacobini, M. (ed.) EvoWorkshops 2007. LNCS, vol. 4448, pp. 291–300. Springer, Heidelberg (2007)
9. Olaguea, G., Cagnoni, S., Lutton, E. (eds.): special issue on evolutionary computer vision and image understanding. Pattern Recognition Letters 27(11) (2006)
10. Hettich, S., Blake, C., Merz, C.: UCI repository of machine learning databases (1998), http://www.ics.uci.edu/~mlearn/MLRepository.html
11. Zhang, M., Ciesielski, V.B., Andreae, P.: A domain-independent window approach to multiclass object detection using genetic programming. EURASIP Journal on Applied Signal Processing 2003(8), 841–859 (2003); special Issue on Genetic and Evolutionary Computation for Signal Processing and Image Analysis
12. Zhang, M., Gao, X., Lou, W.: A new crossover operator in genetic programming for object classification. IEEE Transactions on Systems, Man and Cybernetics, Part B 37(5), 1332–1343 (2007)
13. Zhang, M., Smart, W.: Using gaussian distribution to construct fitness functions in genetic programming for multiclass object classification. Pattern Recognition Letters 27(11), 1266–1274 (2006); evolutionary Computer Vision and Image Understanding

How Far Is It from Here to There?
A Distance That Is Coherent with GP Operators

James McDermott[1], Una-May O'Reilly[1],
Leonardo Vanneschi[2], and Kalyan Veeramachaneni[1]

[1] EvoDesignOpt, CSAIL, MIT
{jmmcd,unamay,kalyan}@csail.mit.edu
[2] D.I.S.Co., University of Milano-Bicocca, Milan, Italy
vanneschi@disco.unimib.it

Abstract. The distance between pairs of individuals is a useful concept in the study of evolutionary algorithms. It is particularly useful to define a distance which is *coherent* with, i.e. related to, the action of a particular operator. We present the first formal, general definition of this operator-distance coherence. We also propose a new distance function, based on the multi-step transition probability (MSTP), that is coherent with any GP operator for which the one-step transition probability (1STP) between individuals can be defined. We give an algorithm for 1STP in the case of subtree mutation. Because MSTP is useful in GP investigations, but impractical to compute, we evaluate a variety of means to approximate it. We show that some syntactic distance measures give good approximations, and attempt to combine them to improve the approximation using a GP symbolic regression method. We conclude that 1STP itself is a sufficient indicator of MSTP for subtree mutation.

1 Introduction

The distance between pairs of individuals is a useful concept for many topics in evolutionary computation (EC). It can be used to compare and control the effects of variation operators [2,8]. It is useful in studying measures of problem difficulty such as locality [12] and fitness distance correlation (FDC) [9,17]. It can be used to calculate and control population diversity and implement fitness sharing [7]. It is an essential component of the representation-independent theory of geometric operators [13].

In some of these applications, a definition of distance which is related to the action of a particular operator is required. However, although several distances have been proposed which are clearly related to actions of particular operators, the desired relationship between operator and distance—often referred to as *operator-distance coherence*—has been left imprecise [7,16,17]. Also, no distance has been shown to be coherent with multiple operators. We provide a precise definition of operator-distance coherence in this paper.

The probability of transition between individuals by a single operator application can be used to define a distance, but it doesn't account for individuals

S. Silva et al. (Eds.): EuroGP 2011, LNCS 6621, pp. 190–202, 2011.

more than one step apart. The minimum number of operator applications required to transition between one individual and another can also be used to define a distance, and accounts for individuals more than one step apart, but doesn't consider the likelihood of each step. In this contribution, we provide a new distance function based on multi-step transition probability (D_{MSTP}) which is coherent with a GP operator from the dual perspectives of quantifying likelihood and accounting for individuals one or more steps apart. It relates the actions of an operator involved in changing one individual to another to their likelihood. D_{MSTP} computes the difference between two individuals by figuring out how one is changed into the other by operator applications and summing the probabilities of these applications.

A version of D_{MSTP} can be defined to be coherent with any given GP operator, including subtree mutation. However, it is impractical to compute. We therefore investigate its approximation. We consider a first order estimation and syntactic distances or combinations of them. An approximation of a distance coherent with subtree mutation is valuable for multiple reasons. It supports an operational definition of population diversity, i.e. the difference between two individuals is expressed in terms of how likely one is to be generated with subtree mutation from the other. It can also be used in fitness distance correlation calculations and to assess locality.

We proceed as follows. Sect. 2 distinguishes between inspection-based and action-based senses of the term *distance—dissimilarity* and *remoteness* respectively. It provides the first formal and general definition of operator-distance coherence. In Sect. 3 we reinterpret previous work in the light of these definitions. Next, we investigate whether an approximation of D_{MSTP} can be found in the case of GP's subtree mutation operator: Sects. 4 and 5 describe our experimental methods and results respectively. Sect. 6 concludes.

2   Definitions

Different distance functions quantify different concepts of distance. We distinguish between *dissimilarity* and *remoteness*.

Definition *Dissimilarity* means the degree to which the syntax or semantics of two objects differ. It does not rely upon an evolutionary computation, stochastic search or meta-heuristics interpretation though it is useful in these contexts. Dissimilarity can be judged by inspection or observation.

Definition *Remoteness* means the relative inaccessibility of one individual when starting from another, via a given operator or set of operators, and in a given representation. Remoteness is grounded in the context of evolutionary computation, stochastic search and meta-heuristics. It depends on the operators and representation in use. Remoteness is an action-based distance, in contrast to the inspection basis of dissimilarity.

Since remoteness is based on inaccessibility via operators, a distance function useful in the context of remoteness requires the key property of *operator-distance coherence*. We propose the first formal definition, which generalises the concepts of coherence used in previous work.

Definition *Operator-distance coherence.* In a discrete space, a distance function is coherent with an operator if the distance from one individual to another is a monotonically decreasing function of the probability that one will be transformed to the other by one or more operator applications.

Remoteness can be helpfully conceptualized via a graph of the problem search space, as discussed by [9]. Each operator induces a different graph. The nodes of the graph represent individuals. A directed edge in the graph represents that the operator applied to the parent (the source node) can generate the child (the destination node). A probability is associated with each such application of the operator. This transition probability is well-known but has not previously been used to define a distance.

Definition *One-step transition probability* (1STP). The 1STP between individuals x and y is the probability that when the operator is applied to x it will generate y. For notational convenience we define $a_e = 1\mathrm{STP}(x, y)$ for an edge e from x to y.

Between any two individuals x and y in the graph there may exist *multiple* paths, each consisting of one or more edges. The probability of traversing a particular path is the product of the probabilities (the 1STP values) of these edges. The overall probability of reaching y from x is the sum of the probabilities of all paths from x to y (excluding paths containing cycles):

Definition *Multi-step transition probability* (MSTP).

$$\mathrm{MSTP}(x, y) = \sum_{p \in \mathrm{paths}(x,y)} \left(\prod_{e \in p} a_e \right)$$

This quantity is a probability: the proof is available in supplementary materials online[1]. This probability will be high when y is easy to reach, and low when y is hard to reach. To express remoteness, then, we take the negative log probability.

Definition $D_{\mathrm{MSTP}} = -\log(\mathrm{MSTP})$.

This negative log transform also avoids underflow and allows shortest-path algorithms to be run (see Sect. 4.3). By definition, D_{MSTP} is coherent with any operator for which 1STP is defined. It satisfies the requirements of coherence because it quantifies the likelihood of an operator generating one individual from the other in one or more steps. Note that D_{MSTP} does *not* conform to the metric axioms of symmetry and the triangle inequality. This does not prevent it from being useful in practice [7].

---

[1] http://www.skynet.ie/~jmmcd/gpdistances.html

3   Discussion of Definitions w.r.t. Previous Work

Dissimilarity. Many authors have used *dissimilarity* and related ideas in GP. The dissimilarity of pairs of individuals allows study of the effects of different operators. For example, it is intuitive to say that subtree mutation [10] makes larger changes to an individual than inflate/deflate mutation [18], or that 90/10 subtree mutation [10] makes larger changes than standard subtree mutation. This is equivalent to saying that one operator makes a child more dissimilar to its parent than another. To quantify such statements, a general purpose distance function of tree dissimilarity is required. A typical method is to measure the genotypic step-size caused by an average application of each operator [5,8,12]. In other applications phenotypic distance is of interest [11,15]. In the context of dissimilarity, operator-distance coherence is not useful because it would equate step sizes between operators and make them indistinguishable.

Dissimilarity-based distances, i.e. distances that depend on observation of syntax or features, will not necessarily be coherent with any operator. However, some dissimilarity-based distances have been shown to be coherent with particular operators [4,14,18].

Remoteness. The concept of *remoteness* of pairs of individuals is needed to study problem difficulty, diversity control, and fitness sharing. It is also essential to theories of landscapes [9] and the representation independent theory of geometric operators [13]. Such studies require that the distance between a pair of individuals reflects their relative inaccessibility via operator applications.

Coherence. One framework for associating operators with distances is proposed by Moraglio [13]. The idea of operator-distance coherence is also common in previous work, though the concept has been defined only implicitly. Some distances have been shown to be coherent with particular operators, and used successfully in applications. This work divides into two cases.

Case 1: number of operator applications. When an operator is not capable of transforming any individual into any other in a single step, it is natural to think of remoteness as meaning the minimum *number of operator applications* required to make the transformation. Two studies have used this approach.

The tree-edit distance computes the minimum-cost set of edits required to transform one tree into another. The allowable edits are insertion, deletion, and substitution, each with cost 1. O'Reilly's hierarchical variable-length (HVL) mutation works by making a single edit of the same type, though node deletion can cause subtree deletion. It was designed to be closely related (or "aligned") to the tree-edit distance. HVL' (HVL-prime) improves the relationship by preventing subtree deletion [4].

The tree-alignment distance recursively computes the cost of transforming each node in a tree into its corresponding node in the other under alignment. The cost of transforming a single node is given by a cost function. Vanneschi *et al.*'s inflate/deflate mutation [17] operates by choosing a single node one level above the leaves, incrementing or decrementing its arity, and adding or deleting a leaf as appropriate. Like HVL mutation, inflate/deflate mutation was motivated by the existence of a corresponding distance: it was proven that the tree-alignment distance of two trees is linearly related to the number of inflate/deflate mutations required to move between them [18].

Case 2: one-step transition probability. By contrast, when an operator is capable of transforming any individual into any other (or almost any other) in a single step, it is natural to think of remoteness as relating to *the one-step transition probability* (1STP) [9]. This idea of remoteness has been used to give a distance function strongly related to the action of subtree crossover [7]. Subtree crossover distance computes the probability of transforming one tree into another in a single step with subtree crossover, in the presence of a particular population. That is, it calculates the 1STP. It finds the required crossover points and for each, calculates the frequency in the population of the subtree required to be pasted in at that point.

The two implied definitions of coherence given in this previous work are incompatible. In **Case 1**, an individual which can only be produced after multiple applications (never after only one) will be seen as infinitely distant in terms of 1STP. The **one-step transition probability** concept thus fails to distinguish between individuals which require 2 applications and those which require 100. In **Case 2**, the graph representing the space is *complete*: there is an edge between every pair of individuals. An individual which is quite likely to be produced, and one which could be produced but is very unlikely, are seen as equally accessible, in terms of **number of operator-applications** required. The new definition of coherence, given in the previous section, resolves the incompatibility by unifying and generalising the two cases and being applicable to any operator.

4   Methods

We turn next to an experimental study. Since the calculation of D_{MSTP} is very computationally intensive, a method of approximating it using other distance functions would be useful. Therefore the aim of our experiments is to collect samples of D_{MSTP} values and syntactic distances between many pairs of individuals, investigate whether the syntactic distances individually approximate D_{MSTP} well, and then attempt to learn D_{MSTP} as a function of some or all of them. All code is available for download.

4.1   Representation

The GP language for our experiments consists of the four arithmetic operators (functions) and the variables x and y (terminals). We consider only subtree mutation, since it is common in GP and has not previously been studied in the context of operator-distance coherence. Other operators are left for future work. In subtree mutation, a node is selected with uniform probability, and replaced with a subtree created by the "grow" method. The maximum depth of trees is 6, counting the root as depth 0.

4.2   Calculation of 1STP for Subtree Mutation

It is possible to calculate $1STP(x, y)$ exactly for subtree mutation using a procedure reminiscent of subtree crossover distance [7]. Consider each node m in
the original tree x. If there are differences between x and y outside the subtree
rooted at m, then a mutation at this point can not lead to y. Otherwise, we find
the required subtree s_m which, if pasted in at m, would result in y. The probability of generating s_m can be calculated as shown in Algorithm 1, GROWPROB.
The probability of choosing a particular node as the mutation point in a tree of
n nodes is $1/n$. Thus $1STP(x, y) = \sum_{m \in x} GROWPROB(s_m)/n$.

Algorithm 1. GROWPROB: Calculate probability of generating tree t using
"grow" method. "Grow" chooses nodes randomly from functions and terminals
until depth $= d$, then chooses only from terminals.

Require: function set F, terminal set T, maximum depth d, tree t
 1: **if** $d = 0$ **then**
 2:     return $1/|T|$
 3: **end if**
 4: **if** t is a single terminal **then**
 5:     return $1/(|T| + |F|)$
 6: **end if**
 7: $p \leftarrow 1/(|T| + |F|)$
 8: **for** $i = 1$ to arity(root(t)) **do**
 9:     $p \leftarrow p \times GROWPROB(F, T, d - 1, child_i(root(t)))$
10: **end for**
11: return p

4.3   Calculation of Graph-Based Distances

Recall that the search space is represented by a graph in which nodes are individuals and directed edges are possible one-step transitions between individuals,
and a_e is the 1STP for an edge e. We define a distance $D_{1STP} = -\log(a_e)$ and
assign to edge e a weight $w_e = D_{1STP}$. The sum of edge weights for a path p is
then $s = \sum_{e \in p} w_e = \sum_{e \in p} -\log(a_e) = -\log(\prod_{e \in p} a_e)$. Therefore, the product
of the probabilities $\prod_{e \in p} a_e$ can be recovered as e^{-s}.

This negative-log transform between multiplicative probabilities and additive
costs allows us to run shortest-path algorithms, e.g. Dijkstra's algorithm, to find
the most probable paths. By analogy with MSTP, we define the highest path probability (HPP) as the probability of traversing the graph from one individual to an-
other via the single most probable path. $HPP(x, y) = \max_{p \in paths(x,y)} \left(\prod_{e \in p} a_e \right)$.
This leads to distance function $D_{HPP} = -\log(HPP)$.

In theory HPP and MSTP are calculated over the graph representing the
entire space. In practice they are calculated over sub-graphs, as explained next.

4.4 Creating and Sampling from Sub-graphs

It is not computationally feasible to create the graph of the entire search space. We proceed by creating sub-graphs, each representing some individuals and transitions of interest. We assume that path probabilities in such sub-graphs will give results which do not differ greatly from those which would be obtained in the graph of the entire space.

The procedure begins by sampling a random individual. It creates a path of many new individuals from it using a variant of the Metropolis-Hastings algorithm, as customised for EC sampling [18]. It performs mutation on an individual until a fitness improvement of sufficient size is found, then updating so that the next mutation is performed on the new individual. Fitness is measured here as the sum of squared errors against a target polynomial $x^2 + xy$ over 100 fitness cases in $[0, 1]^2$. This simple problem was chosen arbitrarily and its purpose is purely to allow the effect of selection to be studied.

After running Metropolis-Hastings multiple times (each starting from the original), edges are added in both directions between every pair of individuals in the graph. Since shortest-path algorithms scale poorly with the number of edges n^2, the number of nodes is limited to at most $n = 100$ (slightly less when paths coincide). Distances are then calculated from the original to all other nodes. Multiple runs of the algorithm, each beginning with new random individuals, are performed. The fitness criterion can be turned off for study of populations without selection, e.g. at initialisation.

Several other methods of creating sub-graphs by mutation have also been investigated. They generally lead to similar results as that described above but are not described to save space. Sampling methods which do not use mutation may give different results, but are left for future work.

4.5 Calculation of Syntactic Distances

Fifteen syntactic distances are calculated, as follows.

Tree-edit distance (TED) [16] is a generalisation of the Levenshtein string-edit distance. Three primitive editing operations—node insertion, deletion, and substitution—each have an associated cost. TED is defined as the minimum-cost sequence of edits transforming one tree into another.

Tree-alignment distance (TAD) was studied extensively in the context of GP by Vanneschi [18] and colleagues. Different configuration of TAD are possible: the six used in this paper are termed TAD 0-5. TAD 0-2 have depth weighting factors 1, 2, and 10 respectively, with the discrete metric on symbols. TAD 3-5 use the same depth weighting factors but the metric on symbols uses a coding function as in [18].

Normalised compression distance (NCD) is an approximation to the *universal similarity metric*, using compression to approximate Kolmogorov complexity [3]. It can be applied to GP trees by encoding them as s-expression strings.

Feature vector distance (FVD) is the mean of the normalised differences between six features of the two trees: node count, minimum leaf depth, mean leaf depth, maximum leaf depth, mean fan-out, and symmetry. Symmetry is calculated as the sum of abs(size of left subtree - size of right subtree) over all nodes. A small constant is added to FVD when the trees are syntactically distinct, to ensure that non-identical trees are not at distance zero. The six normalised differences are also available individually.

5   Results: Relationships among Distance Functions

Results on the relationships among the distances are presented next.

5.1   Distribution of D_{MSTP}

The distribution of D_{MSTP} is shown in Fig. 1. Fewer steps per restart (a) biases the distribution towards individuals near the original. Many steps gives the opposite (b). Selection also tends to drive individuals farther away (c).

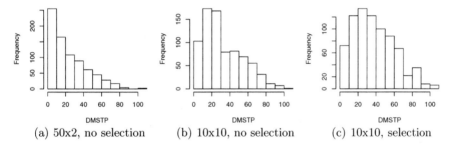

(a) 50x2, no selection   (b) 10x10, no selection   (c) 10x10, selection

Fig. 1. Distribution of D_{MSTP}. The figures NxM give the number of restarts (N) and number of steps per restart (M) in the sampling algorithm of Sect. 4.4.

We next consider how TAD, TED, FVD, and NCD (described in Sect. 4.5) relate to D_{1STP}, D_{HPP}, and D_{MSTP} (described in Sect. 2). Fig. 2 shows that TED, FVD, and NCD are somewhat related to D_{MSTP}. The relationships are quantified in Table 1. It gives the Pearson correlations among the three graph-based distances, shown in the columns, and the syntactic distances, shown in the rows (rank-based correlations are broadly similar: we report Pearson correlation for comparison with the ParetoGP method in the next section).

The most important results are for the correlation between syntactic distances and D_{MSTP}: in all cases, a statistically significant correlation is achieved, at the $p < 0.05$ level after a Bonferroni correction for running 45 tests (that is, the individual confidence levels are at $p < 0.001$). This gives grounds for hypothesizing that the syntactic distances can be used to estimate D_{MSTP}. However, since many of the syntactic distances are correlated *with each other* (not shown), they may not all be giving useful, independent information concerning D_{MSTP}.

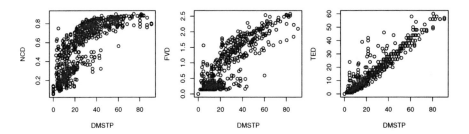

Fig. 2. Scatter plots showing the relationship of D_{MSTP} to three syntactic distances. TAD0 (not shown) is similar to TED.

Table 1. Correlations between distance measures. The four highest correlations in each column occur in the same four rows, highlighted.

Distance	D_{1STP} Corr [95% conf]	D_{HPP} Corr [95% conf]	D_{MSTP} Corr [95% conf]
NCD	0.62 [0.54, 0.69]	0.78 [0.73, 0.83]	0.78 [0.73, 0.83]
FVD	0.69 [0.62, 0.75]	0.85 [0.81, 0.88]	0.85 [0.81, 0.88]
TED	**0.86** [0.82, 0.89]	**0.95** [0.93, 0.96]	**0.95** [0.93, 0.96]
TAD0	**0.87** [0.83, 0.89]	**0.95** [0.93, 0.96]	**0.94** [0.93, 0.96]
TAD1	0.58 [0.49, 0.65]	0.74 [0.68, 0.79]	0.74 [0.68, 0.79]
TAD2	0.45 [0.35, 0.55]	0.62 [0.54, 0.69]	0.62 [0.54, 0.69]
TAD3	**0.86** [0.82, 0.89]	**0.94** [0.92, 0.95]	**0.94** [0.92, 0.95]
TAD4	0.37 [0.26, 0.47]	0.51 [0.42, 0.60]	0.51 [0.42, 0.60]
TAD5	0.21 [0.09, 0.32]	0.33 [0.22, 0.43]	0.33 [0.22, 0.43]
NodeCount	**0.81** [0.77, 0.85]	**0.92** [0.90, 0.94]	**0.92** [0.90, 0.94]
MinDepth	0.51 [0.41, 0.59]	0.59 [0.50, 0.66]	0.59 [0.50, 0.66]
MeanDepth	0.61 [0.53, 0.68]	0.79 [0.74, 0.83]	0.79 [0.74, 0.83]
MaxDepth	0.53 [0.44, 0.61]	0.71 [0.65, 0.77]	0.71 [0.65, 0.77]
Symmetry	0.39 [0.28, 0.48]	0.56 [0.47, 0.63]	0.56 [0.47, 0.64]
MeanFanout	0.40 [0.30, 0.50]	0.55 [0.46, 0.63]	0.55 [0.46, 0.63]

Syntactic distances are also correlated with D_{1STP} and D_{HPP}. The NodeCount distance is surprisingly well-correlated with the graph-based distances, though it is easy to construct examples in which it fails to predict them.

5.2   Learning D_{MSTP} as a Function of Syntactic Distances

We next attempt to learn D_{MSTP} as a function of the various syntactic distances. The aim is not only to approximate D_{MSTP}, but also to understand whether it is in a sense composed of other distances. The method is an example of using GP symbolic regression to investigate GP. Somewhat similar problems have been addressed in previous work, for example using EC to learn edit cost parameters to give the correct tree-edit distance values for pairs of points whose distance is determined by some domain knowledge [1].

We used a total of 15 distance measures as described in Sect. 4.5. These terminals were to be combined using the four arithmetic operators $(+, -, *, /)$. We used the ParetoGP symbolic regression method described fully in [19,6]. ParetoGP uses a GP engine to generate multiple models and evaluates these models along two objectives: (1) *error* $(= 1 - R^2)$, where R is the Pearson correlation coefficient, and (2) *size* (a surrogate for generalization), measured as the number of nodes per subtree summed over all complete subtrees (thus most nodes are counted multiple times). ParetoGP attempts to evolve models to minimize both objectives, generating a Pareto front. The algorithm is run multiple times to create a super-Pareto front as shown in Fig. 3.

We divided the data into training and testing subsets. Out of 744 data points (sampled as in Sect. 4.4 and calculated as in Sect. 4.5) we used approximately 60% to build our models. One of the best and simplest models was: $4.45583 + 0.143304 * \text{TAD0} * (7 + 0.586852 * \text{TAD2})$. It has size 32 and error $(1 - R^2) = 0.091 \Rightarrow R = 0.95$ on training data. On test data, it achieves an error $(1 - R^2) = 0.11 \Rightarrow R = 0.94$. This result is comparable to the very high R values achieved by some individual syntactic distances (see Table 1). Therefore, the best individual distances may be a better choice for approximating D_{MSTP}.

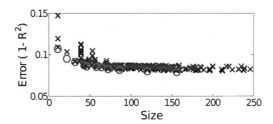

Fig. 3. ParetoGP results. D_{MSTP} is modelled as a function of syntactic distances, also minimising solution size. Crosses represent solutions on the Pareto front after single runs; circles represent solutions on the super-Pareto front of multiple runs.

5.3   Relationships among Graph-Based Distances

Two previous models of remoteness can be seen as approximations to the new D_{MSTP} model. It is natural to ask how accurate these approximations are. Under the *number of operator applications* model, all individuals are equally remote under subtree mutation, and so there is no correlation between number of operator applications and D_{MSTP}.

However, there is a strong relationship between the *one-step transition probability* model and D_{MSTP}. Fig. 4 shows the relationships among D_{1STP}, D_{HPP}, and D_{MSTP}. The path between two individuals consisting of a single edge is usually the best, since paths of multiple edges tend to be unlikely (Fig. 4 (a)). The single-edge path also dominates the overall remoteness (b). The single best path is an almost perfect predictor of the sum of paths (c), but suffers from the same problem of computational infeasibility.

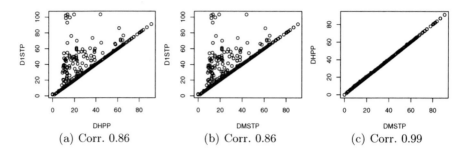

Fig. 4. Scatter plots relating $D_{1\text{STP}}$, D_{MSTP}, and D_{HPP}

The relationship between $D_{1\text{STP}}$ and D_{MSTP} is not perfect, however. We can construct examples in which a path of multiple edges between a pair of individuals will be *more* likely than a path of a single edge. In such cases, the $D_{1\text{STP}}$ will be an inaccurate approximation of D_{MSTP}. We also emphasise that for operators other than subtree mutation, especially where every individual is not accessible in a single mutation from every other, neither $D_{1\text{STP}}$ nor D_{HPP} will be so strongly correlated with D_{MSTP}.

We conclude that $D_{1\text{STP}}$ is a good estimator of (subtree mutation) D_{MSTP} for applications. This result strongly supports previous approaches [7] where (subtree crossover) $D_{1\text{STP}}$ was the central concept of distance.

6   Conclusions

This contribution gives the first formal definition of operator-distance coherence which is general enough to apply to any operator. The definition was used to study operator-distance coherence for subtree mutation for the first time. It motivates a new distance function, D_{MSTP}, which is coherent with subtree mutation (and with any operator for which $D_{1\text{STP}}$ can be defined).

We have also investigated which of several syntactic distances is most closely coherent with subtree mutation (best results $R > 0.92$). Experiments attempted to learn, using GP, non-linear combinations of these distances which are closely coherent with subtree mutation (best result $R = 0.94$). The one-step transition probability distance $D_{1\text{STP}}$, which can be calculated quickly and exactly for subtree mutation, also approximates D_{MSTP} well ($R = 0.86$) and thus approaches operator-distance coherence. According to the results in this initial study, D_{MSTP} seems to provide a good compromise between accuracy and computation time.

In future work, we hope to extend the D_{MSTP} framework to consider other mutation operators, crossover operators, and the use of multiple operators simultaneously, as in typical EC algorithms. For operators where 1STP is difficult or impossible to calculate exactly, a sampling method can be used to estimate it. The remainder of our methods apply equally well to other operators. We hope also to report on other sampling methods.

Acknowledgements

JMcD acknowledges the support of the Irish Research Council for Science, Engineering and Technology under the Inspire scheme. UMO'R gratefully acknowledges the support of VMWare and USA D.O.E. Grant DE-SC0005288. LV acknowledges project PTDC/EIA-CCO/103363/2008 from FCT, Portugal.

References

1. Bernard, M., Habrard, A., Sebban, M.: Learning stochastic tree edit distance. In: Fürnkranz, J., Scheffer, T., Spiliopoulou, M. (eds.) ECML 2006. LNCS (LNAI), vol. 4212, pp. 42–53. Springer, Heidelberg (2006)
2. Brameier, M., Banzhaf, W.: Explicit control of diversity and effective variation distance in linear genetic programming. In: Foster, J., Lutton, E., Miller, J., Ryan, C., Tettamanzi, A. (eds.) EuroGP 2002. LNCS, vol. 2278, pp. 37–49. Springer, Heidelberg (2002)
3. Cilibrasi, R., Vitanyi, P.M.B.: Clustering by compression. IEEE Transactions on Information Theory 51(4), 1523–1545 (2005)
4. Durrett, G., Neumann, F., O'Reilly, U.M.: Computational complexity analysis of simple genetic programming on two problems modeling isolated program semantics. In: Foundations of Genetic Algorithms (2010)
5. Ekárt, A., Németh, S.Z.: A metric for genetic programs and fitness sharing. In: Poli, R., Banzhaf, W., Langdon, W.B., Miller, J., Nordin, P., Fogarty, T.C. (eds.) EuroGP 2000. LNCS, vol. 1802, pp. 259–270. Springer, Heidelberg (2000)
6. Evolved Analytics LLC: DataModeler Release 1.0. Evolved Analytics LLC (2010)
7. Gustafson, S., Vanneschi, L.: Crossover-based tree distance in genetic programming. IEEE Transactions on Evolutionary Computation 12(4), 506–524 (2008)
8. Igel, C., Chellapilla, K.: Investigating the influence of depth and degree of genotypic change on fitness in genetic programming. In: Proceedings of the Genetic and Evolutionary Computation Conference, vol. 2, pp. 1061–1068 (1999)
9. Jones, T.: Evolutionary Algorithms, Fitness Landscapes and Search. Ph.D. thesis, University of New Mexico, Albuquerque (1995)
10. Koza, J.R.: Genetic Programming: On the Programming of Computers by Means of Natural Selection. MIT Press, Cambridge (1992)
11. Krawiec, K., Lichocki, P.: Approximating geometric crossover in semantic space. In: GECCO 2009: Proceedings of the 11th Annual Conference on Genetic and Evolutionary Computation, pp. 987–994. ACM, New York (2009)
12. McDermott, J., Galván-López, E., O'Neill, M.: A fine-grained view of GP locality with binary decision diagrams as ant phenotypes. In: Schaefer, R., Cotta, C., Kołodziej, J., Rudolph, G. (eds.) PPSN XI. LNCS, vol. 6238, pp. 164–173. Springer, Heidelberg (2010)
13. Moraglio, A.: Towards a geometric unification of evolutionary algorithms. Ph.D. thesis, University of Essex (November 2007), http://eden.dei.uc.pt/~moraglio/
14. Moraglio, A., Poli, R.: Geometric landscape of homologous crossover for syntactic trees. In: CEC, vol. 1, pp. 427–434. IEEE, Los Alamitos (2005)
15. Nguyen, Q.U., Nguyen, X.H., O'Neill, M.: Semantic aware crossover for genetic programming: The case for real-valued function regression. In: Vanneschi, L., Gustafson, S., Moraglio, A., De Falco, I., Ebner, M. (eds.) EuroGP 2009. LNCS, vol. 5481, pp. 292–302. Springer, Heidelberg (2009)

16. O'Reilly, U.M.: Using a distance metric on genetic programs to understand genetic operators. In: IEEE International Conference on Systems, Man, and Cybernetics: Computational Cybernetics and Simulation, vol. 5 (1997)
17. Tomassini, M., Vanneschi, L., Collard, P., Clergue, M.: A study of fitness distance correlation as a difficulty measure in genetic programming. Evolutionary Computation 13(2), 213–239 (2005)
18. Vanneschi, L.: Theory and Practice for Efficient Genetic Programming. Ph.D. thesis, Université de Lausanne (2004)
19. Vladislavleva, E., Smits, G., Kotanchek, M.: Better solutions faster: Soft evolution of robust regression models in Pareto genetic programming. In: Riolo, R.L., Soule, T., Worzel, B. (eds.) Genetic Programming Theory and Practice V, ch. 2, pp. 13–32. Springer, Ann Arbor (2007)

Evolution of a Brain-Computer Interface Mouse via Genetic Programming

Riccardo Poli, Mathew Salvaris, and Caterina Cinel

School of Computer Science and Electronic Engineering,
University of Essex,
Wivenhoe Park, Colchester, CO4 3SQ, UK
{rpoli,mssalv,ccinel}@essex.ac.uk

Abstract. We propose the use of genetic programming as a means to evolve brain-computer interfaces for mouse control. Our objective is to synthesise complete systems, which analyse electroencephalographic signals and directly transform them into pointer movements, almost from scratch, the only input provided by us in the process being the set of visual stimuli to be used to generate recognisable brain activity. Experimental results with our GP approach are very promising and compare favourably with those produced by support vector machines.

Keywords: Genetic Programming, Brain-Computer Interfaces, Mouse, Support-vector Machines.

1 Introduction

Over the past few years an increasing number of studies (e.g., [5,12,7,2,11]) have shown that Brain-Computer Interfaces (BCIs) are possible. These convert signals generated from the brain into control commands for devices such as computers, wheel chairs or prostheses. Such systems are often based on the analysis of brain electrical activity obtained via electroencephalography (EEG).

Within EEG-based BCIs, the analysis of P300 waves and other event related potentials (ERPs) has been particularly effective [5,9,1,3]. ERPs are electrical components of brain activity produced in response to external stimuli. The P300 is a positive ERP which, in experimental conditions, can be generated by the brain when an observer attends to rare and/or significant stimuli, around 300 ms after the presentation of such stimuli. P300 waves can be used to determine user intentions in BCIs by relating their presence (or an increase in their amplitude) to specific external stimuli.

Over the years, there have been some attempts to develop BCI systems aimed at controlling 2–D pointer movements. The most successful of these, to date, are those based on the detection of μ or β rhythms in EEG [11], which require subjects to spend weeks or months training their brain to produce such rhythms, and those using invasive cortical interfaces (e.g., [4]), which require surgery.

Much less troublesome are systems based on P300 waves since they require neither user training nor invasive procedures. Some initial success in developing a P300-based mouse has been reported in [1], where rather long intervals between the end of a stimulus and the beginning of the following one were used. This resulted in the pointer moving at very slow rate (one movement every 10 seconds). Slightly better performance

S. Silva et al. (Eds.): EuroGP 2011, LNCS 6621, pp. 203–214, 2011.

(one cursor movement every 4 seconds) was achieved in [9], but accuracy in detecting P300s was only about 50% leading to many incorrect movements.

A more responsive P300-based system for the 2–D control of a cursor on a computer screen was presented in [3]. In this system four randomly-flashing squares are displayed on the screen to represent four directions of movement. Users focus their attention on the flashes of the square representing the desired direction of motion for the cursor. Flashing of the attended square produces ERPs — including P300 components — which are larger in amplitude compared to ERPs produced by the flashing of non-attended squares. Through the analyses of such brain potentials, the BCI system can then infer the user's intentions and move the cursor. The system presents three important features: it completely dispenses with the problem of detecting P300s (a notoriously difficult task) by logically behaving as an *analogue* device (i.e., a device where the output is a continuous function of the input, as opposed to a binary classifier), it uses a single trial approach where the mouse performs an action after every trial (once per second), and it *heavily relies on a genetic algorithm* for the selection of the EEG best channels, time steps and wavelet scales to use as features in the control of the mouse. The use of an analogues approach provides the system with more information about the brain state, which, in turn, makes it a more accurate, gradual and controllable mouse. We will use similar ingredients in the work presented here.

A variety of alternatives to this scheme were explored in [10] where 8 different stimuli (4 for up, down, left and right, and 4 for the 45 degree diagonal directions) were used. A linear support-vector machine (SVM) was trained to classify the ERPs produced in response to stimuli into two classes: target and non-target. After training, the raw (continuous) output produced by the SVM was used to score ERPs (the higher the score, the higher the "targetness" of an ERP). The SVM's score for each flash was then turned into a vector pointing in the direction of the stimulus and with an length proportional to the score. This was then used together with the vectors associated with other directions to determined the pointer's motion.

Despite these successes, the trajectories of the BCI-mouse pointer tend to be very convoluted and indirect. This is mainly because of the noise present in EEG signals (which is often bigger than the signal itself), the presence of eye-blinks and other muscle contractions (which produce artifacts up to two orders of magnitude bigger than the signal) and the difficulty of maintaining a user's mind focused on the target stimulus. However, the success of an analogue approach to BCI mouse design and the benefits accrued in it through the use a machine learning techniques suggest that there is more mileage in this approach. In particular, we should note that, effectively, although both in [3] and [10] feature selection was performed using powerful algorithms, only semi-linear transformations were used to transform EEG signals into mouse movements. Linear systems have obvious limitations, particularly in relation to noise and artefact rejection. So, we wondered if genetic programming (GP) [8], with its ability to explore the huge space of computer programs, could produce even more powerful transformations while also performing feature selection and artefact handling at the same time.

In this paper we report the results of our efforts to use GP as a means to evolve complete and effective BCI mouse controllers. More specifically, in Section 2 we describe the stimuli, procedure, participants and analysis performed in our BCI mouse. Section 3 describes the GP system used, its primitives, parameter settings and fitness function. In Section 4 we report our experimental results, while we provide some conclusions in Section 5.

2   BCI Mouse

Our BCI mouse uses a flashing-stimuli protocol with some similarities to the P300-based BCI mice described in the previous section. More specifically, we used visual displays showing 8 circles (with a diameter of 1.5 cm) arranged around a circle at the centre of the display as in Figure 1(far left). Each circle represents a direction of movement for the mouse cursor. Circles temporarily changed colour – from grey to either red or green – for a fraction of a second. We will call this a *flash*. The aim was to obtain mouse control by mentally focusing on the flashes of the stimulus representing the desired direction and mentally naming the colour of the flash. Flashes lasted for 100 ms and the inter-stimulus interval was 0 ms. Stimuli flashed in clockwise order. This meant that the interval between two target events for the protocol was 800 ms. We used a black background, grey neutral stimuli and either red or green flashing stimuli.

Data from 2 healthy participants aged 29 (male, normal eyesight) and 40 (female, corrected to normal eyesight) were used for GP training. The ERPs of these subjects present significant morphological differences and are affected by different types of artifacts. Each session was divided into runs, which we will call *direction epochs*. Each participant carried out 16 direction epochs, this resulted in the 8 possible directions being carried out twice.

Each direction epoch started with a blank screen and after 2 seconds the eight circles appeared near the centre of the screen. A red arrow then appeared for 1 second pointing to the target (representing the direction for that epoch). Subjects were instructed to mentally name the colour of the flashes for that target. After 2 seconds the flashing of the stimuli started. This stopped after 20 to 24 trials, with a trial consisting of the sequential activation of each of the 8 circles. In other words each direction epoch involves between $20 \times 8 = 160$ and $24 \times 8 = 192$ flashes. After the direction epoch had been completed, subjects were requested to verbally communicate the colour of the last target flash.

Participants were seated comfortably at approximately 80 cm from an LCD screen. Data were collected from 64 electrode sites using a BioSemi ActiveTwo EEG system. The EEG channels were referenced to the mean of the electrodes placed on either earlobe. The data were initially sampled at 2048 Hz, the filtered between 0.15 and 30 Hz and finally sub-sampled to 128 samples per second. Then, from each channel an 800 ms

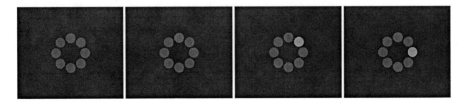

Fig. 1. Stimuli used in our BCI experiments: initial display and three sequentially flashed stimuli

epoch following each flash was extracted and further sub-sampled to 32 samples per second. This resulted in 26 data points per channel. Thus each epoch is described by a feature vector of $26 \times 64 = 1,664$ elements.

Our training set for GP included approximately 2,800 such feature vectors per subject (16 direction epochs per subject, 160–192 trials per direction epoch). Note that we didn't have a target output for each feature vector: we have only have a target direction for each of the 16 direction epochs.

3   GP System and Parameter Settings

We used a strongly-typed GP system implemented in Python with all numerical calculations done using the Numpy library (which is implemented in C). Since fitness evaluation in our domain of application is extremely computationally intensive, we created a parallel implementation which performs fitness evaluations across multiple CPU cores (via farming).

The system uses a steady-state update policy. It evolves a population of 10,000 individuals with tournament selection with a tournament size of 5, a strongly-typed version of the grow method with a maximum initial depth of 4, and strongly-typed versions of sub-tree crossover and sub-tree mutation. Both are applied with a 50% rate and use a uniform selection of crossover/mutation points. The system uses the primitive set shown in Table 1. Program trees were required to have a `Float` return type.

Table 1. Primitive set used in our application

Primitive	Output Type	Input Type(s)	Functionality
`0.5, -0.5, 0.1, -0.1, 0, 1, ···, 25`	Float	None	Floating point constants used for numeric calculations and as array indexes (see below)
`Fp1, AF7, AF3, F1, ...` (60 more channel names)	Array	None	Returns an array of 26 samples following a flash from one of the channels. The samples are of type Float.
`+, -, *, min, max`	Float	(Float, Float)	Standard arithmetic operations plus maximum and minimum on floats.
`>, <`	Bool	(Float, Float)	Standard relational operations on floats
`if`	Float	(Bool, Float, Float)	If-then-else function. If the first argument evaluates to True, then the result of evaluating its second argument is returned. Otherwise the result of evaluating the third argument is returned.
`abs`	Float	Float	Returns the absolute value of a Float.
`mean, median, std, Amin, Amax`	Float	(Float, Float, Array)	Given an 26-sample Array and two floats, treat the floats as indices for the array by casting them to integer via truncation and then applying a modulus 26 operation (if the indices are identical, one is increment by 1). Then compute the mean, median, standard deviation, minimum or maximum, respectively, of the samples in the Array falling between such indices (inclusive).

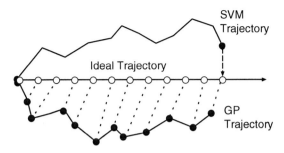

Fig. 2. Ideal and actual trajectories used in the fitness calculation. Dashed lines indicate pairs of matching points. Fitness is the average distance between such points across 16 trajectories. The end point of the ideal trajectory is computed by projecting the end point of the trajectory produced by an SVM-based flash scorer.

With this setup we performed runs of up to 50 generations, manually stopping them whenever we felt they were unlikely to make further significant progress. Because of the extreme computational load required by our fitness evaluation and the complexity of the problem (which forced us to use a relatively large population), here we only report the results of one run per subject.[1] We feel this is reasonable since we are really interested in the output produced by GP — as is the case in many practical applications of GP — rather than in optimising the process leading to such output. Each run took approximately 40 *CPU days* to complete.

Let us now turn to the fitness function we used to guide evolution. Fitness is the dissimilarity between the ideal trajectory and the actual trajectory produced by a program averaged over the 16 direction epochs. Measuring this requires executing each program for over 2,800 times. Being an error measure, fitness is, naturally, minimised in our system. We describe its elements below.

The actual trajectory produced by a program on a training epoch is obtained by iteratively evaluating the program, each time feeding the samples relating to a new flash into the Fp1, AF7, etc. terminals (which effectively act as a sliding window on the EEG). The output of the program, which, as noted above, is of type Float, is multiplied by a unit vector representing the direction corresponding to the stimulus that flashed on the screen. This produces a result of the form $(\Delta x, \Delta y)$ which is used as a displacement to be applied to the current mouse position.

As illustrated in Figure 2, the ideal trajectory for each direction epoch is obtained by sampling at regular intervals the line segment connecting the origin to a point along the desired direction. The point is chosen by projecting the end-point of the trajectory produced by a linear SVM trained on the same data as GP in the same way as in [10]. In this way, when comparing the results obtained by GP to those produced by an SVM, we ensure both had the same ideal trajectories. The ideal trajectory is sampled in such a way to have the same number of samples as the actual trajectory. The comparison between actual and ideal trajectory is then a matter of measuring the Euclidean distance between pairs of corresponding points in the two trajectories and taking an average. Notice that any detours from the ideal line and any slow-downs in the march along it in the actual trajectory are strongly penalised with our fitness measure.

---

[1] Naturally, we performed a number of smaller runs while developing the fitness function.

4   Experimental Results

Figure 3 shows the dynamics of the median and best program's fitness in our runs. The best evolved programs are presented in simplified tree form in Figure 6.[2] To evaluate their performance we will compare their output to the output produced by the SVM on the same ERP data.

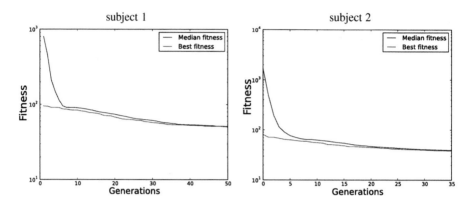

Fig. 3. Plots of the median and best fitness vs generation number in our runs

Let us start from a qualitative analysis. Figure 4(top) shows the output produced by SVM for each of the direction epochs in the training set of each subject (after the transformation of the SVM scores into $(\Delta x, \Delta y)$ displacements). Ignoring the small scales wiggles in the trajectories (which are due to the periodicity of the flashing stimuli), we see that the SVMs do a reasonably good job at producing straight trajectories. This feature is the result of the SVM being trained to respond only when the target stimulus is flashed, and to produce much smaller outputs for all other (seven) stimuli. A close inspection of the trajectories, however, reveals that while they are often straight, often they do not point exactly in the desired direction. For example, one of the two trajectories labelled as $(-0.7, 0.7)$ in subject 1 points almost directly upwards instead of pointing towards the top-left corner of the plot, one of the two trajectories labelled as $(0.7, -0.7)$ in subject 2 points almost directly downwards instead of pointing towards the bottom-right corner of the plot, and one of the two trajectories labelled as $(1.0, 0.0)$ in subject 2 points towards the top-right corner of the plot instead of pointing to the right. These biases are the result of the SVM training algorithm not being able to consider an important fact: in order for overall trajectories to point in the correct direction, the reduced (but often non-zero) output produced in the presence of flashes of stimuli that are one (two) before and one (two) after a target stimulus must be symmetric (at least on average). Overall, because of this phenomenon, SVM produces trajectories which show unsatisfactory clustering towards the 8 prescribed directions of motion.

---

[2] The programs in Figure 6 were first syntactically simplified replacing expressions involving only constants with their value and expressions of the form (if True ExpA ExpB) or (if False ExpA ExpB) with ExpA and ExpB, respectively. Then, starting from the leaves, we replaced sub-trees with their median output if the replacement influenced fitness by less than 2%.

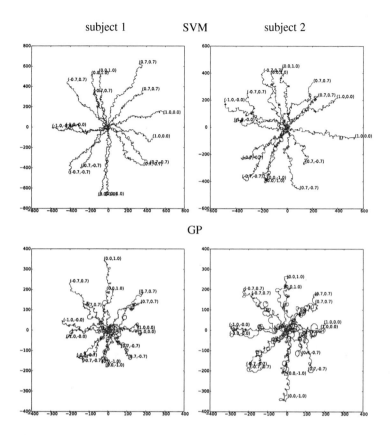

Fig. 4. Graphical representation of the 16 sequences of SVM scores (top) and the evolved-program scores (bottom) for each of our two subjects. The tuples of numbers labelling trajectory endpoints are unit vectors indicating the desired direction of motion.

Figure 4(bottom) shows the corresponding trajectories produced by our best evolved programs. Qualitatively it is clear that these trajectories are more convoluted. This has to be expected since, unlike SVM, GP does not try to classify flashes as targets or non-targets. So, there is no explicit bias towards suppressing the output in the presence of non-target flashes. There is, however, a strong pressure in the fitness measure towards ensuring that there is overall motion in the target direction. Also, there is pressure to ensure that the outputs produced for non-target flashes are such that they either cancel out or add up in such a way to contribute to the motion in the target direction, with minimum deviations from the ideal trajectory. As a result, the trajectories produced by the best programs evolved by GP are quite close to the ideal line in each of the prescribed directions of motion. Note, for example, how close the labels at the end of each trajectory in Figure 4(bottom) are to the corresponding target directions.

To quantitatively verify these observations, Tables 2(a)–(c) show a statistical comparison between the trajectories produced by GP and those produced by SVM. More specifically: Tables 2(a) shows the mean, median, standard deviation and standard

Table 2. Statistical comparison between SVM and evolved solutions: (a) basic statistics of the distribution of distances between ideal and actual mouse trajectories, (b) *p*-values for the Kolmogorov-Smirnov one-sided two-sample test for pairwise comparison of distributions, and (c) *p*-values for the one-sided Wilcoxon signed rank test for paired data

Subject	Program	Mean	Median	Standard Deviation	Standard Error
1	GP	50.813	44.367	17.440	4.503
	SVM	52.638	43.709	30.550	7.888
2	GP	38.424	37.797	10.030	2.590
	SVM	69.766	59.997	34.518	8.913

(a)

Subject		GP	SVM
1	GP	—	0.210
	SVM	0.779	—
2	GP	—	1.000
	SVM	0.018	—

(b)

Subject		GP	SVM
1	GP	—	0.590
	SVM	0.430	—
2	GP	—	1.000
	SVM	0.000	—

(c)

error of the mean of the distances between the ideal trajectory and the actual trajectory recorded in each of the 16 direction trials of each subject; Table 2(b) reports the *p*-values for the Kolmogorov-Smirnov one-sided two-sample test for pairwise comparison of distributions; and, Table 2(c) shows the one-sided Wilcoxon signed rank test for paired data. For both subjects the evolved programs produce trajectories which are on average closer to the ideal line than the corresponding trajectories produced by SVM. In addition, GP programs shows a much more consistent behaviour having much smaller standard deviations from the ideal line than SVM. Statistically, the evolved programs produces trajectories that are significantly better than those produced by SVM for subject 2 and are on par with SVM in the case of subject 1.

Naturally, the more pronounced swirls present in the trajectories produced by the evolved programs may be a distraction for a user. It is, however, easy to remove them by post-processing the mouse movements $(\Delta x_i, \Delta y_i)$ produced by the system via a smoothing function. To illustrate the benefits of this, we have used an exponential IIR filter of the form: $(\Delta x_i^s, \Delta y_i^s) = \alpha \cdot (\Delta x_i, \Delta y_i) + (1 - \alpha) \cdot (\Delta x_{i-1}^s, \Delta y_{i-1}^s)$, with $\alpha = 1/30$ and initialisation $(\Delta x_0^s, \Delta y_0^s) = (\Delta x_0, \Delta y_0)$, which is effectively equivalent to associating a momentum to the mouse cursor. Figure 5 shows the resulting trajectories.

Clearly smoothing improves significantly the straightness of the trajectories produced by evolved programs, thereby removing the potential negative effects of the swirls in the raw trajectories. Smoothing, of course, does not improve the directional biases in the SVM-produced trajectories, although it makes them more rectilinear.

While all this is very encouraging, it is also important to see what lessons can be learnt from the evolved programs themselves. Looking at the program evolved for subject 1, we see that its output is simply the mean of a consecutive block of samples taken from channel P8. The choice of P8 is good, since this and other parietal channels are often the locations where the strongest P300 are recorded. The choice of sample 15 (which corresponds to approximately 470 ms after the presentation of the stimulus) as one of the two extremes of the averaging block is also very appropriate, since this

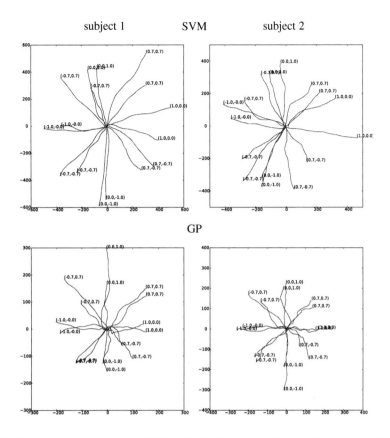

Fig. 5. Trajectories produced by SVM and GP after the application of a low-pass filter

sample falls right in the middle of typical P300 waves. The sample marking the other end of the averaging block is the result of a more elaborate choice determined by three cascaded if statements. Effectively this sample can take four different values: 0 (0.5 is truncated to 0), 4 (125 ms), 10 (310 ms) or 20 (625 ms). Sample 0 is chosen when the leftmost shaded sub-tree in Figure 6(left) returns True. Careful analysis has clarified that this tree is specialised in detecting near targets, i.e., flashes that immediately precede or follow the flashing of a target. When not in the presence of a near target, sample 4 is chosen if the second shaded sub-tree returns True. Analysis suggests that this sub-tree is a negated detector of eye-blinks in the early part of the data. If an eye-blink is detected, then control passes to the last if in the chain which moves away the averaging window from the early samples, forcing the average to cover either the range 10–15 or the range 15–20 depending on the likelihood of finding a target related P300 in the corresponding time windows (310–470 ms and 470–625 ms, respectively). This decision is taken by the right-most shaded sub-tree.

Turning now to the program evolved for subject 2, we see that it too uses the strategy of returning the average of a particular channel, FC2 in this case, where the average is between a fixed sample (13) and a second sample which is computed based on a number

subject 1

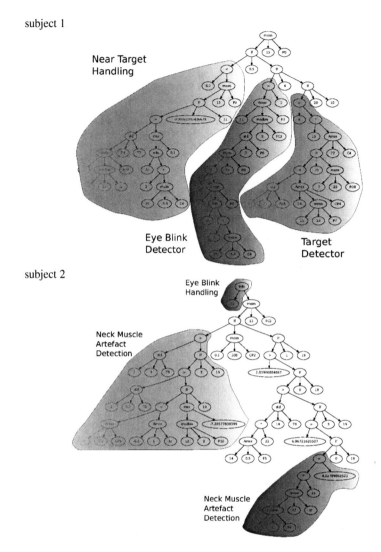

Fig. 6. Best programs evolved in our runs after syntactic and semantic simplification (see text) along with the approximate interpretation of the functionality of their sub-trees

of conditions. In this case, the output is clipped to 13. We think this clipping is an eye-blink handling strategy. Since channel FC2 is relatively frontal, the system needed to prevent eye-blinks from producing disastrous off course mouse movements. By clipping the output to 13, eye-blinks (which typically produce huge voltages for many hundreds of milliseconds, i.e., many stimulus flashes) will cause the pointer trajectory to loop, thereby never going too far off course. Note that the choice of sample 13 (406 ms) as one extreme for the windows where the mean is computed is quite appropriate since this sample, too, falls in the middle of typical P300s.

The values used as a second extreme of the averaging window depends on the result of the comparison carried out in the left-most shaded sub-tree in Figure 6(right). In this subtree electrode Iz (which is at the very base of the scalp near the neck) has a prominent role. The subtree is effectively a detector for neck movements and other muscular artifacts in the early samples of an epoch. If one such artefact is detected the second extreme for the mean is based on the calculation (mean 0.1 308 CP2) which is internally transformed into (mean 0 22 CP2). The stronger the effect of the artefact on centro-parietal areas, the more the resulting sample moves away from the beginning of the epoch, thereby avoiding the influence of spurious data in the determination of the program output. If no early muscle artefact is detected then the second extreme of the averaging block is either 1 (32 ms) or 19 (590 ms).

The decision about which sample to use is essentially made by a second artefact detection subtree (right-most shaded tree in Figure 6(right)). When activated this checks for muscle artifacts over a wider range of samples (including the end of the epoch). If none is detected, this causes a domino effect involving the five if statements connecting the subtree to the mean over FC2 instruction near the top of the program, with all such if statements returning 19. Sample 19 is then used as the second extreme of the averaging window for FC2. In the absence of eye-blinks, the output of the program is thus the average of FC2 over the range 406–590 ms. This makes complete sense since the range effectively covers most of the energy in the P300 wave.

5 Conclusions

Brain-computer interfaces are an exciting research area which one day will hopefully turn into reality the dream of controlling computers hands-free through intelligent interfaces capable of interpreting users' commands directly from electrical brain signals. Progress is constantly made in BCI but it is slowed down by many factors including the noise present in brain signals, muscular artefacts and the inconsistency and variability of user attention and intentions.

Recent research has shown that genetic algorithms and support vector machines can be of great help in the selection of the best channels, time steps and/or filtering kernels to use for the control of an ERP-based BCI mouse. In this paper we propose the use of genetic programming as a means to evolve BCI mouse controllers — a very difficult task that has never been attempted before.

Our objective was to synthesise complete systems, which analyse electroencephalographic signals and directly transform them into pointer movements. The only input we provided in this process was the set of visual stimuli to be used to generate recognisable brain activity. Note that there is no high-performance human-designed system for ERP-based mouse pointer control in BCI. There simply aren't any design guidelines for a domain such as this. This is why the very few systems of this kind reported in the literature all use some form of machine learning (either a GA or an SVM).

Experimental results with our approach show that GP can produce very effective BCI mouse controllers which include clever mechanisms for artefact reduction. The picture emerging from our experiments is that not only GP has been successful in the automated design of a control system for a BCI mouse, but it has also been able to perform better than SVM — which until now has been considered perhaps the best machine-learning technology available for BCI. Additionally, GP produces controllers that can be understood and from which we can learn new effective strategies for BCI.

All this suggests that our evolved systems are at or above state-of-the-art, indicating that, perhaps, they qualify for the attribute of human-competitive, in the sense indicated by Koza (e.g., see [6]).

Acknowledgements

We would like to thank EPSRC (grant EP/F033818/1) for financial support.

References

1. Beverina, F., Palmas, G., Silvoni, S., Piccione, F., Giove, S.: User adaptive BCIs: SSVEP and P300 based interfaces. PsychNology Journal 1(4), 331–354 (2003)
2. Birbaumer, N., Ghanayim, N., Hinterberger, T., Iversen, I., Kotchoubey, B., Kübler, A., Perelmouter, J., Taub, E., Flor, H.: A spelling device for the paralysed. Nature 398(6725), 297–298 (1999)
3. Citi, L., Poli, R., Cinel, C., Sepulveda, F.: P300-based BCI mouse with genetically-optimized analogue control. IEEE Transactions on Neural Systems and Rehabilitation Egineering 16(1), 51–61 (2008)
4. Donoghue, J.: Connecting cortex to machines: recent advances in brain interfaces. Nature Neuroscience 5, 1085–1088 (2002)
5. Farwell, L.A., Donchin, E.: Talking off the top of your head: toward a mental prosthesis utilizing event-related brain potentials. Electroencephalography and Clinical Neurophysiology 70(6), 510–523 (1988)
6. Koza, J.R.: Human-competitive results produced by genetic programming. Genetic Programming and Evolvable Machines 11(3/4), 251–284 (2010)
7. Pfurtscheller, G., Flotzinger, D., Kalcher, J.: Brain-computer interface: a new communication device for handicapped persons. Journal of Microcomputer Applications 16(3), 293–299 (1993)
8. Poli, R., Langdon, W.B., McPhee, N.F.: A field guide to genetic programming (2008), Published via http://lulu.com and freely available at http://www.gp-field-guide.org.uk (With contributions by J. R. Koza)
9. Polikoff, J.B., Bunnell, H.T., Borkowski Jr., W.J.: Toward a P300-based computer interface. In: Proc. Rehab. Eng. and Assistive Technology Society of North America (RESNA 1995), pp. 178–180. Resna Press, Arlington (1995)
10. Salvaris, M., Cinel, C., Poli, R., Citi, L., Sepulveda, F.: Exploring multiple protocols for a brain-computer interface mouse. In: Proceedings of the 32nd Annual International Conference of the IEEE Engineering in Medicine and Biology Society (EMBS), Buenos Aires, September 2010, pp. 4189–4192 (2010)
11. Wolpaw, J.R., McFarland, D.J.: Control of a two-dimensional movement signal by a non-invasive brain-computer interface in humans. Proceedings of the National Academy of Sciences 101(51), 17849–17854 (2004)
12. Wolpaw, J.R., McFarland, D.J., Neat, G.W., Forneris, C.A.: An EEG-based brain-computer interface for cursor control. Electroencephalography and Clinical Neurophysiology 78(3), 252–259 (1991)

Operator Self-adaptation in Genetic Programming

Min Hyeok Kim[1], Robert Ian (Bob) McKay[1], Nguyen Xuan Hoai[2], and Kangil Kim[1]

[1] Seoul National University, Korea
[2] Hanoi University, Vietnam
{rniritz,rimsnucse,nxhoai,kangil.kim.01}@gmail.com
http://sc.snu.ac.kr

Abstract. We investigate the application of adaptive operator selection rates to Genetic Programming. Results confirm those from other areas of evolutionary algorithms: adaptive rate selection out-performs non-adaptive methods, and among adaptive methods, adaptive pursuit out-performs probability matching. Adaptive pursuit combined with a reward policy that rewards the overall fitness change in the elite worked best of the strategies tested, though not uniformly on all problems.

Keywords: Genetic Programming, Adaptive Operator Selection, Adaptive Pursuit, Probability Matching, Evolutionary Algorithm, Tree Adjoining Grammar, Grammar Guided Genetic Programming.

1 Introduction

Evolutionary algorithms (EAs) have many parameters, so that it is not surprising that parameter setting has been a major focus of study in evolutionary computation. In the early days, the focus was making good static choices of parameter values [1], but it quickly became recognised that good parameter settings might change from problem to problem, or even vary within a run, so that the focus shifted to adaptive parameter setting [2].

Recently, there has been renewed interest in mechanisms for adaptive operator rates in particular, with a flurry of significant advances in understanding [3,4,5]. These advances have focused on fixed-complexity evolutionary algorithms (Genetic Algorithms and similar). Genetic Programming (GP) uses a variable-size genome with more complex structure. While it resembles other EAs in many ways, it also exhibits substantial differences. For example, in most EAs, there is only one reservoir of entropy, the differences between individuals, while in GP there is a second, the regularity (or irregularity) of the structure of individuals. This results in differences in desirable operator settings, with good static crossover rates for GP often being quite high (over 50%), while such rates would be considered extreme in most EAs. As a consequence, we cannot simply apply general EA results about operator selection strategies – adaptive or not – willy-nilly to GP. The aim of this work is to investigate whether some recent results [6] on adaptive operator selection carry over to GP.

S. Silva et al. (Eds.): EuroGP 2011, LNCS 6621, pp. 215–226, 2011.

The paper is organised as follows. Section 2 discusses previous work on operator selection, and introduces the 'adaptive pursuit' and 'probability matching' rate adaptation algorithms. Section 3 describes our experimental framework for comparing different adaptation mechanisms, while section 4 presents the results. In section 5, we discuss the meaning and implications of these results, concluding in section 6 with the assumptions and limitations of the work, a summary of the conclusions, and pointers for further work.

2 Background

2.1 Operator Parameter Adaptation

The parameters that control evolutionary operators have long been an issue in EAs. Two kinds of parameters are involved: parameters that control the rates of application of specific operators, and parameters that control the scale of the operators. The former were of particular concern in the Genetic Algorithms literature, and the latter in Evolution Strategies. In both cases, the rates were originally set based on preliminary experiments, and rules of thumb were developed to suggest reasonable general values [1]. It did not take long before it was recognised that fixed rates were undesirable, and that they should change during the search. A simple approach is to pre-schedule the change in the parameters, usually linearly. More sophisticated methods use the preceding performance of operators as a guide to their likely future performance, beginning with Schwefel's one-fifth success rule [2]. More recently, two approaches have come to the fore: Probability Matching (PM [7,8,9]) and Adaptive Pursuit (AP [3,10]).

Probability Matching aims to match the operator probabilities to their relative reward from the environment. A detailed algorithm is given in table 1(left). The parameters passed to the algorithm are the number of operators, K, a probability vector P giving the probability of applying the corresponding operator, a quality vector Q measuring the reward obtained from previous use of the operator, a minimum allowable probability P_{min}, and a decay parameter α determining the responsiveness to changes in reward.

Adaptive Pursuit – table 1(right) – differs in applying a greater increase in probability to the most effective operator, and correspondingly decreased probability to the other operators. It aims to provide faster response to changes in the reward profile. It adds to the parameters of probability matching another, β, which determines the rate of increase of probability of the best operator.

Adaptive Pursuit vs. Probability Matching in GA. In [6], Thierens showed that adaptive pursuit substantially out-performed probability matching in adapting GA operator rates. The aim of this study is to confirm whether these results carry over to GP. In studying this, we need to choose a suitable form of GP.

Table 1. Algorithms for Probability Matching (left) and Adaptive Pursuit (right)[1]

ProbabilityMatching(P,Q,K,P_{min},α)

for $i \leftarrow 1$ **to** K **do**
  $P(i) \leftarrow \frac{1}{K}$
  $Q(i) \leftarrow 1.0$
end for
while NotTerminated?() **do**
  $a^s \leftarrow$ ProportionalSelectOperator(P)
  $R_{a^s}(t) \leftarrow GetReward(a^s)$
  $Q_{a^s}(t+1) \leftarrow Q_{a^s}(t)$
    $+\alpha(R_{a^s}(t) - Q_{a^s}(t))$

  for $a \leftarrow 1$ **to** K **do**

    $P_a(t+1) \leftarrow P_{min}$
      $+(1 - K.P_{min})\frac{Q_a(t)}{\sum_{i=1}^{K} Q_i(t)}$

  end for
end while

AdaptivePursuit(P,Q,K,P_{min},P_{max},α,β)
$P_{max} \leftarrow 1 - (K - 1)P_{min}$
for $i \leftarrow 1$ **to** K **do**
  $P(i) \leftarrow \frac{1}{K}$
  $Q(i) \leftarrow 1.0$
end for
while NotTerminated?() **do**
  $a^s \leftarrow$ ProportionalSelectOperator(P)
  $R_{a^s} \leftarrow$ GetReward(a^s)
  $Q_{a^s}(t+1) \leftarrow Q_{a^s}(t)$
    $+\alpha(R_{a^s}(t) - Q_{a^s}(t))$
  $a^* \leftarrow$ ArgMax($Q_{a^s}(t+1)$)
  $P_{a^*}(t+1) \leftarrow P_{a^*}(t) + \beta(P_{max} - P_{a^*}(t))$
  for $a \leftarrow 1$ **to** K **do**
    if $a \neq a^*$ **then**
      $P_a(t+1) \leftarrow P_a(t)$
        $+\beta(P_{min} - P_a(t))$
    end if
  end for
end while

2.2 Tree Adjoining Grammars and Genetic Programming

Tree adjoining grammars (TAGs) are tree-generating systems for Natural Language Processing [11]. They have also been used as the basis for a number of GP systems [12,13,14]. In this work, we use a linear form closest in spirit to that described in [12]; following the nomenclature of previous authors, we refer to it as Linear Tree Adjoining Grammar Guided Genetic Programming (LTAG3P).

3 Methods

3.1 Problems

To compare the different operator selection mechanisms, we used a number of symbolic regression problems [15]. Three of the functions were from the well-known family $F_n(x) = \sum_{i=1}^{n} x^n$; we used F_6, F_9 and F_{18}. For each of these three problems, we used two different versions, with $s = 50$ and $s = 20$ sample points over the range [-1,1], denoted by F_n^s. In addition, we used two other functions, $Q(x) = x^5 - 2x^3 + x$ and $S(x) = x^6 - 2x^4 + x$; for these, we used 50 sample points over the same range. To reduce the complexity of the problems, we used a reduced function set, $\{+, -, \times, \div\}$.[2] Parameters are detailed in table 2.

---

[1] As presented by D. Thierens in [3].
[2] Commonly, the function set also includes $\sin, \cos, \log, \exp$.

Table 2. Problem Definitions

Objective	Minimise MAE		Cases	Random sample, 50(20) points from [-1..+1]
Targets	$F_n : n = 6, 9, 18, Q, S$		Fitness	Sum of absolute errors of fitness cases
Terminals	X		Hit	Mean absolute error over fitness cases $< \epsilon$
Functions	+, -, ×, ÷		Success	Evolved function scores 50(20) hits

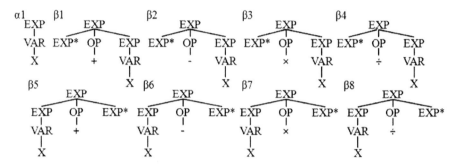

Fig. 1. TAG ElementaryTrees for Reduced Symbolic Regression Search Space

3.2   Defining the LTAG3P System

Search Space. As a grammar-based GP system, LTAG3P relies on a user-defined grammar specifying the search space. For these problems, we used the grammar defined by the elementary trees in figure 1.

Initialisation. We randomly grow an individual upward from the leaf using β trees to the specified minimum depth, then grow further upward sampling from all elementary trees (α and β) until an α tree is sampled (automatically the root) or we reach the specified maximum, when an α tree is inserted.

Operators. In these experiments, we use three LTAG3P operators: crossover, mutation and reproduction.

One-point Crossover chooses random points in the two parent chromosomes, and exchanges the chromosome segments beyond them. if a child exceeds the maximum depth, it is cut at that point and an α tree inserted.

Mutation selects a random point in the chromosome, deletes the chromosome above that point, and replaces it with a new segment generated using the initialisation algorithm.

Reproduction copies the parent to the child population unchanged.

3.3   Adaptive Mechanisms

We compared four adaptive mechanisms: fixed operator rates, linearly scheduled rates (changing according to equation 1; t_f is the total number of generations),

Table 3. Operator Rate Parameters for Adaptive Mechanisms

	Operator	Crossover	Mutation	Reproduction
Fixed	Fixed Rate P	0.5	0.3	0.2
Scheduled	Initial Rate P_i	0.7	0.2	0.1
	Final Rate P_f	0.2	0.6	0.2
AP and PM	Initial Rate P_i	0.333	0.333	0.333
	Min. Rate P_{min}	0.167	0.167	0.167
AP only	Max. Rate P_{max}	0.667	0.667	0.667

probability matching (PM) and adaptive pursuit (AP). The necessary parameters are shown in table 3.

$$P(t) = P_i + \frac{t}{t_f}(P_f - P_i) \tag{1}$$

Reward Policy. Adaptive mechanisms such as PM and AP rely on a reward, reflecting the operator's effectiveness in improving the population. We compared two methods. They are based on the fitness values of children F_c and their corresponding parents F_p. GP focuses resources on the elite, so our reward is calculated over the elite 20% of the children created by that operator.

Reward Policy 1: $R_1 = $ Mean of $\frac{F_p}{F_c}$ over the 20% fittest children.

Reward Policy 2: $R_2 = \frac{\sum F_p}{\sum F_c}$ over the 20% fittest children.

Abbreviations: we denote the fixed mechanism by Fix, the scheduled policy by Scd, and PM and AP with rewards R_1 and R_2 by PM1, PM2, AP1, AP2.

3.4   Evolutionary Parameters

Table 4 shows the evolutionary parameter settings that were used in the experiments. The values of α and β were determined by preliminary experiments.

Table 4. General Evolutionary Parameter Settings

Parameter	Value	Parameter	Value	Parameter	Value
Runs	20	Elite	None	Samples	50(20)
Generations	50	Tournament	Size 8	Hit bound ϵ	0.1
Population	500	Crossover	1-point	PM, AP: α	0.8
Ind. length	2...40	Mutation		AP: β	0.8
		Reproduction			

4 Results

We conducted detailed analysis of our eight problems, but only have space for detailed plots for F_6^{20} and F_{18}^{50}, the easiest and toughest, which most clearly illustrate the trends. Tabular results are given for all eight. For most measurements, we present mean values, but where they are misleading (extreme outliers – small numbers of very poor runs) we present medians.

4.1 Overall Performance

To summarise overall performance, all operator adaptation mechanisms perform well for simple problems such as F_6^{20}, but there is more distinction with harder problems such as F_{18}^{50} (figure 2). In this case, AP1 and PM2 were unable in some runs to ever converge to a good solution. In terms of fitness convergence (error minimisation), AP2 is almost always the best performer. Note that there was no elitism in the algorithm, so the fluctuating behaviour is to be expected.

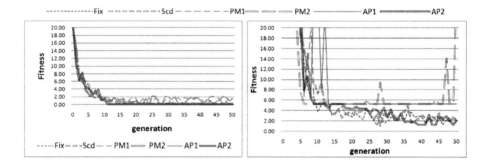

Fig. 2. Median Best Fitness (MAE on Sample Points) Left:F_6^{20}, Right:F_{18}^{50}

In many problem domains, the time taken to first find an acceptable solution (hit) may be as, or more, important than the fitness convergence. Table 5 shows the mean and standard deviation of this first hitting time. We used a one-tailed Mann-Whitney test to check whether there were real improvements over the baseline fixed-ratio treatment.[3] Cases where the performance was better than that of F1 at the 1% significance level are shown in bold (5% level in italic).

In addition to how quickly we find our first hit, we are also interested in how reliably an algorithm hits – that is, if we stop a run at a given point, what is the probability that the best individual in the population is a hit? This is shown for the final (50th) generation for all problems in table 6, and in detail for all generations for F_6^{20} and F_{18}^{50} in figure 3. For the same problems, we also show the overall proportion of hits, across all runs, in figure 4.

---

[3] Some runs did not achieve any hits at all, we assigned a first hitting time of 70 (i.e. 40% beyond the end of the runs) to those cases. This arbitrary choice may slightly distort the mean and standard deviation values, but does not affect the Mann-Whitney results, since they are rank-based.

Table 5. Mean ± Standard Deviation (Failures) of First Hitting Time

Problem	Q	S	F_6^{20}	F_6^{50}
Fix	15.0 ± 5.5 (0)	18.6 ± 8.6 (0)	12.8 ± 5.8 (0)	21.3 ± 12.4 (0)
Scd	17.2 ± 7.9 (0)	27.2 ± 18.1 (2)	12.8 ± 4.7 (0)	26.6 ± 18.3 (1)
PM1	12.6 ± 3.6 (0)	30.5 ± 27.6 (5)	13.9 ± 6.0 (0)	15.7 ± 7.4 (0)
PM2	13.4 ± 8.6 (0)	46.4 ± 30.7 (11)	**10.1 ± 1.7** (0)	*17.2 ± 14.1* (0)
AP1	13.6 ± 4.1 (0)	22.2 ± 18.0 (1)	11.8 ± 3.8 (0)	**8.9 ± 2.2** (0)
AP2	**15.0 ± 19.6** (1)	*17.9 ± 18.5* (2)	**9.4 ± 2.5** (0)	**11.1 ± 4.1** (0)

Problem	F_9^{20}	F_9^{50}	F_{18}^{20}	F_{18}^{50}
Fix	32.4 ± 22.9 (4)	21.9 ± 8.8 (0)	45.8 ± 22.4 (8)	48.1 ± 23.3 (10)
Scd	44.1 ± 23.4 (7)	29.2 ± 16.2 (1)	36.0 ± 19.0 (4)	53.0 ± 22.7 (12)
PM1	35.6 ± 24.5 (5)	15.4 ± 5.3 (0)	53.1 ± 23.2 (11)	58.6 ± 19.6 (14)
PM2	**19.1 ± 18.3** (2)	19.4 ± 18.2 (2)	64.9 ± 16.1 (19)	70.0 ± 0.0 (20)
AP1	*22.4 ± 17.9* (2)	16.1 ± 3.5 (0)	*32.5 ± 21.3* (4)	58.0 ± 20.3 (14)
AP2	**19.4 ± 18.1** (1)	**11.0 ± 2.5** (0)	43.9 ± 27.7 (9)	55.0 ± 20.7 (11)

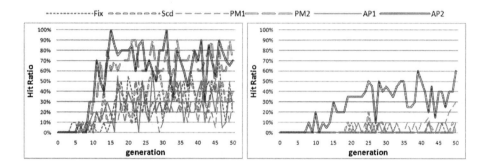

Fig. 3. Proportion of Runs where Best Individual is a Hit

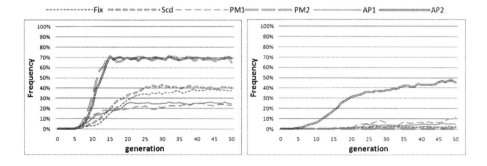

Fig. 4. Overall Proportion of Hits over all Runs

Table 6. Overall Percentage of Hits in Final Generation (Mean ± Standard Deviation)

Problem	Q	S	F_6^{20}	F_6^{50}
Fix	45.4 ± 8.2	30.1 ± 17.2	37.9 ± 8.2	30.1 ± 14.4
Scd	44.0 ± 9.4	18.5 ± 8.5	40.4 ± 10.8	25.3 ± 9.8
PM1	35.1 ± 21.1	19.9 ± 15.6	23.1 ± 11.1	20.4 ± 3.0
PM2	62.7 ± 6.7	29.2 ± 35.0	65.1 ± 9.6	67.8 ± 6.5
AP1	29.0 ± 21.0	25.4 ± 19.0	24.6 ± 16.6	20.3 ± 2.5
AP2	63.3 ± 22.5	62.6 ± 22.2	69.4 ± 2.4	70.3 ± 2.5

Problem	F_9^{20}	F_9^{50}	F_{18}^{20}	F_{18}^{50}
Fix	22.2 ± 13.2	26.1 ± 6.9	7.9 ± 6.0	2.2 ± 1.6
Scd	13.1 ± 12.2	26.6 ± 12.5	11.1 ± 6.5	1.8 ± 1.3
PM1	20.7 ± 22.3	19.3 ± 3.3	19.3 ± 25.3	11.5 ± 20.1
PM2	46.8 ± 32.7	61.3 ± 21.7	4.3 ± 13.7	0.0 ± 0.0
AP1	32.2 ± 26.0	22.2 ± 10.1	55.5 ± 29.3	4.8 ± 8.4
AP2	60.5 ± 21.4	70.3 ± 2.4	35.7 ± 36.1	45.4 ± 8.2

4.2   Operator Application Rates

In addition to observing the performance of the adaptation algorithms, it is important to see their overall effect: what operator rates do they actually select? Figure 5 shows the evolution of these rates for the F_{18}^{50} problem. Space precludes showing such plots for all problems, but inspection of them shows that all problems and treatments may be divided into three regions: up to generation 5, in which the crossover rate either rises or stays steady and the mutation rate either stays steady or falls; generations 5 to 20, during which the crossover rate may remain steady for some time or drop (sometimes precipitously) and the mutation rate generally stays fairly steady; and generations 20 to 50, during which the crossover rate either stabilises or continues to fall, and the mutation rate generally stays fairly steady. Thus we may characterise the behaviour by observing the ratios at generations 5, 20 and 50, which we show for all problems in table 7.

It appears that AP2 gains its good overall performance in most cases by:

- rapidly raising its crossover rate and dropping its reproduction and mutation rates in the first few generations (close to the maximum and minimum rates).
- dropping the crossover rate again after about generation 5 – in most problems reaching close to the minimum by generation 20, but in harder problems dropping more slowly. The mutation rate is also kept low during this period.
- further dropping the crossover rate during the remaining period if it has not already reached the minimum.

The performance of the other algorithms and policies seems to reflect almost directly the extent to which they follow these behaviours. PM and AP1 often behave quite similarly to the fixed rate strategy, though often with a higher rate of crossover in the later generations – these characteristics seem to be associated with poorer performance.

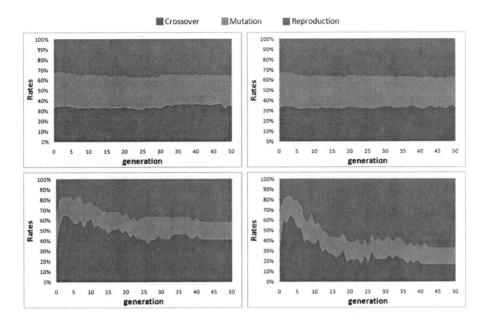

Fig. 5. Evolution of Operator Application Rates for F_{18}^{50}
Top: PM, Bottom: AP; Left: Reward 1, Right: Reward 2
X: Crossover; M: Mutation; R: Reproduction

Table 7. Mean Operator Rates (%) at Generations 5, 20 and 50
X: Crossover; M: Mutation; R: Reproduction

		Q			S			F_6^{20}			F_6^{50}			F_9^{20}			F_9^{50}			F_{18}^{20}			F_{18}^{50}		
	Gen.	5	20	50	5	20	50	5	20	50	5	20	50	5	20	50	5	20	50	5	20	50	5	20	50
X	PM1	44	56	50	34	49	53	60	54	60	39	56	61	36	35	39	39	53	67	37	35	34	35	34	35
	PM2	33	24	24	33	27	27	36	17	18	34	22	17	35	23	21	35	22	18	37	33	31	33	34	33
	AP1	58	57	56	60	47	57	63	48	62	63	61	53	67	47	47	67	61	60	67	22	17	58	51	42
	AP2	43	22	17	46	25	18	63	17	17	45	17	17	58	17	17	67	17	17	67	17	17	54	21	17
M	PM1	30	22	28	36	28	22	20	27	21	33	27	23	32	30	25	31	26	17	32	26	25	32	30	29
	PM2	33	27	21	34	26	26	32	17	17	33	21	21	32	21	19	32	21	18	32	25	24	33	31	29
	AP1	26	22	23	23	27	21	21	30	17	21	22	30	17	17	17	17	22	23	17	17	17	21	17	17
	AP2	24	17	17	28	18	17	21	17	17	30	17	17	17	17	17	17	17	17	17	17	17	25	17	17
R	PM1	27	22	23	30	23	24	20	19	19	28	17	17	32	35	35	30	21	17	31	39	42	33	36	36
	PM2	33	49	55	33	47	47	32	67	65	33	57	61	33	56	59	33	58	64	32	42	44	34	35	38
	AP1	17	21	22	17	27	22	17	22	22	17	17	17	17	37	37	17	17	17	17	61	67	21	32	42
	AP2	33	61	67	26	57	66	17	67	67	25	67	67	25	67	67	17	67	67	17	67	67	21	61	67

5   Discussion

If one's aim is to reliably find good solutions, our results strongly favour adaptive pursuit with reward policy 2. It finds solutions for most problems, and when it is beaten, it is generally by only a couple of other treatments, and by small margins relative to the dispersion of the results. If the focus is on finding solutions quickly but unreliably, then AP2 is still a good overall performer, though it may not be quite as fast, especially with tougher problems, as a fixed operator rate strategy.

Overall, the results show of a general trend that adaptive pursuit out-performs probability matching. Reward policy 2 clearly out-performs policy 1 for adaptive pursuit, but the results are more equivocal for probability matching, with policy 1 sometimes the best.

The importance of adaptive operator rates is clear from these results: a linearly-scheduled change of operator rates performed, if anything, worse than fixed rates. Perhaps this indicates that we chose the schedule poorly. The results confirm 'received wisdom' in GP, that high crossover rates are generally desirable in early search, but undesirable later. In this GP differs substantially from other EAs. This provides additional confirmation of the effectiveness of adaptive pursuit operator in particular: although the actual profiles it needs to track in GP are very different from those in other EAs, it nevertheless tracks them accurately, generating good performance.

Generally, good performance is associated with a division of the search regime into four phases:

1. a rapid initial ramp-up of crossover and ramp-down of both other operators.
2. an intermediate regime during which this high crossover rate is maintained.
3. a transition phase during which the crossover rate drops.
4. a low final rate of both mutation and crossover.

The major variation in this behaviour is the length of the two intermediate phases. The second phase varies from one or two generations up to ten or so (for F_{18}^{20}). The third phase varies from quite abrupt to a gradual decline over the remainder of the evolution (with easier problems being associated with a more abrupt drop). The difference from this profile is a quite reasonable predictor of performance; this may explain the relatively poor performance of scheduled rates: a linear schedule cannot approximate the highly nonlinear preferred profile.

6   Conclusions

6.1   Summary

The main conclusion we can draw from this work is that, as in other evolutionary algorithms, GP performance benefits substantially from adaptive operator rates. Given that they are relatively easy to implement, they are worth considering for most GP systems. We can also confirm Thierens' results [6], that adaptive pursuit out-performs probability matching; of the methods we tried, adaptive pursuit combined with reward policy 2 was generally the most reliable.

6.2   Limitations

The study was conducted with a specific linear GP system, LTAG3P. Since the results are consistent with those for other EAs, we expect that they will extend to other GP systems as well, but this requires confirmation. Only symbolic regression problems were used in this study. Symbolic regression, as well as being an important application in its own right, is a good surrogate for many more complex real-world applications, especially in continuous domains. But GP results from symbolic regression don't always extend well, especially to discrete domains. In this case, we expect them to, but this needs to be verified.

6.3   Further Work

Fialho and colleagues [16,5] have recently shown that more exploratory reward policies – emphasising rare but very beneficial changes – work better, especially for rough fitness landscapes. This is likely to be the case for GP as well, and is certainly worthy of further study.

Our interest in operator adaptation stemmed from work with systems with larger numbers of evolutionary operators, where rate tuning can become complex. Our next step in this work will be to undertake such an extension. However it is clear from this study that rate adaptation can be beneficial in GP, even with only two or three classical operators. We plan to extend our work to cover a range of GP systems, operators and problems, and also to investigate, in more detail, what kinds of reward policies are most beneficial for GP.

Acknowledgments. Seoul National University Institute for Computer Technology provided research facilities for this study, which was supported by the Basic Science Research Program of the National Research Foundation of Korea (NRF) funded by the Ministry of Education, Science and Technology (Project No. 2010-0012546), and the BK21-IT program of MEST.

References

1. De Jong, K.: Parameter setting in EAs: a 30 year perspective. In: Lobo, F.G., Lima, C.F., Michalewicz, Z. (eds.) Parameter Setting in Evolutionary Algorithms. SCI, vol. 54, pp. 1–18. Springer, Heidelberg (2007)
2. Schwefel, H.: Numerical optimization of computer models. John Wiley & Sons, Inc., New York (1981)
3. Thierens, D.: An adaptive pursuit strategy for allocating operator probabilities. In: Proceedings of the 7th Genetic and Evolutionary Computation Conference, pp. 1539–1546 (2005)
4. Lobo, F., Lima, C., Michalewicz, Z.: Parameter setting in evolutionary algorithms. Springer Publishing Company, Incorporated, Heidelberg (2007)
5. Fialho, Á., Da Costa, L., Schoenauer, M., Sebag, M.: Dynamic multi-armed bandits and extreme value-based rewards for adaptive operator selection in evolutionary algorithms. In: Stützle, T. (ed.) LION 3. LNCS, vol. 5851, pp. 176–190. Springer, Heidelberg (2009)

6. Thierens, D.: Adaptive strategies for operator allocation. Parameter Setting in Evolutionary Algorithms, 77–90 (2007)
7. Goldberg, D.: Probability matching, the magnitude of reinforcement, and classifier system bidding. Machine Learning 5(4), 407–425 (1990)
8. Tuson, A., Ross, P.: Adapting operator settings in genetic algorithms. Evolutionary Computation 6(2), 161–184 (1998)
9. Igel, C., Kreutz, M.: Operator adaptation in evolutionary computation and its application to structure optimization of neural networks. Neurocomputing 55, 347–361 (2003)
10. Thathachar, M., Sastry, P.: A class of rapidly converging algorithms for learning automata. IEEE Transactions on Systems, Man and Cybernetics 15, 168–175 (1985)
11. Joshi, A.K., Levy, L.S., Takahashi, M.: Tree adjunct grammars. Journal of Computer and System Sciences 10(1), 136–163 (1975)
12. Hoai, N.X., McKay, R.I., Essam, D.: Some experimental results with tree adjunct grammar guided genetic programming. In: Foster, J.A., Lutton, E., Miller, J., Ryan, C., Tettamanzi, A.G.B. (eds.) EuroGP 2002. LNCS, vol. 2278, pp. 228–237. Springer, Heidelberg (2002)
13. Hoai, N.X., McKay, R.I.B., Essam, D.: Representation and structural difficulty in genetic programming. IEEE Transactions on Evolutionary Computation 10(2), 157–166 (2006)
14. Murphy, E., O'Neill, M., Galván-López, E., Brabazon, A.: Tree-adjunct grammatical evolution. In: 2010 IEEE Congress on Evolutionary Computation (CEC), July 1-8 (2010)
15. Koza, J.R.: Genetic Programming: On the Programming of Computers by Means of Natural Selection. MIT Press, Cambridge (1992)
16. Fialho, Á., Da Costa, L., Schoenauer, M., Sebag, M.: Extreme value based adaptive operator selection. In: Rudolph, G., Jansen, T., Lucas, S., Poloni, C., Beume, N. (eds.) PPSN 2008. LNCS, vol. 5199, pp. 175–184. Springer, Heidelberg (2008)

Evolving Fitness Functions for Mating Selection

Penousal Machado and António Leitão

CISUC, Department of Informatics Engineering, University of Coimbra
Polo II of the University of Coimbra, 3030 Coimbra, Portugal
machado@dei.uc.pt, apleitao@student.dei.uc.pt

Abstract. The tailoring of an evolutionary algorithm to a specific problem is typically a time-consuming and complex process. Over the years, several approaches have been proposed for the automatic adaptation of parameters and components of evolutionary algorithms. We focus on the evolution of mating selection fitness functions and use as case study the Circle Packing in Squares problem. Each individual encodes a potential solution for the circle packing problem and a fitness function, which is used to assess the suitability of its potential mating partners. The experimental results show that by evolving mating selection functions it is possible to surpass the results attained with hardcoded fitness functions. Moreover, they also indicate that genetic programming was able to discover mating selection functions that: use the information regarding potential mates in novel and unforeseen ways; outperform the class of mating functions considered by the authors.

1 Introduction

The choice of an adequate representation, genetic operators and parameters is critical for the performance of an Evolutionary Algorithm (EA). To attain competitive results it is usually necessary to choose or develop problem specific representations, operators and fitness functions, and fine-tune parameters. This can be a complex and time-consuming process. The use of self-adaptive EAs – e.g., EAs that automatically adjust their parameters or components in order to improve the performance in a specific problem or set of problems – has the potential to overcome this problem.

The fitness function has a deep effect on the selection and survival process. In optimization problems the choice of a fitness function may appear to be straightforward. Nevertheless, in the context of sexual reproduction, individuals may have an advantage in choosing mating partners in accordance to criteria other than fitness. For instance, an individual that chooses its mates in accordance to their genetic compatibility may gain an evolutionary advantage.

We focus on the evolution of fitness functions for mating selection, using as test problem the Circle Packing in Squares (CPS). Each individual is composed by two chromosomes: (i) a candidate solution for the CPS problem encoded as a vector (ii) a mating fitness function which is used to evaluate the potential mates of the individual. We test two different approaches for the representation

S. Silva et al. (Eds.): EuroGP 2011, LNCS 6621, pp. 227–238, 2011.

of the evolving mating fitness functions. In the first mating fitness functions are weighted sums, and the weights are under evolutionary pressure. In the second, the mating fitness function is conventional Genetic Programming (GP) tree [1].

We begin by making a brief overview of the state of the art on adaptive EAs focusing on their classification. Next, we introduce the CPS problem and its equivalence with the point scattering problem. In the fourth section, we describe our approach to the evolution of mating fitness functions. The experimental setup is described in the fifth section, while the experimental results are presented and analyzed in the sixth. Finally, we present overall conclusions and future research.

2   State of the Art

EA researchers often face the challenge of designing adequate EAs for specific problems. To automate the task of finding a good design, and set of parameters, one may adapt these components and variables along the evolutionary process. Over the years, several approaches have been proposed for the evolution of EAs. Angeline [2] has presented a classification of these approaches, which was later expanded by Hinterding [3]. He proposes two axis of classification: *adaptation type* and on the *adaptation level*.

Adaptation Type. Evolving EAs may be *static* or *dynamic*. In *static adaptation* the tuning of parameters is made between runs. This is usually accomplished by an external program (or human) that performs a set of test runs and attempts to find a link between settings and performance. The work of De Jong [4] constitutes an example of such an approach. He focused on single-point crossover and bit mutation probabilities in GA, performing a wide number of tests, and empirically establishing recommended mutation and crossover probabilities for several problems.

Dynamic adaptation relies on mechanisms that modify EA parameters during the evolutionary run without external control. They may use, or not, some form of feedback from the evolutionary process to control adaptation. This approach can be further divided into *deterministic, adaptive* and *self-adaptive*.

Deterministic dynamic adaptation uses no feedback from the EA. Instead, it uses a set of deterministic rules that alter the strategy parameters in a predetermined way when a given event takes place. For instance, Hinterding [3] describes an approach where the mutation rate of a Genetic Algorithm (GA) decreases as the number of generations increases.

Adaptive dynamic adaptation approaches use feedback information regarding the progress of the algorithm to determine the direction and magnitude of the changes made to strategy parameters. E.g., Bean and Hadj-Alouane [5] adjust a penalty component for constraints violation on a GA by periodically stopping the run and evaluating the feasibility of the best individuals of the last generations.

Self-adaptive dynamic adaptation relies on evolution to modify EA parameters. The control parameters are encoded into each individual in the population and undergo recombination and mutation as part of the genotype. Eiben et al.

[6] studied the self-adaptation of the tournament size. Each genotype has an extra parameters that encodes the tournament size. On each selection step tournament size is determined by a voting system. Spears [7] proposes the use of an extra gene to encode the type of crossover, two-point or uniform, that should be use when recombining the genetic code of the corresponding individual.

Adaptation Levels. The level within the EA where adaptation takes place is another axis of classification. Adaptation can be defined in four levels: *environment, population, individual* and *component*.

Environment level adaptation takes place when the response of the environment to the individual is changed. Mostly this affects, somehow, the fitness function, either by changing the penalties or weights within it or by changing the fitness of an individual in response to niching. In [8], Angeline argues that competitive fitness functions have many advantages over typical approaches, specially when there is little knowledge about the environment.

Population level adaptation consists of adapting parameters that are shared and used by the entire population. The work of Fogarty [9] where the mutation rate of the entire population evolves through time is an example of such an approach.

Individual level adaptation evolves strategy parameters that affect particular individuals. Often these components are encoded in each individual, e.g., Braught [10] uses a self-adaptive scheme where an extra gene that codifies the mutation rate of each individual.

Component level adaptation acts on strategy parameters that affect specific components or genes of an individual. Fogel's study [11] on the self-adaptation of finite state machines is an example of this approach, exploring the association of a mutability parameter to each component of the finite state machines.

2.1 Other Approaches

Grefenstette [12] suggests conducting a search over a parameterized space of GAs in order to find efficient GAs for a function optimization task (this work was later applied to GP as well). The search uses two levels: a *macro level* representing a population of EAs that undergo evolution; a *micro level* where each represented EA acts on a population of candidate solutions for a given problem. The use of linear variants of GP to evolve EAs has also been an active field of research, consider for instance the work of Oltean on Multi Expression Programming (MEP) [13] and Linear Genetic Programming (LGP) [14]. Spector and Robinsons [15] explore the use of Push – a stack-based programming language able to represent GP systems – to evolve EAs, including the co-evolution of mate selection.

2.2 Evolving the Fitness Function

Evolving the fitness function implies changes in the way the environment responds when evaluating individuals of a population. Darwen and Yao [16] have synthesized some previous work done on *fitness sharing* to tackle a multi optima problem. The fitness sharing technique modifies the search landscape during the evolutionary run by reducing the payoff in niches that become overcrowded. Hillis [17] explores the co-evolution of two populations – candidate solutions and test cases – for the evolution of sorting networks. The fitness of the candidate solutions is determined accordingly to their ability to sort the test cases of the second population; the fitness of a test case is determined by the amount of sorting errors it induces.

3   Circle Packing in Squares

We focus on CPS, a surprisingly challenging geometrical optimization problem relevant for real life storage and transportation issues. It consists in finding a spacial arrangement for a given number circles of unitary radius that minimizes the area of the minimal square that contains them. The circles may not overlap. In Fig. 1 we present the optimal solutions for the packing of 7, 15, and 24 circles into a square. The optimal solutions for sets from 2 to 24 circles are known.

The CPS problem is closely related to that of scattering n points in a unit square such that the minimal distance between any of them is maximized [18]. In order to transform from this model to that of a square containing fixed size circles one must apply the transformation presented in Fig. 2 where r is the radius of the circles, d is the greatest minimal distance between points and S is the side of the resulting square.

We conducted a wide set of experiments using GA and GP approaches [19], which show that, in an evolutionary context, it is advantageous to see the CPS problem as a scattering problem: it avoids the overheads caused by invalid candidate solutions (e.g. where circles overlap) leading to better results. As such, we will adopt this representation in the experiments described in this paper.

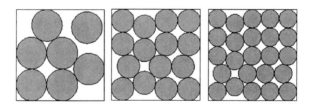

Fig. 1. Optimal solutions for 7, 15, and 24 circle packing in a square

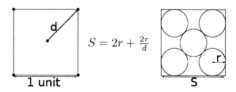

$$S = 2r + \frac{2r}{d}$$

Fig. 2. Calculation of the side of the square for the 5 circles optimal solution

4   Mating Selection

Without loss of generality, considering tournament based selection, the mating selection procedure can be described as follows: (i) a parent is selected from the population using tournament selection based on fitness; (ii) a set of t mating candidates is randomly selected from the population; (iii) the candidate that, according to the parent, is fittest for mating purposes is selected for sexual reproduction with this parent; The process is repeated until sufficient offsprings are generated.

4.1   Mating Selection in the Circle Packing in Squares Problem

In order to better understand the means by which the mating fitness can be evolved, there's a characteristic of the CPS problem that should be mentioned. Usually, optimal solutions can be decomposed in subsets that are in turn optimal (or near optimal) solutions for a different instance of the problem. More formally, the optimal solution for the packing of n circles in a square may also encode optimal solutions for the packing of $n-i$ circles for i in $[1, n-2]$. As Fig. 3 shows the optimal representation for 9 circles also encodes the optimal solution for 5 circles, 4 circles and 2 circles. One may be inclined to think a good performance in a $n - i$ problem promotes a good performance in the n packing problem. However, this is not always the case. For instance, to pack 7 circles optimally one must use an inefficient packing of 4 and 6 circles, and an optimal packing of 4 (see Fig. 1) . In other words, the optimal packing of 7 circles can be created by adding 3 circles to the optimal packing of 4 circles; adding circles to the optimal packing of 5 or 6 circles leads to an inefficient packing of 7 circles. Thus, for evolving solutions for the 7 circle instance it makes sense to value the ability of the candidates to efficiently pack 4 circles and penalize candidates that pack 5 or 6 circles optimally. These observations made us consider the following hypotheses: the performance of an individual on a $n - i$ packing problem may provide information regarding its performance on the target n packing problem; this information may be valuable for mating selection purposes.

As such, when evaluating a mating candidate, the individual has access to information regarding the fitness achieved by the candidate in the packing of $n - i$ circles for all i in $[1, n - 2]$. Our expectation is to evolve mating fitness functions that use this information wisely, leading to a better performance in the n circles instance.

To determine the fitness of an individual packing n circles in all instances of the CPS problem of lower dimensionality one would need to consider a total of $C_2^n + C_3^n + ... + C_{n-2}^n + C_{n-1}^n$ packings of circles.

This becomes computationally intensive. In order to reduce computation and increase the speed of the tests executed we perform the following simplification: For determining the fitness an individual in the packing of k circles (with $k < n$) we only consider the first k first circles represented in the chromosome, instead of C_k^n possible combinations of k circles. With this simplification the computational overhead becomes minimal and negligible.

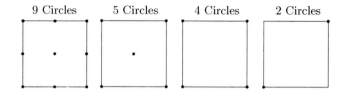

Fig. 3. Composition of optimal solutions by close to optimal subsets

```
evaluate_mating_candidates(mating_candidates,parent) {
    for i = 1 to #(mating_candidates) {
        candidate_mating_fitnesses_i ← eval(mating_candidates_i,parent)
    }
    return candidate_mating_fitnesses
}
```

Fig. 4. Evaluation of the mating candidates [20]

Each parent picks its mate among the candidates and so is bound to evaluate them. In order to do this each individual encodes a mating fitness function that it uses to evaluate its mating candidates. The *evaluate_mating_candidates* step, described in Fig. 4, has been designed to receive the set of mating candidates and the parent which is evaluating them, returning the candidates' mating fitnesses. The *eval* step uses the mating fitness function encoded in the parent to evaluate each candidate.

Each individual is composed of two chromosomes. The first encodes a candidate solution for the CPS problem through a vector of n x, y coordinates. The second encodes a mating fitness function. We explore two different approaches for the encoding of the mating fitness function: one using GA and the other using GP.

In the GA approach the mating fitness functions are weighted sums and the chromosome encodes a set of weights. The *MatingFitness* of candidate mc according to parent p is given by the following formula:

$$MatingFitness^{mc} = \sum_{k=2}^{n} w_k^p F_k^{mc},$$

where w_k^p is the weight given by parent p to the fitness of the candidates in the packing of k circles and F_k^{mc} is the fitness of the candidate mc in the packing of k circles. In the GP approach the chromosome encodes a standard GP tree that defines the mating fitness function. The terminal set includes the variables F_k^{mc}, which allows the individual to use this information to access the mating candidates.

5   Experimental Setup

In order to understand the effect of mating selection and of evolving of the mating fitness function we need to have a basis for comparison. We performed several experiments using a standard GA approach to solve the CPS problem [19]. We considered a wide set of parameters for tournament size, mutation rate and type of mutation, and performed static adaptation over this set of parameters. The parameters that resulted in best overall performance are presented in table 1.

Table 1. Parameter set for the standard GA approach

Representation	$x_1, y_1, ..., x_n, y_n$
Initialization	x_n, y_n in $[0, 1]$
Parents Selection	$Tournament\ size = 5$
Crossover	$1 - point$
Crossover probability	90%
Mutation	Gaussian mutation; $mean = 0.0$; $stdev = 0.08$
Mutation probability	2%
Elitism	1 individual

Table 2. Additional settings used in GA-based mating selection evolution

Representation	$x_1, y_1, ..., x_n, y_n, w_2, ..., w_n$
Initialization	x_n, y_n in $[0, 1]$; $w_{s(n)}$ in $[-1, 1]$
Crossover	$1 - point$
Crossover probability	90%
Mating Candidates Selection	$Tournament\ size = 5$
Mutation	Gaussian mutation; $mean = 0.0$; $stdev = 0.08$
Mutation probability	5%
Evolving Mating Fitness	$MatingFitness^{mc} = \sum_{k=2}^{n} w_k^p F_k^{mc}$

To evaluate the effects of evolving the mating fitness function we conducted experiments using the GA and GP based approaches for mating selection.

In GA-based mating fitness evolution the mating fitness function has been designed by the authors to be a weighted sum of the fitness values of the candidate in smaller instance of the CPS problem. An additional chromosome is added to the standard GA representation to encode the weights. The genetic operators are applied independently to each chromosome. For the chromosome encoding the candidate solutions we use the previously established set of parameters presented in table 1. For the chromosome encoding the mating fitness function we use the parameters presented in table 2. As previously this set of parameters was established empirically using static adaptation (See section 2).

Likewise, in GP-based mating selection evolution we have two chromosomes: a linear chromosome encoding the candidate solution; a mating fitness function

Table 3. Additional settings used in GP-based mating selection evolution

Representation	$x_1, y_1, ..., x_n, y_n, GPtree$
GP terminals	$F_{s(2)}^{mc}, ..., F_{s(n)}^{mc}, 0, 1, 2$
GP functions	$+, -, *, /$
GA Initialization	x_n, y_n in $[0, 1]$
GP Initialization	half-half
Mating Candidates Selection	$Tournament\ size = 5;$
Evolving Fitness Function	Output from the execution of the tree

represented by a GP tree. For the candidate solution chromosome we use the parameters presented in table 1, for the GP chromosome the parameters and operators are those employed by Koza [1]. The additional settings for this approach can be found in table 3.

The fitness of an individual in the CPS problem is used for parent selection in all approaches, and it is given by the size of the minimum square that contains the circles (see Fig. 2). As such this is a minimization problem.

6   Experimental Results

For a given number of n circles, a given set of parameters and a chosen approach, 30 runs are executed. In each run 100 individuals evolve along 5000 generations and the fitness in the CPS problem of the best individual and population mean is saved at each generation. We performed tests for all n in $[2, 24]$. Table 4 shows the results obtained in this study. Column N indicates the instance of the CPS problem being tackled, while column *optimal* presents the optimal values for that instance. Column *Standard* shows the results obtained by the standard GA approach, i.e. both parents chosen by tournament selection based on CPS fitness. Column *Random* present the results attained by choosing the first parent using tournament selection based on CPS fitness and its mating partner randomly from the current population. Column *GA* presents the results attained by using GA-based mating fitness selection – i.e. the first parent is chosen using using tournament selection based on CPS fitness and its mating partner is selected by tournament selection based on the mating fitness of the candidates accordingly to the mating fitness function encoded by the first parent. Likewise, column *GP* presents the results achieved by GP-based mating fitness selection.

For each approach, the *best* column presents the CPS fitness of the best individual found in the 30 runs, while the *avg* column presents the average fitness of the best individuals over the 30 runs. Underlined values indicate the result is better than the one attained using the Standard approach. Bold values indicate that the value is better and that a statistically significant difference exists (obviously, this only applies to average values).

A comparison of the results attained by the Standard and Random mating selection approaches reveals that for small instances of the CPS it is often advantageous to select mates randomly. It is important to notice that this is not

Table 4. Comparison of the results attained by the different approaches. Results are averages of 30 runs. Lower values indicate better performance. Underlined values signal results better than the ones attained using the Standard approach. Bold values indicate that a statistically significant difference exists. Statistically significance determined through the Wilcoxon–Mann–Whitney test. Confidence level of 0.95.

		Static Mating Selection Function				Evolved Mating Selection Function			
		Standard		Random		GA		GP	
N	optimal	best	avg	best	avg	best	avg	best	avg
2	3.4142	3.4142	3.4142	3.4142	3.4142	3.4142	3.4142	3.4142	3.4142
3	3.9319	3.9319	3.9320	3.9319	3.9320	3.9319	3.9319	3.9319	3.9319
4	4.0000	4.0000	4.0266	4.0000	4.0001	4.0000	4.0255	4.0000	4.0001
5	4.8284	4.8288	5.0056	4.8287	4.9911	4.8285	**4.9250**	4.8286	**4.9475**
6	5.3282	5.3296	5.3669	5.3299	5.3674	5.3306	5.3685	5.3303	5.3804
7	5.7321	5.7426	5.8227	5.7379	5.8081	5.7353	5.8296	5.7348	5.8098
8	5.8637	5.8665	6.0212	5.8714	5.9615	5.8693	5.9913	5.8643	5.9898
9	6.0000	6.0072	6.5184	6.0086	6.4907	6.0042	6.5401	6.0018	6.5154
10	6.7474	6.7564	6.8936	6.7804	6.8854	6.7642	6.9110	6.7581	**6.8536**
11	7.0225	7.0323	7.1619	7.0822	7.1764	7.0600	7.2232	7.0418	7.1564
12	7.1450	7.1540	7.3966	7.2416	7.3565	7.1966	7.4809	7.1682	7.3438
13	7.4630	7.4977	7.8088	7.6036	7.8167	7.5663	7.8355	7.4816	**7.7147**
14	7.7305	7.8059	8.0705	7.8859	8.0950	7.9190	8.1509	7.8498	**8.0048**
15	7.8637	8.0332	8.3324	8.1102	8.4173	8.0296	8.4345	7.9677	**8.2581**
16	8.0000	8.4015	8.7014	8.4542	8.8632	8.3030	8.8153	8.3980	**8.6012**
17	8.5327	8.6688	8.8765	9.0022	9.2345	8.7143	9.0836	8.7065	8.8665
18	8.6564	8.8566	9.0996	9.1902	9.4966	8.9189	9.2724	8.8582	9.0984
19	8.9075	9.1482	9.4442	9.4789	9.9422	9.2049	9.6036	9.0178	**9.3511**
20	8.9781	9.3889	9.7212	9.9433	10.2839	9.2951	9.7641	9.1795	**9.6030**
21	9.3580	9.6980	9.9788	10.2998	10.7402	9.7305	10.1307	9.6730	9.9425
22	9.4638	9.9210	10.2610	10.6887	11.0512	9.9546	10.3705	9.9969	10.2693
23	9.7274	10.0625	10.5201	10.9262	11.5476	10.0631	10.6498	10.0943	10.5892
24	9.8637	10.3198	10.7725	11.2717	11.8382	10.5232	10.8163	10.4678	10.8034

equivalent to performing random search, the first parent is chosen by tournament selection using CPS fitness, only its mate is selected randomly. By choosing mating partners randomly one lowers the selection pressure. Furthermore, it may promote population diversity. The combination of these factors may avoid premature convergence, explaining the best performance of the Random mating selection approach in small instances. The analysis of the evolution of CPS fitness throughout the evolutionary runs supports this hypothesis. For larger instances the lack of selection pressure penalizes the Random mating selection approaches.

The performance of the GA-based mating selection approach is disappointing. Although it was able to find better solutions than the standard approach for six instances of the problem, the average of the best solutions is only better than the one attained by the Standard approach in four instances, and the difference is only statistically significant in two of them. This suggests that the overhead inflicted on the evolution process by the necessity of evolving adequate weights is not rewarding. The approach should be able to reproduce the original fitness function, and by doing so it would be expectable for it to achieve closer results to those of the Standard approach. An analysis of the runs indicates that the approach is unstable, in the sense that some runs produce significantly better results than others, which partially explains the worst average fitness. Despite this, comparing the results with the ones attained by the Random mating selection approach, revels that better results are attained for large instances of the CPS problem. This indicates that the individuals are able to evolve mating fitness functions that take advantage of the information provided to them, although the reward is not enough to compete with the Standard approach.

Considering these results, the performance of the GP-based mating selection approach is surprising. It attains better averages than the Standard approach in eighteen instances and the differences are statistically significant for thirteen of them. Moreover, the best solutions surpass those found using the Standard approach in ten of the instances. For small instances of the CPS problem, where Random mating performed better than the Standard approach, GP-based mating selection is also competitive, the averages are comparable to those attained by Random mating, and the best individuals found tend to be better. Overall, these results indicate that: the GP-based approach is able to evolve mating selection functions that take advantage of the information regarding the performance of the candidate mates in smaller instances of the CPS problem in a useful way, leading to better overall performance in spite of the overheads caused by the need to evolve the mating selection functions.

The results attained by GP-based mating selection contrast with the ones of the GA-based approach. This was a totally unexpected result, that demands further analysis.

In GA-based mating, the mating fitness function was designed by the system developers in a way that appeared adequate to the problem at hand, and that is focused on the selection of the mating candidate that seem most fit for the role. Moreover, the evolution process only evolves the weights used in the mating fitness function. In GP-based mating selection the system must evolve the entire mating fitness function. Thus, the search space of GP-based contains all the mating fitness functions that can be evolved by the GA-based and a vast number of functions of a different kind. It is, therefore, significantly larger. By these reasons, the task of the GP-based approach appeared to be significantly harder than the one of the GA-based approach. We find two, non-exclusive, hypotheses for the better performance of the GP-based approach: (i) The "ideal" mating fitness function varies during the course of the run – e.g., the best way to choose mating partners in the beginning of the run may be useless in later generations – and GP is able to adapt faster to these changing requirements than GA. (ii) GP was able to evolve fitness selection functions that outperform the ones constructible using the weighted sum template designed by us, and based on our knowledge on the regularities and irregularities of the CPS problem.

Although, the experimental results are insufficient to draw definitive conclusions, the analysis of the evolution of the weights during individual runs appears to support the first hypothesis. The analysis of the fitness mating functions evolved by the GP-based approach is extremely difficult and the bottom line is that we are unable to understand exactly how mating candidates are being evaluated. Nevertheless, it is clear that what the fitness functions evolved by this approach are doing is radically different from a weighted sum, which supports the second hypothesis. Thus, the analysis points towards a combination of both factors.

Our inability to explain how the GP-based approach is determining mating fitness makes it impossible to fully understand the results. It does, however, also indicate that the GP-based approach is using fitness mating functions that we would be unlikely to design by hand, and surpassing the performance attained by the ones we designed.

7 Conclusions

We focus on the self-adaptive evolution of mating fitness functions using the CPS problem for test purposes. We perform an overview of previous work in the area, and introduce the CPS problem, describing its regularities and irregularities, and proposing ways to explore them for the purpose of developing mating fitness functions that improve the performance of the EA. Two approaches, based on GA and GP, are presented for the evolution of mating fitness functions.

The experimental results in the packing of two to 24 circles are presented and discussed. These results reveal that the GA-based mating fitness evolution approach – which evolves parameters for a mating fitness function designed by the authors based on their knowledge of the CPS problem – is unable to surpass the performance of a conventional approach. Contrastingly, the GP-based mating fitness evolution approach, which explores a significantly larger search space of mating fitness functions, is able to outperform all of the considered approaches. Although the nature of the mating fitness evolved by the GP-based approach was not fully understood, it is safe to say that they are very different from the ones designed by the authors. This indicates that the GP-based approach is not only able to evolve fitness mating functions that outperform hand coded ones, but also to discover mating fitness function designs that would be unlikely to be created by hand.

Acknowledgment

This work has been partially supported by the project PTDC/EIA-EIA/102212/-2008, High-Performance Computing over the Large-Scale Internet.

References

1. Koza, J.R., Poli, R.: Genetic programming. In: Search Methodologies, pp. 127–164. Springer, Heidelberg (2005)
2. Angeline, P.J.: Adaptive and self-adaptive evolutionary computations. In: Computational Intelligence: A Dynamic Systems Perspective, pp. 152–163. IEEE Press, Los Alamitos (1995)

3. Hinterding, R., Michalewicz, Z., Eiben, A.E.: Adaptation in evolutionary computation: A survey. In: Proc. of the 4th International Conference on Evolutionary Computation, pp. 65–69 (1997)
4. De Jong, K.A.: An analysis of the behavior of a class of genetic adaptive systems. PhD thesis, University of Michigan, Ann Arbor, MI, USA (1975)
5. Bean, J., Hadj-Alouane, A.: A dual genetic algorithm for bounded integer programs. Technical Report 92-53, University of Michigan (1993)
6. Eiben, A., Schut, M., de Wilde, A.: Boosting genetic algorithms with self-adaptive selection. In: IEEE Congress on Evolutionary Computation, pp. 477–482 (2006)
7. Spears, W.M.: Adapting crossover in a genetic algorithm. In: Proc. of 4th Annual Conference on Evolutionary Programming, pp. 367–384 (1995)
8. Angeline, P.J., Pollack, J.B.: Competitive environments evolve better solutions for complex tasks. In: Proc. 5th International Conference on GAs, pp. 264–270 (1994)
9. Fogarty, T.C.: Varying the probability of mutation in the genetic algorithm. In: Proc. of the 3rd International Conference on Genetic Algorithms, pp. 104–109 (1989)
10. Braught, G.: Evolving evolvability: Evolving both representations and operators. In: Adaptive and Natural Computing Algorithms, pp. 185–188. Springer, Heidelberg (2005)
11. Fogel, L., Angeline, P., Fogel, D.: An evolutionary programming approach to self-adaptation on finite state machines. In: Evolutionary Programming, pp. 355–365 (1995)
12. Grefenstette, J.: Optimization of control parameters for genetic algorithms. IEEE Transactions on Systems, Man and Cybernetics 16(1), 122–128 (1986)
13. Oltean, M.: Evolving evolutionary algorithms with patterns. Soft Computing - A Fusion of Foundations, Methodologies and Applications 11, 503–518 (2007)
14. Oltean, M.: Evolving evolutionary algorithms using linear genetic programming. Evolutionary Computation 13, 387–410 (2005)
15. Spector, L., Robinson, A.: Genetic programming and autoconstructive evolution with the push programming language. Genetic Programming and Evolvable Machines 3, 7–40 (2002)
16. Darwen, P., Yao, X.: Every niching method has its niche. In: Ebeling, W., Rechenberg, I., Voigt, H.-M., Schwefel, H.-P. (eds.) PPSN 1996. LNCS, vol. 1141, pp. 398–407. Springer, Heidelberg (1996)
17. Hillis, W.D.: Co-evolving parasites improve simulated evolution as an optimization procedure. In: Emergent Computation, pp. 228–234. MIT Press, Cambridge (1991)
18. Hifi, M., M'Hallah, R.: A literature review on circle and sphere packing problems: Models and methodologies. Advances in Operations Research (2009)
19. Leitão, A.: Evolving components of evolutionary algorithms. MSc Thesis, Faculty of Science and Technology, University of Coimbra (2010)
20. Tavares, J., Machado, P., Cardoso, A., Pereira, F.B., Costa, E.: On the evolution of evolutionary algorithms. In: Keijzer, M., O'Reilly, U.-M., Lucas, S., Costa, E., Soule, T. (eds.) EuroGP 2004. LNCS, vol. 3003, pp. 389–398. Springer, Heidelberg (2004)

Experiments on Islands

Michael Harwerth

Miq1@gmx.de

Abstract. The use of segmented populations (Islands) has proven to be advantageous for Genetic Programming (GP). This paper discusses the application of segmentation and migration strategies to a system for Linear Genetic Programming (LGP). Besides revisiting migration topologies, a modification for migration strategies is proposed — migration delay. It is found that highly connected topologies yield better results than those with segments coupled more loosely, and that migration delays can further improve the effect of migration.

Keywords: Linear Genetic Programming, Island Model, Migration Delay, Topologies.

1 Introduction

The Island Model, here referred to as Segmented Population, has been extensively analyzed in GA (Genetic Algorithms [4]) and GP (Genetic Programming [1]). It is in common use as it has proven to be advantageous to several problems to be solved by evolutionary computing. Migration of individuals between segments is governed by migration interval, number of migrating individuals and topology. With linear GP, migration topologies have been discussed before mainly for parallelization [6] or with few selected topologies only [2,7].

This paper provides results from extensive experiments on different segmentation topologies for three different data sets, attempting to find a topology that reduces the time spent to find a sufficiently fit solution. The topologies given here are representatives for a wide range of possible candidates with differing degree of connectedness. We found promising topologies, and, based on more experiments on the torus topology (that has been found performing well), give a proposal to further improve the classical migration scheme by a new parameter controlling migration delay.

Section 2 gives an overview on the LGP system used in these analyses, section 3 describes the three test case sets used. In section 4 eight different population topologies are compared, one of those found performing better is further developed in section 5 with the new proposed migration delay parameter. Finally a short conclusion is given in section 6.

S. Silva et al. (Eds.): EuroGP 2011, LNCS 6621, pp. 239–249, 2011.

2   LGP System Used

Originally started as a fun experiment in 1995, the 'e6[1] system has grown to a useful tool to examine the LGP landscape over the years. In the meantime it has been used for several applications, as seafloor habitat classification [3], prediction of air traffic flight arrival times or the evaluation of error correction functions for approximations of the Prime counting function $\pi(x)$.

The core of e6 is a virtual machine evaluating byte code, generated as phenotype of an individual. The byte code set includes most math operations, conditionals, logical operators and branching instructions. If required, numerical constants and variables can be activated; input is restricted to reading test case values, output to writing result values. Fitness is determined by an error function to be minimized. Individuals with equal fitness are ranked by resources used — less steps taken are considered better. Operations resulting in infinite or undefined values (NaN etc.) force termination of the individual run.

A multitude of options allows the configuration of instruction sets and runtime conditions that are tailored for any application. In the experiments described here, the relevant settings were:

population size	1960
population segments	49
selection	elitist[2]
mutation rate	0.1
crossover rate	0.9
mating pool	10%
mating pool survives	yes
offspring	90%
error function	sum of errors
generations in a run	3000
segment connections	directed
migration source segment selection	elitist[3]
migration rate	10%

The values are common in literature, except migration source segment selection and number of segments. The former was chosen to maximize distribution speed of fit intermediate solutions, the latter was found empirically to be large enough to show the effects described in this paper (smaller numbers tend to have the macroscopic effect being blurred by local noise) and small enough to be able to limit the population size and necessary run times. The parameters for migration topology, interval and migration delays are subject of this paper and will be specified below.

---

[1] e6 stands for 'evolution, version 6'. Its very first incarnation was of course called 'e2'...'

[2] The best 10% of all individuals in a segment are taken for mating – see 'mating pool'.

[3] The segment with best fitness among all connected to the current is allowed to migrate into it.

3   The Test Cases

The test cases have been selected to cover regression as well as classification problems. The problems must be difficult enough to require longer evaluation runs (several thousands of generations) to be able to detect long-term effects. All experiments have been carried out with all three test cases.

3.1   Intertwined Spirals

The Intertwined Spirals problem is well-known [5]; it is renowned as being difficult to solve by GP. Each of 194 given points has to be classified as belonging to one of two intertwined spirals in the XY plane. The Intertwined Spirals data set is referred to as 'spiral' further on.

In the experiments the instruction set was left unrestricted (only subroutine calls were disabled), code size limited to 30 instructions and the maximum number of instructions executed in an individual's run was set to 30.

3.2   Regression

A "classic" regression test function seems to be the the $f(x) = 4x^3 - 3x^2 + 2x - 1$. This function is quite easy to detect — with the setup as described before it normally takes less than 100 generations to find it. So a slightly more complex function was chosen: $f(x,y) = xy^2 - x^2y + \frac{x}{y} - \frac{y}{x}$. This function (Fig. 1, named 'regress' in the following) is special as it has two pole planes at $x = 0$ and $y = 0$, respectively, that prevent smooth transitions in the solution landscape.

Code size and maximum steps taken were set to 30 each, and the instruction set was unrestricted as before. Restricting the instruction set to math and stack manipulating commands made the problem too easy to solve (most runs prematurely achieved optimal solutions), and made the results difficult to compare with the others.

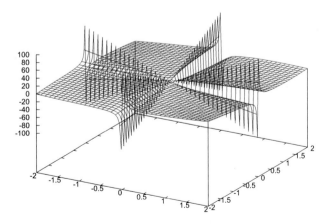

Fig. 1. $f(x,y) = xy^2 - x^2y + \frac{x}{y} - \frac{y}{x}$

3.3 8-Bit Parity

The 8-Bit Parity data set gives the following task: for all 256 combinations of 8 bits, the parity bit has to be determined. The parity problems given in the literature normally present all single bit values to find the parity – which is quite easy to solve. To make it more difficult, here the data set does not give the single bit values, but only the resulting number between 0 and 255. This test is very hard for the system, as currently no bit operations are implemented and the standard floating point data type used by the system is not well suited for manipulating binary values. The name used in this paper to refer to this data set is 'parity8'.

As with the Intertwined Spirals problem, the instruction set was left almost unrestricted (only subroutine calls were removed), while code length was limited to 30 and steps were limited to 60 instructions.

4 Topology Experiments

In [2], migration topologies were examined, namely the ring, grid and random topologies. The conclusion based on those experiments was that topologies play a minor role influencing fitness.

We had made different observations and wanted to try other topologies. We chose eight representatives for different levels of segment connectedness (see Fig. 2, that shows the six possible to be displayed as connection patterns. Panmictic and completely isolated populations have no connections at all) – the 'Name' column gives the terms used further on as reference names:

Name	Description
panmictic	unsegmented population – all 1960 individuals in one
isolated	49 completely isolated segments with 40 individuals each
buckets	linear bucket brigade of 49 segments. Each segment has only one connection to the next in sequence
bin.tree	hierarchical binary tree. Each segment is connected to exactly two others
layered	groups of 7 segments all connected within one layer. 7 layers are hierarchically stacked
torus	a 7×7 toroidal structure with connections from each segment to the NE, SE, SW and NW neighbours
scattered	a layout with 1 to 3 connections from each segment, randomly selected at initialization
fully conn.	a completely interconnected aggregate, where each segment has connections to all the others

The diagrams can be read as follows: for each segment on the Y axis, a black square right of it signals a connection to the corresponding segment on the X axis below.

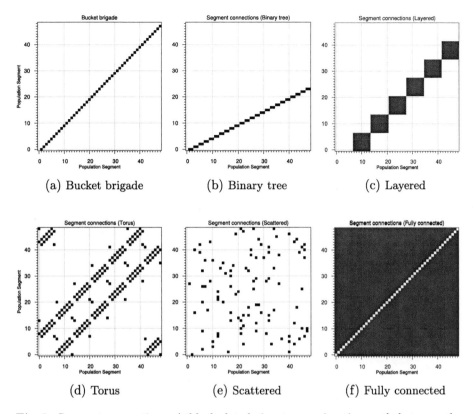

Fig. 2. Segment connections. A black dot designates a migration path between the segment numbered on the Y axis to the corresponding on the X axis.

Each data set was evaluated in 100 runs for each parameter combination analyzed. Only the best performing of all segments connected to the one to migrate into was allowed to attempt migration (elitist selection among segments). An alternative could be migration tournaments of segments, but is not considered in this paper. Migration only takes place if the migrating segment outperformed the target segment.

4.1   Test Results

The results for the three data sets are shown in Fig. 3 to 5 as box plots on the left side, where outliers are drawn as small circles 'o'.

The proposition from [2] regarding the topology influence can not completely be confirmed for the LGP system used here, as there are significant (ANOVA tests were applied to all results reported here and proved a 99.9% significance level) differences within the tested topologies. The 'regress' data set had the best results with the 'buckets', 'torus', 'scattered' and 'fully conn.' topologies. The 'spiral' and 'parity8' data sets showed a similarity; the 'torus', 'scattered' and 'fully conn.' topologies had better results overall than the others tested.

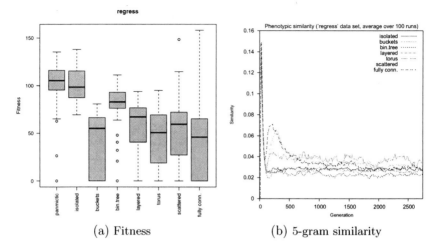

(a) Fitness   (b) 5-gram similarity

Fig. 3. 'regress' data set, topology results

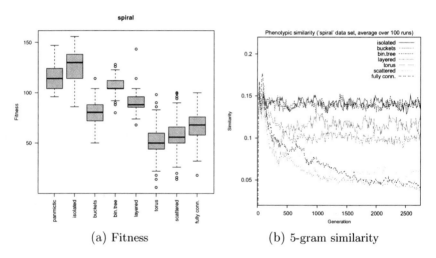

(a) Fitness   (b) 5-gram similarity

Fig. 4. 'spiral' data set, topology results

With the exception of the 'buckets' structure with the 'regress' data set, the topologies yielding the best results were those directly or in a few steps connecting all segments — the 'torus' has a maximum of four "hops" to get to any segment, 'scattered' takes six steps on average, whereas 'fully conn.' needs one only. It would be interesting to further experiment with different torus dimensions and connection variants between torus segments, as the four diagonal connections used here have been chosen arbitrarily.

Being able to quickly spread into less performing segments seems to be the clue for good solutions to evolve. Moreover, as crossover between very differently structured individuals is destructive most of the time, keeping the segments'

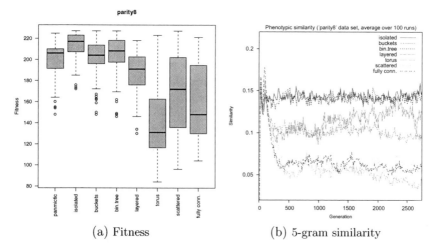

(a) Fitness   (b) 5-gram similarity

Fig. 5. 'parity8' data set, topology results

genetic distance low by frequent contacts between them makes crossover less destructive. The down side of it could be loss of diversity, driving the population towards one solution only.

To evaluate this effect, we calculated the average code similarities between the respective best individuals in each segment. The number of code 5-grams (consecutive blocks of 5 code instructions) was counted, that were used in more than one individual. 5-grams, or generally n-grams, are used in linguistics to characterize and compare texts. As linear GP has a similar, linear structure as written language, n-grams seem to be suitable to compare code fragments as well. The found number of matching 5-grams is scaled to the maximum number of identical 5-grams possible — $\frac{segments(segments-1)}{2}(max.codesize - 5)^2$ — giving a value between 0 (all 5-grams are unique) and 1 (only one 5-gram used throughout the population). The results for the three data sets can be found on the right side of Fig. 3 to 5[2].

In the diagrams, lower values mean higher genetic diversity between segments. Surprisingly, the topologies connected more closely nevertheless are maintaining a higher degree of diversity. The 'regress' data set is an exception again, as all similarity levels are not visibly distinguishable at the end of the runs – the number of representations for the perfect solution found in numerous runs is not as large –, but the 'spiral' and 'parity8' data sets are showing a distinctive pattern. The topologies performing best as seen from the fitness, have a higher degree of genetic diversity as well. First analyses revealed separate levels of diversity: the highly connected topologies had more, but smaller differences, whereas the less connected had a fewer but structural changes between segments. This effect again needs to be further evaluated.

---

[2] Please note the graphics have been drawn in color for better visibility - B&W display may be harder to read.

5   Delaying Migrations

For all experiments so far, the migration interval was left fixed at 10 generations. Given the observations from the previous section, what will happen to
e.g. the well-performing torus topology if we change intervals? Fig. 7 to 8 shows
the results — here only the 'spiral' and 'regress' test results are shown, since
'parity8' has an identical pattern as 'spiral'. The migration intervals of 2, 3 and
5 generations are clearly favoured to one generation and the longer ones with
the 'spiral' and 'parity8' data sets, and are considerably improving the results.
The 3 and 5 intervals are maintaining a slightly higher genetic diversity than
the 2-generation interval.

The 'regress' figures on the other hand are suggesting larger intervals, as
fitness was better on average with the 5, 10 and 20 generation intervals. This is
an indication that the optimal migration interval could be problem dependent.

Yet the problem remains that highly connected topologies will promote a
singular solution, crowding out less fit alternatives. The idea to prevent this is
the introduction of migration delays. Whenever a segment has been improving
locally, migration into it is suspended for one or even more generations. This
gives room for a fresh solution to develop. Once it stops improving, it has to
compete with migrating individuals from other segments again.

Again 100 tests were run on the 'spiral', 'regress' and 'parity8' data sets for
each of the 30 combinations of migration intervals from 1 to 6 and migration
delays from 0 to 4. As can be seen from Fig. 9, some combinations have better average fitnesses (darker areas) than others. The effect to be expected was,
that delaying migrations cures the negative influence of migration interval 1, i.e.,
migration is applied every single generation. Without delays, this effectively prevents promising structures to evolve, since they always are endangered to be run

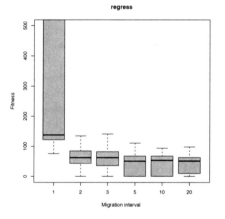
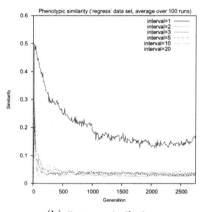

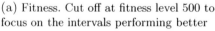

(a) Fitness. Cut off at fitness level 500 to
focus on the intervals performing better

(b) 5-gram similarity

Fig. 6. Effect of different migration intervals, 'regress' data set

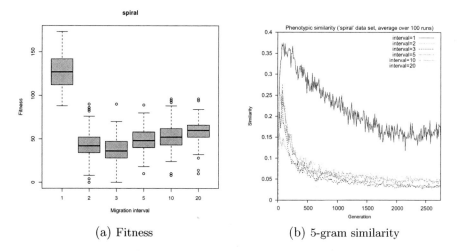

(a) Fitness   (b) 5-gram similarity

Fig. 7. Effect of different migration intervals, 'spiral' data set

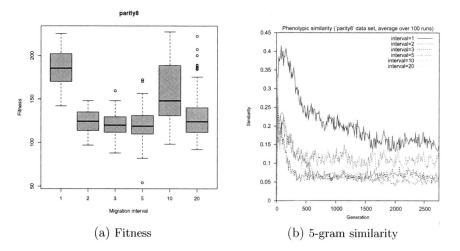

(a) Fitness   (b) 5-gram similarity

Fig. 8. Effect of different migration intervals, 'parity8' data set

over by a neighbouring segment accidentially being better at the moment. Delays will keep the evolving candidates in relative isolation, protecting their progress.

The more interesting areas of better fitness are the "trough" at migration interval 3 and delays of one and two generations with the 'spiral' data set, the top right area (intervals 2 and above and delays of 2 and above) for the 'regress' data set and again the area around delays 1 and 2 and intervals 3 to 6 with the 'parity8' data set. Here an improved fitness is observed. The successful interval/delay combinations seem to be problem-dependent, on one hand protecting evolving solutions sufficiently to survive, on the other hand allowing promising structures to spread through the population to give them more room to further develop.

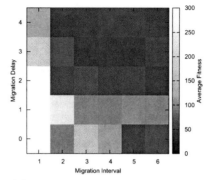

(a) 'regress' data set, average fitness

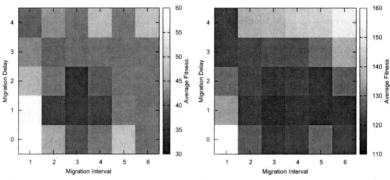

(b) 'spiral' data set, average fitness   (c) 'parity8' data set, average fitness

Fig. 9. Effect of migration delays vs. migration intervals, torus topology

Tests were run on other migration interval/delay combinations on the data sets singularly, but not complete enough to determine potential effects outside the area analysed here. A thorough test over migration intervals 1 to 25 and migration delays 0 to 20 (52.500 test runs in total) is being currently done. The basic pattern evolving is a 'sweet spot' favouring small delays versus higher intervals.

6   Conclusion

It was shown for three different data sets, that the effect of different migration topologies on the performance of a linear GP system is not as subordinate as was concluded in [2]. Instead, highly connected topologies of population segments yielded better results, at the same time maintaining a higher genetic diversity among the population. Prerequisite is to select the best segment among all competing to migrate, and only let migration happen if it outperforms the one to migrate into. A further improvement was achieved for the torus topology by delaying the migration if the target segment had improved in the generation before.

More work will be necessary to explain the contraintuitive higher level of genetic diversity observed. Moreover, all except the last experiment took the migration parameters as mutually independent; one was chosen and modified with all others fixed. A next step will be to analyze the interdependencies between parameters to find combinations performing even better. There are indications, that for instance the migration interval and migration delay parameters may have other useful combinations for other topologies than the tuples found here for the torus, outside the combinations area examined.

Acknowledgement

I would like to thank Sara Silva for helping me with the statistics and insisting on a consistent structure of this paper — and for the zillion other hints!

References

1. Tomassini, M., Vanneschi, L., Fernandez, F., Galeano, G.: A Study of Diversity in Multipopulation Genetic Programming. In: Liardet, P., Collet, P., Fonlupt, C., Lutton, E., Schoenauer, M. (eds.) EA 2003. LNCS, vol. 2936, pp. 243–255. Springer, Heidelberg (2004)
2. Fernandez, F., Tomassini, M., Vanneschi, L.: Studying the Influence of Communication Topology and Migration on Distributed Genetic Programming. In: Miller, J., Tomassini, M., Lanzi, P.L., Ryan, C., Tetamanzi, A.G.B., Langdon, W.B. (eds.) EuroGP 2001. LNCS, vol. 2038, pp. 51–63. Springer, Heidelberg (2001)
3. Silva, S., Tseng, Y.-T.: Classification of Seafloor Habitats Using Genetic Programming. In: Giacobini, M., Brabazon, A., Cagnoni, S., Di Caro, G.A., Drechsler, R., Ekárt, A., Esparcia-Alcázar, A.I., Farooq, M., Fink, A., McCormack, J., O'Neill, M., Romero, J., Rothlauf, F., Squillero, G., Uyar, A.Ş., Yang, S. (eds.) EvoWorkshops 2008. LNCS, vol. 4974, pp. 315–324. Springer, Heidelberg (2008)
4. Cantú-Paz, E.: Topologies, Migration Rates, and Multi-Population Parallel Genetic Algorithms. In: Proceedings of the Genetic and Evolutionary Computation Conference, July 13-17, vol. 1, pp. 91–98. Morgan Kaufmann, San Francisco (1999)
5. Koza, J.R.: A Genetic Approach to the Truck Backer Upper Problem and the Inter-Twined Spiral Problem. In: Proceedings of IJCNN International Joint Conference on Neural Networks, vol. IV, pp. 310–318. IEEE Press, Los Alamitos (1992)
6. Eklund, S.E.: A Massively Parallel GP Engine in VLSI. In: Proceedings of the 2002 Congress on Evolutionary Computation CEC 2002, May 12-17, pp. 629–633. IEEE Press, Los Alamitos (2002)
7. Brameier, M., Banzhaf, W.: A Comparison of Linear Genetic Programming and Neural Networks in Medical Data Mining. IEEE Transactions on Evolutionary Computation 5(1), 17–26 (2001)

A New Approach to Solving 0-1 Multiconstraint Knapsack Problems Using Attribute Grammar with Lookahead

Muhammad Rezaul Karim and Conor Ryan

Biocomputing and Developmental Systems Group
Department of Computer Science and Information Systems
University of Limerick, Ireland
{Rezaul.Karim,Conor.Ryan}@ul.ie

Abstract. In this paper, we introduce a new approach to genotype-phenotype mapping for Grammatical Evolution (GE) using an attribute grammar (AG) to solve 0-1 multiconstraint knapsack problems.

Previous work on AGs dealt with constraint violations through repeated remapping of non-terminals, which generated many introns, thus decreasing the power of the evolutionary search.

Our approach incorporates a form of lookahead into the mapping process using AG to focus only on feasible solutions and so avoid repeated remapping and introns. The results presented in this paper show that the proposed approach is capable of obtaining high quality solutions for the tested problem instances using fewer evaluations than existing methods.

1   Introduction

For many NP-hard combinatorial optimization problems, it is difficult to find an optimal solution. For these problems the only reasonable way is to look for heuristics algorithms that quickly produce good, although not necessarily optimal solutions. Another characteristic of these problems is that the solutions must meet all the problem-specific constraints. The Multiconstraint Knapsack Problem (MKP) belongs to this class of problems.

The MKP has a wide range of real world applications such as cargo loading, selecting projects to fund, budget management, stock cutting etc. [17]. There has been extensive research [3,4,5,7,8,9,10,11,14,18,19] into the application of Evolutionary Algorithms (EAs) to MKP, and they have been found to be most appropriate in solving large instances of MKP for which exact methods have been found inefficient. The EA approaches that currently provide best results for MKP require extra machinery (typically repair) and problem specific information to provide improvements [3,18,19]. These approaches are explored in more detail in Section 3.1.

Due to the power of representation, AGs [12] have been applied in diverse fields of computer science such as semantic analysis and intermediate code generation in compiler construction [1] and programming language specification [15].

S. Silva et al. (Eds.): EuroGP 2011, LNCS 6621, pp. 250–261, 2011.
© Springer-Verlag Berlin Heidelberg 2011

In terms of their power of representation, AGs are more powerful than context-free grammars, and have the same formal power as Turing machines [12].

GE [21] is a genotype-phenotype mapping grammar based system that uses CFGs to evolve structures. There has been some work [6] that extended it to use AG to provide context sensitive information, although these are all problem specific approaches. One approach that extended GE with AG for MKP problems was [14]. We argue that the problem with this approach is that it fails to fully utilize the power of AG. In this approach, many introns are generated due to problem constraint violation, as codons that violate constraints are skipped. To get better results, this approach needs to rely on computationally expensive operation like splicing to remove generated introns [4] or on phenotypic duplicate elimination [4]. In this paper we explore the implementation of a form of lookahead mechanism using AG to solve MKP without generating a single intron.

The rest of the paper is organized as follows: Section 2 introduces AG. Section 3 gives an introduction to MKP and an overview of the existing EA approaches for MKP. The AG used is described in section 4, while the experimental results are reported and analyzed in section 5. The paper concludes in section 6 with a summary and scope of future work.

2   AGs

Knuth [12] first proposed AGs to specify the semantics of programming languages whose syntax is defined by CFGs. An AG is based upon a CFG and consists of a tuple $G=< N, T, P, S >$ where N is the set of non-terminal symbols, T is the set of terminal symbols, P is the set of productions (syntactic rules), and S is the start symbol where $(S \in N)$. Each symbol X in the set of all symbols V ($V = N \cup T$) of the grammar G has associated attributes $A(X)$. Each attribute represents a specific context-sensitive property of the corresponding symbol.

The purpose of these attributes is to transmit information between nodes in the derivation tree for a given sentential form. The notation $X.a$ is used to indicate that a is an element in $A(X)$ or a is an attribute of symbol X. $A(X)$ is partitioned into two disjoint sets; $SA(X)$, the set of *synthesized* attributes and $IA(X)$, the set of *inherited* attributes.

Synthesized attributes $X.s$ are defined in terms of attributes at descendant nodes of node X (meaning the information is passed back up the tree) of the corresponding annotated derivation tree. Inherited attributes $X.i$, on the other hand, are defined in terms of attributes at the parent of X, X itself or the sibling nodes of node X of the corresponding annotated derivation tree (and so are passed down the tree).

Each of the productions $p \in P(p = X_0 \rightarrow X_1 X_2 ... X_n)$ is augmented by a set of semantic functions. All synthesized and inherited attributes have a semantic function which defines the value of the attribute occurrence in terms of the values of other attribute occurrences appearing in the same production. If a grammar uses only synthesized attribute, it is S-attributed. A grammar is called L-attributed if, for a typical production like $p \in P(p = X_0 \rightarrow X_1 X_2 ... X_n)$ of the grammar, the following properties hold for each attribute [1]:

1. Each inherited attribute of a right-hand-side symbol X_i depends only on inherited attributes of X_0 and any arbitrary attributes of the symbols $X_1 X_2 ... X_{i-1}$.
2. Each synthesized attribute of X_0 depends only on its inherited attributes and arbitrary attributes of the right-hand side symbols: $X_1 X_2 ... X_n$.

AGs follow a nonprocedural formalism. Therefore AGs do not impose any sequencing order in the process of evaluating the attributes. Instead, they simply describe the dependencies between the syntactic structures and among the attribute values. An attribute can be evaluated as soon as the value for the attributes on which it depends are available. When a grammar is S-attributed, we can evaluate its attribute in any bottom-up order of the nodes of the derivation tree. It is simple to evaluate the attributes by performing the post-order traversal of the derivation tree and evaluate the attributes at a node N when the traversal leaves the node N for the last time [1]. When evaluating the attributes of an L-attributed production we first evaluate the inherited attributes of X_0 (left-hand-side of production), then evaluate the inherited followed by the synthesized attributes of each X_i on the right side of the production from left to right and finally evaluate the synthesized attribute of X_0.

3   Multiconstraint Knapsack Problem

The Multiconstraint knapsack problem (MKP) is a generalization of the standard 0-1 knapsack problem. The objective of MKP is to fill a series of m knapsacks each of capacity c_j with any number of n items maximizing the items' combined profit while ensuring that the sum of weights of all items considered for inclusion does not exceed any of the capacities. We can formulate MKP as:

$$\text{maximize} \quad \sum_{i=1}^{n} p_i x_i$$

$$\text{subject to} \quad \sum_{i=1}^{n} w_{ij} x_i \leq c_j \quad j = 1, ..., m \tag{1}$$

$$x_i \in \{0, 1\} \quad i = 1, ..., n$$

$$\text{with} \quad p_i > 0, \quad w_{ij} \geq 0, \quad c_j \geq 0$$

Each item $i\{i = 1, ..., n\}$ has profit p and weight w_{ij} with respect to knapsack $j\{j = 1, ..., m\}$.

3.1   Previous Work

Gottlieb [7] has classified EA approaches to solving MKP into two categories based on the search space which is explored by the population: *direct* and *indirect*.

EAs operating in the original search space are called direct search whereas indirect search techniques search on an arbitrary search space which is later mapped to the original search space using a decoder. Some direct search techniques focus on both feasible and infeasible solutions and rely on penalty functions to penalize infeasible solutions [9,11].

One of the earliest penalty-based approaches is Khuri et al. [11] who suggest the use of a graded penalty term to penalize infeasible solutions upon the number of constraint violations. Other direct search techniques searching the feasible space rely on repair techniques and specialized operators to alter infeasible solutions (produced by crossover and mutation) into feasible ones [3,18]. Indirect search methods performance is dependent on the decoder which builds legal solutions. All decoders work in the same way to generate feasible solution but differs in the way they interpret the genome and incorporates problem specific heuristics [4,7,8,10,14,19].

Among the repair based techniques, very good results were reported in [3,18]. Chu et al. in [3] propose a repair operator based on problem specific-pseudo-utility ratio. Raidl [18] reported further improvement by applying repair technique and local improvement operators along with specialized initialization operator. These two techniques suffer from large computational effort spent in repairing and local improvements to solutions. Of the decoder-based techniques, the most successful one is the approach given in [19] and this decoder employs weight-coding representation. In [19] several biasing and decoding heuristics are proposed and compared. The drawback of this approach is extra computational cost involved in biasing and decoding phases.

3.2 Grammatical Evolution Approaches for the MKP

Cleary et al. [14] showed the importance of encoding context-sensitive information for solving MKP and extended Grammatical Evolution with AG to solve MKP. They proposed *AG(Full)*, a mapping approach based on AG to encode context-sensitive information. In their approach, if a codon violates a constraint either by inserting an item that puts the knapsack over the weight limit or by selecting an item too many times, it is skipped and considered an intron.

Remapping of the offending non-terminal is then tried with the next codon. This process is repeated until a new terminal (item) is found that is previously not selected and that does not violate any weight constraint. One of the problems with AG(Full) is the computational effort spent for repeated remapping. For a heavily constrained problem like MKP, many introns are also generated by AG(Full) due to constraint violation, in a similar manner to [16], the ancestor of GE. The performance reported was moderate for AG(Full), with slightly better results reported in [4] for AG(Full) by increasing limit of evaluated individuals and further improvements were reported by applying intron removal techniques like splicing (AG(Full)+Splice). Later, phenotypic duplicate elimination was integrated with splicing and further improvement were reported [4].

4 AG with Lookahead for the MKP

The AG described in this section is used to show the effectiveness of implementing lookahead with AG to derive any string from the set of available items. The CFG is identical to that used in [14] except that a new production is added to the rule for terminal-generating (item-generating) non-terminal I so that a special purpose item ϵ can be generated to indicate that derivation must stop. Unlike the attributes in [14], a set of new attributes are defined to incorporate lookahead concept to eliminate the drawbacks of previous AG based approach (described in Sect. 3.2).

Lookahead mechanism has successfully been used in constraint satisfaction algorithms [13]. When solving a constraint satisfaction problem using simple backtracking, whenever a new variable is considered for instantiation, any of its values that are inconsistent with any previous instantiations cause immediate failure and backtracking occurs. Forward checking is another algorithm for solving constraint satisfaction problem that eliminates this kind of unnecessary backtracking: whenever a new variable instantiation is made, the domains of all as-yet-uninstantiated variables are filtered to contain only those values that are consistent with this instantiation. Our proposed AG is inspired by forward checking. Our solutions for MKP do not have fixed length and we are interested to find a solution which can be represented by a vector $\boldsymbol{x} = \{x_1, x_2, ..., x_k\}$ where $1 \leq k \leq n$ and each x_i represents any distinct item from the set of n available items. In the proposed approach, after generating i^{th} item in the solution vector using grammar, we filter out the domain of the next item in the solution vector with the help of special attributes in the grammar. Thus repeated remapping is avoided during genotype to phenotype mapping of GE like [22].

The proposed AG (AG+LA) is given below:

$$S \rightarrow K \qquad S.alloweditems = initialAllowedItems()$$
$$S.limit = initialWeightConstraints()$$
$$S.items = K.items$$
$$S.weight = K.weight$$
$$K.alloweditems = S.alloweditems$$
$$K.limit = S.limit$$

$$K \rightarrow I \qquad I.alloweditems = K.alloweditems$$
$$K.items = I.item$$
$$K.weight = I.weight$$

$$K_1 \rightarrow IK_2 \qquad K_1.selecteditem = I.item$$
$$K_1.selecteditemweight = I.weight$$
$$K_1.usage = K_1.usage + K_1.selecteditemweight$$
$$K_1.useditems = K_1.useditems + K_1.selecteditem$$

$$K_1.items = K_2.items + I.item$$
$$K_1.weight = K_2.weight + I.weight$$
$$I.alloweditems = K_1.alloweditems$$
$$K_2.usage = K_1.usage$$
$$K_2.useditems = K_1.useditems$$
$$K_2.limit = K_1.limit$$
$$K_2.alloweditems = genAllowedItems(K_2.useditems,$$
$$K_2.usage, K_2.limit)$$

$I \rightarrow i_1$ $I.item = \text{``}i_1''\text{''}$

      $I.weight = getweight(i_1)$

....

$I \rightarrow i_n$ $I.item = \text{``}i_n''\text{''}$

      $I.weight = getweight(i_n)$

$I \rightarrow \epsilon$ $I.item = \text{``}\epsilon''\text{''}$

      $I.weight = 0$

As each symbol in the grammar has its own set of attributes, subscript notation is used in the above grammar to differentiate between attribute occurrences of similar non-terminals. But during the mapping of a non-terminal, all subscripts are ignored, and all productions in the grammar with this non-terminal as left hand symbol form the mapping rule. Thus attribute evaluation rules are different for symbol K_1 and K_2 as shown in above grammar but mapping rule is identical. The above AG has following attributes:

limit: is an inherited attribute for non-terminal symbol K and S of the grammar. This is a special attribute for non-terminal symbol S, as this value is not inherited from any other symbol. At the start of derivation, the value of the attribute *limit* for start symbol S is computed by the semantic function *initialWeightConstraints*. Later this is passed down the nodes of the derivation tree to facilitate weight constraints violation check. This attribute is implemented by a data structure and stores the individual capacity of each of the knapsacks.

selecteditem and **selecteditemweight:** synthesized attributes for non-terminal symbol K. These two attributes are used to record the item derived by the descendant I to this K and the weight of the derived item, respectively.

usage and **useditems:** both are inherited attributes for non-terminal symbol K. *useditems* records the list of all items derived so far. The attribute, *usage*, on the other hand, records the total weight of all items derived so far for each knapsack. These attributes are used for passing the usage and previously used items down the derivation tree to aid in enforcing weight constraint violation check.

items and ***weight:*** the attribute *items* is a synthesized attribute for non-terminal symbol S and K, whereas *weight* is a synthesized attribute of non-terminal symbols S, K and I. *items* is used to keep track of all items selected by the sub-tree rooted at the node represented by any of the symbols S and K. *weight*, on the other hand, is used to compute the total weight of all items represented by the sub-tree. The special terminal ϵ is assigned zero weight for all knapsacks, but for the attribute *weight* of non-terminal I, the semantic function *getweight* is used to return the knapsack specific physical weight of the item. The attribute *weight* stores one record for each knapsack.

item: is a synthesized attribute for non-terminal symbol I. This attribute represents the item that is being derived by this non-terminal.

alloweditems: is an inherited attribute for non-terminal symbol S, K and I. Like the attribute *limit*, *alloweditems* is a special inherited attribute for I. This attribute indicates which items are available for selection at this point of derivation by the next item-generating rule I, without violating weight constraint of any knapsack.

At the start of derivation, the allowed item list is generated by the semantic function *initialAllowedItems* for start symbol S based on the knapsacks capacity constraints and item weights. For non-terminal K, depending on the current production, either it inherits the value from start symbol S or the value is computed with the help of the semantic function *genAllowedItems*. This semantic function includes an item in the allowable item list if that item is not previously selected and if inclusion of the item does not violate weight constraints of knapsacks.

If the domain of the attribute *alloweditems* for any symbol becomes null at any point of derivation, only the special item ϵ is included in the allowed item list and the derivation stops. If a derivation for any non-terminal K is not affected due to inclusion of ϵ, the computed allowed item list is passed down the derivation tree to the immediate descendant node I and the rule for non-terminal I is modified in such a way that only those item-generating productions that represent an item in the allowed item list is made available for mapping by the current codon. Thus not a single codon is skipped or treated as intron during mapping and there is no need for remapping due to selection of previously selected item or due to weight constraint violation of any of the knapsacks.

5   Experiments

The experimental results reported here have been carried out using standard experimental parameters for GE, changing only the population size to 1500 to compare with the results in [4,11]. We compare performances based on the information available in the literature and do not implement the other approaches to generate unavailable data. As our goal is to extend standard GE with AG to solve MKP without employing problem specific heuristics or repair operations, we do not compare our approach with any other approach that employs problem specific knowledge or repair operations using items' value-weight ratio.

Table 1. Properties of the test problems used and problem specific parameter settings

Problem	Items	Knapsacks	Optimal Value	Individual Processed	Max Depth
knap15	15	10	4015	20000	11
knap20	20	10	6120	20000	15
knap28	28	10	12400	20000	21
knap39	39	5	10618	100000	29
knap50	50	5	16537	100000	37
Sento1	60	30	7772	150000	45
Sento2	60	30	8722	150000	45
Weing7	105	2	1095445	200000	78
Weing8	105	2	624319	200000	78

We adopt a variable length one-point crossover probability of 0.9, bit mutation probability of 0.01 and roulette wheel selection. A steady state GA is used
where replacement occurs if the generated child is better than the worst individual in the population. Sensible initialization [20] is used for initializing the
population with grow probability set to 0.2. Grow probability is set to 0.2 to bias
the grammar to produce right-skewed derivation trees 80% of the time. This is
necessary to produce individuals that tend to have more rather than less items.

The minimum depth parameter for sensible initialization is 8. The maximum
depth, on the other hand, is problem specific and is set to three quarters of the
number of items in the problem instance. This proportion is consistent across
all experiments. A high value of maximum depth helps to introduce diversity in
the initial population. We employ standard 8-bit codons with no wrapping. For
comparison purposes, we use nine instances from OR-Library by Beasley [2]. A
total of 30 runs are conducted for each test problem. The basic characteristics
of these problems and various problem specific settings are given in Table 1.

5.1   Results

Table 2 shows the results for all problem instances. For each problem instance,
we note the average fitness of the best individual in the final generation, standard deviation, the percentage of runs yielding optimal solution and the average
percentage gap. The percentage gap is is the difference between the discovered
solutions and the optimal solution (if known) and is calculated as in (2).

$$\%gap = 100 * (opt - opt^{LA})/opt \qquad (2)$$

where opt is the optimal value of the test problem and opt^{LA} is the best solution
found by AG+LA.

Our algorithm, AG+LA, outperforms AG(Full)[4] across all instances not only
in terms of averages but also in terms of the percentage of runs yielding optimal solution. It is also worth noting that for most of the problems, the results
are achieved evaluating fewer individuals than AG(Full). AG(Full) uses 200000
individual evaluations for each problem instance but we use 200000 individual

Table 2. Comparing AG with lookahead with other relevant evolutionary approaches in the literature. All approaches are compared in terms of average fitness of the best individual in the final generation as well as percentage of runs yielding optimal solution. '-' indicates data not available in the literature. Where available, we also report the standard deviation and average percentage gap.The highest mean best fitness for each problem is highlighted in bold. Notice that AG+LA scores the highest in seven out of nine cases.

Problem	AG(Full)			AG(Full)+Splice			Khuri			AG+LA			
	Avg.	Std.	% Opt.	Avg.	Std.	% Opt.	Avg	Std	% Opt.	Avg.	Std.	% Opt.	%Gap
knap15	4011.33	-	83.33%	-	-	86.66%	4012.7	-	83%	**4015**	0.00	100%	0.000
knap20	6111.0	-	76.66%	-	-	80%	6102.3	-	33%	**6120**	0.00	100%	0.000
knap28	12377.66	-	40%	-	-	30%	12374.7	-	33%	**12399.67**	1.83	96.67%	0.003
knap39	10457.63	-	36.66%	-	-	33.33%	10536.9	-	4%	**10610.83**	10.01	60%	0.067
knap50	16266.13	-	3.33%	-	-	3.33%	16378.0	-	1%	**16501.33**	27.81	9.09%	0.216
Sento1	7693.2	-	10%	-	-	40%	7626	-	5%	**7723.20**	55.89	23.33%	0.628
Sento2	8670.06	-	3.33%	-	-	13.33%	**8685**	-	2%	8661.23	53.71	10%	0.697
Weing7	1075950.83	-	0%	-	-	0%	**1093897**	-	0%	1087377.40	2111.49	0%	0.736
Weing8	605993.63	-	0%	-	-	33.33%	613383	-	6%	**615615.30**	8936.77	9.09%	1.394

evaluations only for two instances, the rest of them are tested with fewer individual evaluations (see Table 1).

It is difficult to make a fair comparison with AG(Full)+splice as data on mean fitness is not reported in [4]. Considering the stochastic nature of evolutionary algorithm, comparisons based only the percentage of runs yielding optimal solution can lead us to wrong conclusion if that percentage is not significant. For example, an algorithm that reaches an optimal solution 13.33% time and generates very poor results rest of the time, cannot be considered as a good algorithm. But, for completeness, we mention the results in table 2. For a very difficult problem instance where no algorithm is able to reach optimal solution, or if the optimal solution is unknown which is common for many difficult test problem instances [2], the average percentage gap or mean fitness along with standard deviation can be used to compare one algorithm with others to measure how well the algorithm performs as an approximation algorithm.

AG+LA also produces superior results than Khuri et al. [11] for all instances but Sento2 and Weing7. Even though Weing7 has same number of items and knapsacks as Weing8, AG+LA produces worse result than Khuri for Weing7. For Sento2, AG+LA produces slightly worse mean best fitness but reaches the optimal solution more often than Khuri. The results also indicate that the performance of AG+LA is dependent on the specific characteristics of the problem search space rather than on the number of items. This suggests that the performance of the system is more related to the tightness of the constraints rather than the number of them. However, at this point, this is purely conjecture and further research and testing is required to validate.

Even though the results presented in Table 2 is based on population size 1500, it is also possible to obtain good results with small population size for some problem instances such as knap15, knap20 and knap28. Figure 1 shows the effect of different population size on the quality of solution achieved by AG+LA. It is clear from Fig. 1 that population size 500 is sufficient to reach

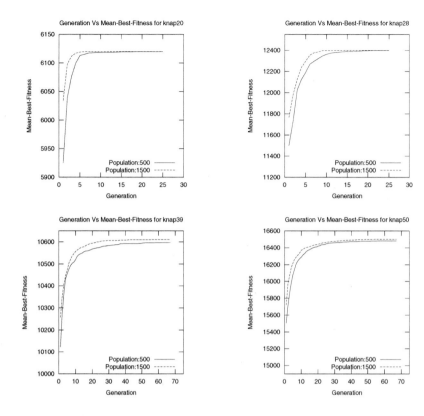

Fig. 1. A comparison of the quality of solutions obtained by AG+LA for different problem instances with different population size

optimal solution for problem instances knap20 and knap28 within maximum individual evaluation limit. It also shows that the quality of solutions greatly improves with larger population size for problems with large number of items such as knap39 and knap50. For the sake of a standard population size, we adopt the population size 1500 as the standard so that the algorithm AG+LA solves all problem instances with a single population size.

In summary, the results show that AG+LA, for all test problems, improves the performance of existing AG based solution for MKP and achieves average fitness value that turns out to be reasonably close to the global optimum. There is still room for improvement and further experiments and research are needed.

6   Conclusions and Future Work

In this paper, we have introduced an approach to solving MKP based on AG that incorporates lookahead to generate feasible solutions. Unlike the existing

AG based approaches, this approach eliminates the need for repeated remapping of item-generating non-terminal due to constraints violation and thus avoids introns.

This is crucial when trying to solve large, highly constrained problems like MKP as it greatly reduces the search space and computational effort as well as eliminates the need for applying intron removal techniques to improve performance. The results show that our technique improves the performance of AG based system on all nine test problems. Since this is an early investigation, further testing is required to establish how well the algorithm will scale and also to investigate the reason for performance drop on some instances, most notably weing7. Additional future work should focus on to determine whether the proposed approach is suitable for other constrained optimization problems especially ordering problems like Traveling Salesperson Problem. Our analysis does not report the effect of important aspects of Grammatical Evolution, such as genetic degeneracy, effect of genetic operators, various methods of population initialization and genetic diversity. Future work will look into that.

References

1. Aho, A.V., Lam, M.S., Sethi, R., Ullman, J.D.: Compilers: Principles, Techniques, and Tools, 2nd edn. Addison-Wesley, Reading (2006)
2. Beasley, J.E.: Or-library: distributing test problems by electronic mail. Journal of the Operational Research Society 41(11), 1069–1072 (1990)
3. Chu, P.C., Beasley, J.E.: A genetic algorithm for the multidimensional knapsack problem. Journal of Heuristics 4(1), 63–86 (1998)
4. Cleary, R.: Extending Grammatical Evolution with Attribute Grammars: An Application to Knapsack Problems. Master of science thesis in computer science, University of Limerick, Ireland (2005)
5. Cotta, C., Troya, J.M.: A hybrid genetic algorithm for the 0-1 multiple knapsack problem. In: Artificial Neural Nets and Genetic Algorithms 3, pp. 250–254. Springer, New York (1998)
6. de la Cruz, M., Ortega de la Puente, A., Alfonseca, M.: Attribute grammar evolution. In: Mira, J., Álvarez, J.R. (eds.) IWINAC 2005. LNCS, vol. 3562, pp. 182–191. Springer, Heidelberg (2005)
7. Gottlieb, J.: On the effectivity of evolutionary algorithms for the multidimensional knapsack problem. In: Fonlupt, C., Hao, J.-K., Lutton, E., Schoenauer, M., Ronald, E. (eds.) AE 1999. LNCS, vol. 1829, pp. 23–37. Springer, Heidelberg (2000)
8. Gottlieb, J.: Permutation-based evolutionary algorithms for multidimensional knapsack problems. In: Proceedings of the 2000 ACM Symposium on Applied Computing, pp. 408–414. ACM, New York (2000)
9. Gottlieb, J.: On the feasibility problem of penalty-based evolutionary algorithms for knapsack problems. In: Boers, E.J.W., Gottlieb, J., Lanzi, P.L., Smith, R.E., Cagnoni, S., Hart, E., Raidl, G.R., Tijink, H. (eds.) EvoIASP 2001, EvoWorkshops 2001, EvoFlight 2001, EvoSTIM 2001, EvoCOP 2001, and EvoLearn 2001. LNCS, vol. 2037, pp. 50–59. Springer, Heidelberg (2001)
10. Gottlieb, J., Raidl, G.R.: The effects of locality on the dynamics of decoder-based evolutionary search. In: Proceedings of the Genetic and Evolutionary Computation Conference 2000, pp. 283–290. Morgan Kaufmann Publishers, San Francisco (2000)

11. Khuri, S., Back, T., Heitkotter, J.: The zero/one multiple knapsack problem and genetic algorithms. In: Proceedings of the 1994 ACM Symposium on Applied Computing, pp. 188–193. ACM Press, New York (1994)
12. Knuth, D.E.: Semantics of context-free languages. Theory of Computing Systems 2(2), 127–145 (1968)
13. Kumar, V.: Algorithms for constraint satisfaction problems: A survey. AI Magazine 13(1), 32–44 (1992)
14. O'Neill, M., Cleary, R., Nikolov, N.: Solving knapsack problems with attribute grammars. In: Proceedings of the Third Grammatical Evolution Workshop (2004)
15. Paakki, J.: Attribute grammar paradigms–a high-level methodology in language implementation. ACM Comput. Surv. 27(2), 196–255 (1995)
16. Paterson, N., Livesey, M.: Evolving caching algorithms in C by genetic programming. In: Koza, J.R., Deb, K., Dorigo, M., Fogel, D.B., Garzon, M., Iba, H., Riolo, R.L. (eds.) Proceedings of the Second Annual Conference on Genetic Programming, pp. 262–267. Morgan Kaufmann, San Francisco (1997)
17. Pisinger, D.: Algorithms for knapsack problems. Ph.D. thesis, University of Copenhagen (1995)
18. Raidl, G.R.: An improved genetic algorithm for the multiconstrained 0-1 knapsack problem. In: Proceeding of the 1998 IEEE International Conference on Evolutionary Computation, pp. 207–211 (1998)
19. Raidl, G.R.: Weight-codings in a genetic algorithm for the multi-constraint knapsack problem. In: Proceedings of the 1999 Congress on Evolutionary Computation, pp. 596–603 (1999)
20. Ryan, C., Azad, R.M.A.: Sensible initialisation in grammatical evolution. In: Barry, A.M. (ed.) Proceedings of the Bird of a Feather Workshops, Genetic and Evolutionary Computation Conference, Chigaco, pp. 142–145 (2003)
21. Ryan, C., Collins, J., O'Neill, M.: Grammatical evolution: Evolving programs for an arbitrary language. In: Proceedings of the First European Workshop on Genetic Programming, pp. 83–95. Springer, Heidelberg (1998)
22. Ryan, C., Nicolau, M., O'Neill, M.: Genetic algorithms using grammatical evolution. In: Foster, J.A., Lutton, E., Miller, J., Ryan, C., Tettamanzi, A.G.B. (eds.) EuroGP 2002. LNCS, vol. 2278, pp. 278–287. Springer, Heidelberg (2002)

An Empirical Study of Functional Complexity as an Indicator of Overfitting in Genetic Programming

Leonardo Trujillo[1], Sara Silva[2,3],
Pierrick Legrand[4,5], and Leonardo Vanneschi[2,6]

[1] Instituto Tecnológico de Tijuana, Av. Tecnológico S/N, Tijuana, BC, México
[2] INESC-ID Lisboa, KDBIO group, Lisbon, Portugal
[3] CISUC, ECOS group, University of Coimbra, Portugal
[4] IMB, Institut de Mathématiques de Bordeaux, UMR CNRS 5251, France
[5] ALEA Team at INRIA Bordeaux Sud-Ouest, France
[6] Department of Informatics, Systems and Communication (D.I.S.Co.),
University of Milano-Bicocca, Milan, Italy
leonardo.trujillo.ttl@gmail.com, sara@kdbio.inesc-id.pt,
pierrick.legrand@u-bordeaux2.fr, vanneschi@disco.unimib.it

Abstract. Recently, it has been stated that the complexity of a solution is a good indicator of the amount of overfitting it incurs. However, measuring the complexity of a program, in Genetic Programming, is not a trivial task. In this paper, we study the functional complexity and how it relates with overfitting on symbolic regression problems. We consider two measures of complexity, Slope-based Functional Complexity, inspired by the concept of curvature, and Regularity-based Functional Complexity based on the concept of Hölderian regularity. In general, both complexity measures appear to be poor indicators of program overfitting. However, results suggest that Regularity-based Functional Complexity could provide a good indication of overfitting in extreme cases.

1 Introduction

In the field of Genetic Programming (GP), a substantial amount of research focuses on the bloat phenomenon Bloat is an excess of code growth without a corresponding improvement in fitness [10], and it can cause several types of problems during a GP run. For instance, bloat can stagnate a GP search because when program trees become excessively large then fitness evaluation can turn into a computational bottleneck. Moreover, it is also assumed that bloat is related to overfitting, one of the most important problems in machine learning. It is often stated that simpler solutions will be more robust and will generalize better than complex ones, with the latter being more likely to overfit the training data [6,14]. The GP community has tacitly assumed that *simple* solutions can be equated with small program trees, and that very large programs correspond to *complex* solutions [8]. Therefore, bloat was assumed to be a good indicator of program overfitting. However, recent experimental work suggests

S. Silva et al. (Eds.): EuroGP 2011, LNCS 6621, pp. 262–273, 2011.
© Springer-Verlag Berlin Heidelberg 2011

that this assumption is not reliable [14]. In particular, [14] showed that a causal link between bloat and overfitting did not exist on three test cases. From this it follows that bloated programs should not be a priori regarded as complex. This leads us towards two important questions. First, how can program complexity be measured? And second, can program complexity be used as an indicator of program overfitting? Here, we study the measure of *functional complexity* proposed in [14], which we call Slope-based Functional Complexity (SFC). SFC in inspired by the concept of curvature, and represents the first measure of complexity that is explicitly intended to be an indicator of program overfitting. While SFC is based on a reasonable intuition, the methodological approach it requires can become complicated for multidimensional problems. Therefore, we propose a measure of complexity based on the concept of Hölderian regularity, and we call it Regularity-based Functional Complexity (RFC), which captures the underlying justification of SFC without being hampered by some of its practical difficulties. Both measures are experimentally tested on two real world problems and compared based on their correlation with overfitting [14].

2   Overfitting

Based on [6,14], for a GP search in program space P an individual program $p \in P$ overfits the training data if an alternative program $p' \in P$ exists such that p has a smaller error than p' over the training samples, but p' has a smaller error over the entire distribution of instances. In the above terms, overfitting is practically impossible to measure because it requires an exhaustive search over P and the entire distribution of instances. Nevertheless, [14] argues that a good approximation can be obtained with the relationship between fitness computed on the training set and fitness computed on an independent test set. In our work, we use the measure for overfitting proposed in [14]; which proceeds as follows for a minimization problem. If at generation g test fitness is better than training fitness, then $(overfitting(g) = 0)$; if test fitness is better than the best test fitness obtained thus far then $(overfitting(g) = 0)$; otherwise, overfitting is quantified by the difference of the distance between training and test fitness at generation g and the distance between training and test fitness at the generation when the best test fitness was obtained. In this procedure, test fitness is computed only for the best individual of the population, the individual with the best training fitness. This individual is chosen because we want to measure the overfitting of the best GP solution, since this is the individual that will be returned at the end of the search. If elitism is used then training fitness will be monotonically decreasing, thus $overfitting(g) \geq 0 \ \forall g$.

Therefore, overfitting depends on the performance of a program on the test dataset. In the best case scenario, the test set is a representative sample of the unseen data that a program might encounter during on-line operation. However, it might also be true that an independent test set could be difficult to produce or might be unavailable. Therefore, an indicator of program overfitting would be a useful practical tool. In this sense, [14] suggests that program complexity could indicate if overfitting is present during a GP run.

3   Program Complexity

Program complexity could be measured in different ways. For instance, [15] proposed two measures. The first is to equate complexity with tree size, based on the assumption that bloated programs are also complex. The second addresses the complexity of the program output or functional complexity, measured as the degree of the Chebyshev polynomial that approximates the output. In [15] complexity is studied as an objective that should be optimized, not as an indicator of program overfitting, which could be studied in future work.

3.1   Functional Complexity

If we are describing the complexity of a program, focusing on the functional output seems to be the best approach for several reasons. First, the size and shape of a program might be misleading if many nodes are introns, in which case they have no bearing on program output. Moreover, even a large program might be simplified and expressed as a compact expression. The size of the program influences the dynamics of the evolutionary process, but tells us little with respect to the output of each program. Secondly, the genotype-phenotype distinction is not clear in GP, where an explicit phenotype is normally not defined [5]. Therefore, it is easier to focus on the functional output of a program in order to describe its behavior. Here, we describe two measures of functional complexity. We begin with the SFC measure proposed in [14], and afterwards present our proposed RFC measure. It is important to state that despite the differences between SFC and RFC, it should become clear that both are concerned with measuring the same underlying characteristic of functional behavior.

Slope-based Functional Complexity. The complexity measure proposed in [14] is inspired by the concept of function curvature. Curvature is the amount by which the geometric representation of a function deviates from being flat or straight. In [14], the authors correctly point out that a program tree with a response that tends to be flat, or smooth, should be characterized as having a low functional complexity, and that such a function will tend to have a lower curvature at each sampled point. Conversely, a complex functional output would exhibit a larger amount of curvature at each point. This basic intuition seems reasonable, but computing a measure of curvature is not trivial [7], especially for multidimensional real-world problems. In order to overcome this, [14] proposes a heuristic measure based on the slope of the line segments that join each pair of adjacent points. Consider a bi-dimensional regression problem, for which the graphical representation of the output of a program is a polyline. This polyline is produced by plotting the points that have fitness cases in the abscissa, the corresponding values of the program output as ordinates, and sorting the points based on the fitness cases. Then, these points are joined by straight line segments and the slope of each segment is computed. The SFC is calculated by summing the differences of adjacent line segments. If the slopes are identical, the value of the measure is zero, thus a low complexity score. Conversely, if the sign of the slope of all adjacent segments changes then complexity is high.

The intuition behind SFC seems to be a reasonable first approximation of a complexity measure that could serve as a predictor of GP overfitting. However, the formal definition of SFC can become complicated for multidimensional problems and require a somewhat heuristic implementation for the following reasons [14]. First, the SFC considers each problem dimensions independently by projecting the output of a function onto each dimension and obtaining a complexity measure for each dimension separately. In [14] the final SFC was the average of all such measures, while in this paper we test the average, median, max and min values. The decision to consider each dimension separately might be problematic since the complexity of a function in a multidimensional space could be different from the complexity of the function as projected onto each individual dimension. Second, by employing such a strategy, the data might become inconsistent. Namely, if we order the data based on a single dimension then the case might arise in which multiple values at the ordinates (program output) correspond with a single value on the abscissa (fitness cases). In such a case a slope measure cannot be computed, and an ad-hoc heuristic is required.

Regularity-based Functional Complexity. Following the basic intuition behind the SFC measure, we propose a similar measure of complexity that focuses on the same functional behavior. Another way to interpret the output we expect from an overfitted function, is to use the concept of Hölderian regularity. In the areas of signal processing and image analysis it is quite clear that the most prominent and informative regions of a signal correspond with those portions where signal variation is highest [13]. It is therefore useful to be able to characterize the amount of variation, or regularity, that is present at each point on signal. One way to accomplish this is to use the concept of Hölderian regularity, which characterizes the singular or irregular signal patterns [11]. The regularity of a signal at each point can be quantified using the pointwise Hölder exponent.

Definition 1: Let $f : \mathbb{R} \to \mathbb{R}$, $s \in \mathbb{R}^{+*} \setminus \mathbb{N}$ and $x_0 \in \mathbb{R}$. $f \in C^s(x_0)$ if and only if $\exists \eta \in \mathbb{R}^{+*}$, and a polynomial P_n of degree $n < s$ and a constant c such that

$$\forall x \in B(x_0, \eta), |f(x) - P_n(x - x_0)| \leq c|x - x_0|^s , \qquad (1)$$

where $B(x_0, \eta)$ is the local neighborhood around x_0 with a radius η. The pointwise Hölder exponent of f at x_0 is $\alpha_p(x_0) = \sup_s \{f \in C^s(x_0)\}$.

In the above definition, P_n represents the Taylor series approximation of function f. When a singularity (or irregularity) is present at x_0 then f is non-differentiable at that point, and P_n represents the function itself. In this case, Eq. 1 describes a bound on the amount by which a signal varies, or oscillates, around point x_0 within an arbitrary local neighborhood $B(x_0, \eta)$. Hence, when the singularity is large at x_0, with a large variation of the signal, then $\alpha_p \to 0$ as $x \to x_0$. Conversely, $\alpha_p \to 1$ when the variation of the signal $(f(x) - P_n) \to 0$ as $x \to x_0$, the point is smoother or more regular. Figure 1 shows the envelope that bounds the oscillations of f expressed by the Hölder exponent α_p at x_0.

The basic intuition of SFC is that a complex functional output should exhibit an irregular behavior (high curvature). Conversely, a simple function should

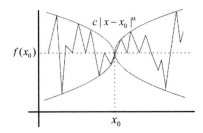

Fig. 1. Hölderian envelope of signal f at point x_0

produce a smoother, or more regular, output. Therefore, we propose to use the Hölder exponent as the basis for a complexity measure which we call Regularity-based Functional Complexity, or RFC. The idea is to compute the Hölder exponent at each fitness case and use these measures to characterize the overall regularity of the functional output. The final RFC measure could be a variety of statistics computed from the set of Hölder exponents obtained from each fitness case; here, as for SFC, we test the average, median, max and min values. There are several estimators for the Hölder exponent [1,12]; however, here we use the oscillations method, which can be derived directly from Definition 1 [11].

Estimation of the pointwise Hölder exponent through oscillations. The Hölder exponent of function $f(x)$ at point x_0 is the $sup(\alpha_p) \in [0,1]$, for which a constant c exists such that $\forall x'$ in a neighborhood of x_0,

$$|f(x_0) - f(x')| \leq c|x_0 - x'|^{\alpha_p} . \tag{2}$$

In terms of signal oscillations, a function $f(x)$ is Hölderian with exponent $\alpha_p \in [0,1]$ at x_0 if $\exists c \; \forall \tau$ such that $osc_\tau(t) \leq c\tau^{\alpha_p}$, with

$$osc_\tau(x_0) = \sup_{x',x'' \in [x_0-\tau, x_0+\tau]} |f(x') - f(x'')| . \tag{3}$$

Now, since $x' = x_0 + h$ in Eq. 2, we can also write that

$$\alpha_p(x_0) = \liminf_{h \to 0} \frac{\log |f(x_0 + h) - f(x_0)|}{\log |h|} . \tag{4}$$

Therefore, the problem is that of finding an α_p that satisfies 2 and 3, and in order to simplify this process we can set $\tau = \beta^r$. Then, we can write $osc_\tau \approx c\tau_p^\alpha = \beta^{(\alpha_p r + b)}$, which is equivalent to $log_\beta(osc_\tau) \approx \alpha_p r + b$.

Therefore, an estimation of the regularity can be built at each point by computing the slope of the regression between the logarithm of the oscillations osc_τ and the logarithm of the dimension of the neighborhood at which the oscillations τ are computed. Here, we use least squares regression to compute the slope, with $\beta = 2.1$ and $r = 0.1, 0.2, \ldots, 9$. Also, it is preferable not to use all sizes of neighborhoods between two values τ_{min} and τ_{max}. Hence, we calculate the

Table 1. GP parameters used for real world experiments

Parameter	Description
Population size	500 individuals
Iterations	100 generations
Initialization	Ramped Half-and-Half
Crossover probability	$p_c = 0.5$
Mutation probability	$p_\mu = 0.5$
Initial max. depth	Six levels
Selection	Lexicographic Parsimony Tournament
Survival	Elitism
Runs	30

oscillation at point x_0 only on intervals of the form $[x_0 - \tau_r : x_0 + \tau_r]$. For a function in $\mathbb{R}^D$, τ_r is defines a hyper-volume around x_0, such that $d(x', x_0) \leq \tau_r$ and $d(x'', x_0) \leq \tau_r$, where $d(a, b)$ is the Euclidean distance between a and b.

The implicit difficulties of applying SFC are not an issue for RFC, since it is not necessary to consider each problem dimension separately. Moreover, unlike the concept of curvature, estimating signal regularity does not require derivatives in a multidimensional space. Therefore, the RFC provides a straightforward measure of the intuitive functional behavior that SFC attempts to capture.

4 Experiments

To test both SFC and RFC, we use the same two real world problems as [14]. They consist on predicting the value of two pharmacokinetic parameters for a set of candidate drug compounds based on their molecular structure. The first data set relates to the parameter of human oral bioavailability, and contains 260 samples in a 241 dimensional space. The second data set relates to the median lethal those of a drug, or toxicity, and contains 234 samples from a 626 dimensional space. The experimental setup for our GP is summarized in Table 1. Most parameters are the same as those in [14], except for the probabilities of each genetic operator. The reason for this is that we want a GP system that does not bloat, however the Operator Equalisation technique used in [14] was considered to be too computationally costly. Therefore, we combine simple one-point crossover with standard subtree mutation in order to avoid code bloat and still achieve good fitness scores. A bloat-free GP is used because we want to confirm that the causal relationship between bloat and overfitting is not a general one.

In Section 3 we presented SFC and RFC, however there is one more important consideration that must be accounted for before computing these scores. Since we are to compute a complexity score for any program tree, it will sometimes be the case that a particular program will use only a fraction of the total dimensions (expressed as terminals) of the problem. Therefore, a decision must be made regarding which dimensions to include when either SFC or RFC are computed. In this case we only use those dimensions that are part of a programs computations. Finally, we point out that all of our code was developed using Matlab 2009a, using the GPLAB toolbox for our GP system [9], and the FracLab toolbox to compute the Hölder exponents of the RFC measure [4].

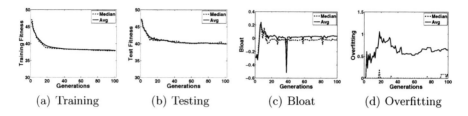

(a) Training   (b) Testing   (c) Bloat   (d) Overfitting

Fig. 2. Evolution of the bioavailability problem

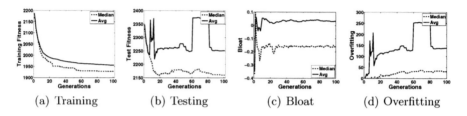

(a) Training   (b) Testing   (c) Bloat   (d) Overfitting

Fig. 3. Evolution of the toxicity problem

Results. We begin by presenting the results obtained for training fitness, test fitness, bloat and overfitting for the bioavailability and toxicity problems, shown in figures 2 and 3 respectively. In each case we present the median and average performance over all runs. There are several noticeable patterns in these plots. First, in both cases we can see that our GP systems converges to good fitness scores [14]. Second, the GP system does not incur bloat, in fact the median values in both problems fall below zero, which means that the average population size actually decreased with respect to the initial population. This is an interesting result, where bloat is eliminated from our runs using a simple strategy. Third, we see that in both problems there is only a small amount of overfitting based on the median scores, with values much lower than those published in [14]. However, the average overfitting is substantially larger than the median value, this indicates that our runs produced outliers.

The evolution of the complexity measures are presented in figures 4 and 5 for the bioavailability and toxicity problems respectively. Each plot considers a different representative statistics for each measure, as stated in Section 3, and shows the median value of this statistics over all the runs. However, it is not possible to draw a clear conclusion regarding the relationship, or correlation, between complexity and the overfitting curves shown in figures 2(d) and 3(d).

Therefore, scatter plots of complexity and overfitting are shown in figures 6 and 7 for the bioavailability and toxicity problems respectively. These plots take the best individual from each generation and each run, and their overfitting and complexity scores as ordered pairs. Additionally, each plot shows the Pearson's correlation coefficient ρ between each complexity score and overfitting. For analysis, we follow the empirical estimates of [2] and consider values of $\rho > 0.15$

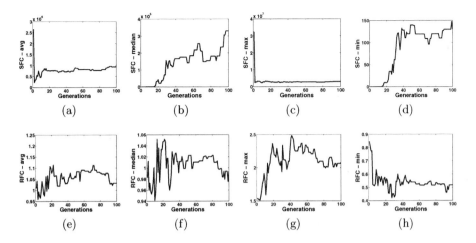

Fig. 4. Evolution of both complexity measures for the bioavailabilty problem. (a)-(d) correspond to the SFC measure and show the average, median, max and min of the SFC values of all problem dimensions. (a)-(d) correspond to the RFC measure and show the average, median, max and min of all the Hölder exponents computed for each fitness case. All plots show the median value obtained over all 30 runs.

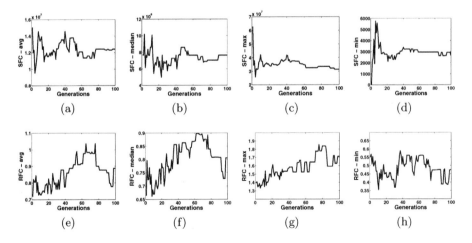

Fig. 5. Evolution of both complexity measures for the toxicity problem. (a)-(d) correspond to the SFC measure and show the average, median, max and min of the SFC values of all problem dimensions. (a)-(d) correspond to the RFC measure and show the average, median, max and min of all the Hölder exponents computed for each fitness case. All plots show the median value obtained over all 30 runs.

as an indication of positive correlation, $\rho < -0.15$ as negative correlation, and $-0.15 \leq \rho \leq 0.15$ as an indication of no correlation. It is important to point out that SFC is expected to have a positive correlation with overfitting while RFC should have a negative one given the definition of each. Most tests show

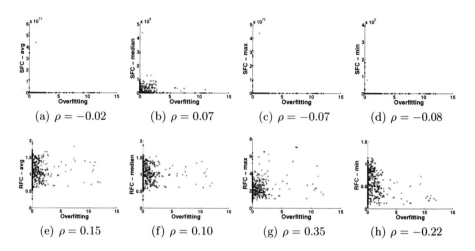

Fig. 6. Scatter plots of both complexity measures for the bioavailabitly problem. (a)-
(d) correspond to the SFC measure and show the average, median, max and min.
(a)-(d) correspond to the RFC measure and show the average, median, max and min.
All plots also show Pearson's correlation coefficient ρ.

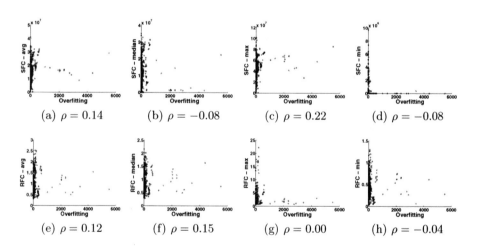

Fig. 7. Scatter plots of both complexity measures for the toxicity problem. (a)-(d)
correspond to the SFC measure and show the average, median, max and min. (a)-(d)
correspond to the RFC measure and show the average, median, max and min. All plots
also show Pearson's correlation coefficient ρ.

and absence of correlation, with only SFC-max achieving the expected result on
toxicity, and RFC-min on bioavailability.

Since the best case scenario would be to have a complexity measure that predicts when a program is overfitted, then it is more informative to discard those
individuals with low overfitting, and focus on individuals with a large amount

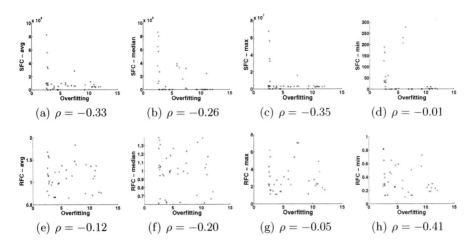

Fig. 8. Scatter plots of both complexity measures for the bioavailabitly problem with the 5% of individuals that have the highest overfitting scores. (a)-(d) SFC measure, showing the average, median, max and min. (a)-(d) RFC measure, showing the average, median, max and min. All plots also show Pearson's correlation coefficient ρ.

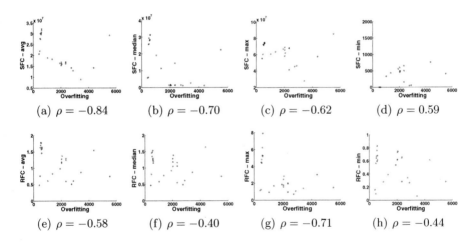

Fig. 9. Scatter plots of both complexity measures for the toxicity problem with the 5% of individuals that have the highest overfitting scores. (a)-(d) SFC measure, showing the average, median, max and min. (a)-(d) RFC measure, showing the average, median, max and min. All plots also show Pearson's correlation coefficient ρ.

of overfitting. In other words, if a complexity measure cannot be used as an indicator of extreme cases of overfitting, then its usefulness would be quite low. Therefore, in figures 8 and 9 we show the same scatter plots as before, however in this case we only present the 5% of individuals with the highest amount of overfitting. First, consider the results for the SFC measure. In most cases, SFC shows a negative correlation with overfitting on both problems, the exact opposite of what we expected. It is only SFC-min that shows the expected correlation on the toxicity problem. Second, for RFC on the bioavailability problem, only RFC-median and RFC-min show the expected negative correlation. However, given the low overfitting on this problem, we should only expect to detect a small amount of correlation in the best case scenario. On the other hand, for the toxicity problem all RFC variants show the expected correlation. Hence, in this case for the higher overfitting values (the top 5%), the RFC measure could provide an indication of program overfitting. It appears that the above results are not very strong, given low correlation values achieved even in the best cases. Nonetheless, we would like to stress the difficulty of the problem at hand. In both test cases we have a small sample of data that lies within a highly multidimensional space. Therefore, extracting a proper description of program behavior, and thus functional complexity, is not trivial. For instance, the RFC measures relies on the pointwise Hölder exponent, and in order to estimate it at any point it is necessary to consider how the function oscillates within a series of progressively larger and concentric local neighborhoods. However, in both of our test cases it becomes extremely difficult to correctly estimate the oscillations of the function, and sometimes it cannot be done, because of the sparseness of the data. One possible way to solve this problem is to generate or interpolate the missing data in a way that preserves the regularity of the function [3], however this is left as future work. In summary, if any measure can produce, even a slight indication of overfitting, this should suggest that we are proceeding in a promising direction.

5   Summary and Concluding Remarks

This work presents an empirical study of functional complexity in GP. It is commonly stated that simpler solutions to a problem should be preferred because they are less likely to be overfitted to the training examples. In this work, we study two complexity measures, Slope-based Functional Complexity [14], and Regularity-based Functional Complexity derived from the concept of Hölderian regularity. We measure the evolution of SFC and RFC on a bloat-free GP system using two real-world problems, in order to evaluate if they can serve as indicators of overfitting during a GP run. Overall, the results show almost no correlation between both complexity measures and program overfitting. However, when we consider only highly overfitted solutions, then the RFC measures does appear to provide a somewhat useful indicator of overfitting. On the other hand, the SFC measure fails to achieve any useful correlation with program overfitting, and in some cases it produces the opposite of what is expected. Nonetheless, it is important to stress that this should be taken as an initial, and partial, empirical study of functional complexity in GP. Therefore, further research should focus on a more comprehensive evaluation of these, and possibly other, measures of complexity as indicators of GP overfitting.

Acknowledgements. This work was partially supported by FCT (INESC-ID multiannual funding) through the PIDDAC Program funds. Sara Silva and Leonardo Vanneschi acknowledge project PTDC/EIA-CCO/103363/2008 from FCT, Portugal.

References

1. Jaffard, S.: Wavelet techniques in multifractal analysis. In: Fractal Geometry and Applications: A Jubilee of Benoit Mandelbrot, Proceedings of Symposium in Pure Mathematics, vol. 72, pp. 91–151 (2004)
2. Jones, T., Forrest, S.: Fitness distance correlation as a measure of problem difficulty for genetic algorithms. In: Proceedings of the Sixth International Conference on Genetic Algorithms, pp. 184–192. Morgan Kaufmann, San Francisco (1995)
3. Legrand, P., Lévy-Véhel, J.: Local regularity-based interpolation. In: WAVELET X, Part of SPIE's Symposium on Optical Science and Technology, vol. 5207 (2003)
4. Lévy-Véhel, J., Legrand, P.: Signal and Image Processing with FRACLAB. In: Thinking in Patterns, pp. 321–322 (2004),
   `http://fraclab.saclay.inria.fr/homepage.html`
5. McDermott, J., Galvan-Lopez, E., O'Neill, M.: A fine-grained view of GP locality with binary decision diagrams as ant phenotypes. In: Schaefer, R., Cotta, C., Kołodziej, J., Rudolph, G. (eds.) PPSN XI. LNCS, vol. 6238, pp. 164–173. Springer, Heidelberg (2010)
6. Mitchell, T.: Machine Learning. McGraw-Hill, New York (1997)
7. Morvan, J.-M.: Generalized Curvatures, 1st edn. Springer, Heidelberg (2008)
8. Rosca, J.: Generality versus size in genetic programming. In: Proceedings of Genetic Programming 1996, pp. 381–387. MIT Press, Cambridge (1996)
9. Silva, S., Almeida, J.: Gplab: A genetic programming toolbox for matlab. In: Proceedings of the Nordic MATLAB Conference, pp. 273–278 (2003)
10. Silva, S., Costa, E.: Dynamic limits for bloat control in genetic programming and a review of past and current bloat theories. Genetic Programming and Evolvable Machines 10(2), 141–179 (2009)
11. Tricot, C.: Curves and Fractal Dimension. Springer, Heidelberg (1995)
12. Trujillo, L., Legrand, P., Lévy-Véhel, J.: The estimation of hölderian regularity using genetic programming. In: Proceedings of GECCO 2010, pp. 861–868. ACM, New York (2010)
13. Trujillo, L., Legrand, P., Olague, G., Pérez, C.: Optimization of the hölder image descriptor using a genetic algorithm. In: Proceedings of GECCO 2010, pp. 1147–1154. ACM, New York (2010)
14. Vanneschi, L., Castelli, M., Silva, S.: Measuring bloat, overfitting and functional complexity in genetic programming. In: Proceedings of GECCO 2010, pp. 877–884. ACM, New York (2010)
15. Vladislavleva, E.J., Smits, G.F., Den Hertog, D.: Order of nonlinearity as a complexity measure for models generated by symbolic regression via pareto genetic programming. IEEE Trans. Evol. Comp. 13, 333–349 (2009)

Estimating Classifier Performance with Genetic Programming

Leonardo Trujillo, Yuliana Martínez, and Patricia Melin

Instituto Tecnológico de Tijuana, Av. Tecnológico S/N, Tijuana, BC, México
{leonardo.trujillo.ttl,ysaraimr}@gmail.com, pmelin@tectijuana.edu.mx

Abstract. A fundamental task that must be addressed before classifying a set of data, is that of choosing the proper classification method. In other words, a researcher must infer which classifier will achieve the best performance on the classification problem in order to make a reasoned choice. This task is not trivial, and it is mostly resolved based on personal experience and individual preferences. This paper presents a methodological approach to produce estimators of classifier performance, based on descriptive measures of the problem data. The proposal is to use Genetic Programming (GP) to evolve mathematical operators that take as input descriptors of the problem data, and output the expected error that a particular classifier might achieve if it is used to classify the data. Experimental tests show that GP can produce accurate estimators of classifier performance, by evaluating our approach on a large set of 500 two-class problems of multimodal data, using a neural network for classification. The results suggest that the GP approach could provide a tool that helps researchers make a reasoned decision regarding the applicability of a classifier to a particular problem.

1 Introduction

Classification is one of the most common problems within many fields of research, such as data mining, pattern recognition and image analysis. When confronted with a classification problem, a researcher faces an obvious methodological decision, he (or she) must choose which classifier to use. In order to do so, it is necessary to assess the main characteristics of the data, and from this infer the expected performance that each classifier at his disposal might achieve. Afterward, it is then possible to make a reasoned choice. However, straightforward algorithmic solutions do not exist for this basic task. Hence, it is normally solved using trial and error, a process that depends on the personal background and preferences of the researcher.

It is clear, however, that an algorithmic solution to this problem is required if truly autonomous learning systems are to be developed. Indeed, several works directly address such issues by analyzing problem complexity [2], proposing meta-learning approaches [7,14] and determining a classifiers domain of competence [5]. In fact, such a problem is not restricted to the domain of data classification,

S. Silva et al. (Eds.): EuroGP 2011, LNCS 6621, pp. 274–285, 2011.
© Springer-Verlag Berlin Heidelberg 2011

with researchers in the field of search and optimization also studying similar issues [3,16,11]. The present work addresses the task of inferring the performance of particular classifier, an artificial neural network (ANN), by analyzing specific features of the data that will be classified.

We formulate this goal as an optimization problem and solve it using Genetic Programming (GP). The system takes as input several descriptors of the problem data, based on statistical and complexity measures [2,14]. From this, the GP searches for programs that are able to estimate the expected error of the ANN classifier on the given data. The experimental results show that GP evolves accurate performance estimators, based on an extensive validation using 500 two-class problems of multi-modal data. Moreover, our approach could be extended to produce estimators for a variety of classifiers, and thus simplify the process by which a researcher chooses a classification method.

The remainder of the paper proceeds as follows. In Section 2 we review related work. Then, Section 3 provides a precise statement of the problem we study. Section 4 outlines the proposed GP solution and Section 5 presents the experimental results. Finally, Section 6 contains our concluding remarks.

2   Related Work

Over the last decade, it has become increasingly clear that every algorithm or computational techniques has a bounded domain of competence; i.e., no paradigm should be considered as an optimal approach for all possible problem instances [17]. This general idea has become firmly grounded on the theoretical arguments of the No-Free-Lunch Theorem (NFL) [18]. Therefore, when confronted with a new problem instance a researcher must choose the best conceptual and computational tools to develop an algorithmic solution based on the particular characteristics of the problem. In order to automate this task, some researchers have turned to a meta-analysis of common computational problems, such as search [9] and classification [7,14]. This perspective offers promising lines of future development because under certain conditions the NFL does not apply to meta-learning approaches, such as GP [9,10].

With respect to classification, the empirical behavior of an individual classifier will depend on the characteristics of data that needs to be classified. However, only a limited amount of current literature attempts to extract the underlying relationship that exists between the performance of the classifier and the characteristics of the data. One of the first attempts was the STATLOG project [7], which developed a meta-level binary rule that determines the applicability of a classifier to a given problem. In [14], on the other hand, a statistical meta-model was developed that attempts to predict the expected classification performance of several common classifiers on a small set of real-world problems. The above examples perform a meta-analysis of classification problems using statistical descriptors of the datasets. However, both use a small set of problems, only 22 in total, which makes any conclusions based on their results quite limited in scope. Moreover, the predictive accuracy of these approaches has been criticized because they do not consider the geometric relationships amongst the data, a more relevant characteristic to classifier accuracy [2]. On the other hand, [2] provides a detailed analysis of several types of complexity measures directly related to classification difficulty. However, [2] does not attempt to use these measures to estimate classifier performance.

A meta-analysis that attempts to predict the performance of an algorithm on a particular problem has also become common in recent GP research. For instance, some have attempted to predict the difficulty of a search problem using fitness-landscape analysis [3,16,11] and locality-based approaches [6].

3   Problem Statement

In this paper, the goal is to estimate the performance of a classifier on a particular problem. In order to bound our study, we focus on ANN classifiers; in particular, a fully connected multi-layer perceptron (MLP), because it has proven to be a powerful classification tool in various domains [1,8].

First, we build a set of synthetic 2-class multimodal problems, call this set P. It is important to realize that we are performing a meta-analysis of classifier performance, where each sample case is an individual classification problem with its respective data samples. Second, from each problem $p \in P$ we extract a vector of descriptive features β, with the goal of capturing problem specific characteristics. For this, we build each β using statistical measures [14] and measures of problem complexity [2] that have proven to be useful in related research (these descriptors are defined in Section 4.1). Third, we perform en exhaustive search in order to determine what is the best performance that a MLP with one hidden layer can achieve on each problem p, characterized by the classification error ϵ. After which, we have a set of problems P, where each $p \in P$ is described by a descriptive vector β, as well as the lowest classification error ϵ achieved by the MLP on p. Therefore, the problem we pose is that of finding an optimal estimator of classifier performance K^o, such that

$$K^o = \arg\min_{K} \{Err[K(\beta), \epsilon]\} \ \forall p \in P \tag{1}$$

where $Err[,]$ is an error function, in this work the root-mean-square error (RMSE).

4   GP Approach

The solution we propose for the above optimization problem is to use a GP system, and in this section we present its search space and fitness function.

4.1   Search Space

The search space is established by the sets of terminals T and functions F. For the terminal set, we use several descriptive measures that can be extracted from the problem data, these measures are concatenated in order to construct the base descriptive vector β for each problem p. We consider two groups of measures, statistical measures and complexity measures. The statistical measures are a subset of those proposed in [14], those that produced the best results in their meta classification problem. These are basic statistics computed for each feature of the problem data; they are:

- The geometric mean ratio of the pooled standard deviations to standard deviations of the individual population (SD).
- The mean absolute correlation coefficients between two features (CORR) and its squared value (CORR2).
- The mean skewness of features (SKEW).
- The mean kurtosis of features (KURT).
- The average entropy of features (HX) and its squared value (HX2).

Additionally, our terminal set also includes various measures of problem complexity that directly consider the geometry of the distribution of data samples over the feature space. All of these features are a subset of those proposed in [2], and are defined as follows.

Fisher's discriminant ratio (FD). The discriminant ratio proposed by Fisher is defined as

$$f = \frac{(\mu_1 - \mu_2)^2}{\sigma_1^2 + \sigma_2^2} , \tag{2}$$

where μ_1, μ_2, σ_1, σ_2 are the means and standard deviations of the two classes, respectively. This measure is computed independently for each feature of the data. Therefore, for a multidimensional problem the maximum f over all feature dimensions is chosen as the representative value FD [2].

Volume of overlap region (VOR). This measure provides an estimate of the amount of overlap between both classes in feature space. This measure is computed by finding, for each feature, the maximum and minimum value of each class and then calculating the length of the overlap region. The length obtained from each feature can then be multiplied in order to obtain a measure of volume overlap, see Figure 1(a). VOR is zero when there is at least one dimension in which the two classes do not overlap.

Feature efficiency (FE). It is a measure of how much each feature contributes to the separation of both classes. When there is a region of overlap between two classes on a feature dimension, then data is considered ambiguous over that region along that dimension. However, it is possible to progressively remove the ambiguity between both classes by separating those points that lie inside the overlapping region, see Figure 1(b). The efficiency of each feature is defined as the fraction of all reaming points separable by that feature, and the maximum feature efficiency (FE) is taken as the representative value.

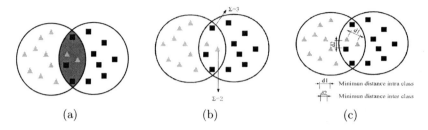

Fig. 1. Graphical representations of measures of problem complexity. (a) Shows the volume of class overlap VOR. (b) Illustrates the concept of feature efficiency FE, with the number of sample points within the overlap region. (c) Illustrates the concept of class distance ratio CDR, with the minimum inter and intra class distance.

Class distance ratio (CDR). It is a measure that compares the dispersion within classes to the gap between classes [2]; it is computed as follows. First, for each data sample the Euclidean distance to its nearest-neighbor is computed, within and outside its own class, see Figure 1(c). Second, the average of all intraclass and interclass nearest-neighbor distances is computed. Finally, CDR is the ratio of these two values.

In summary, the eleven descriptive measures described above are included in the terminal set T,

$$T = \{SD, CORR, CORR2, SKEW, KURT, HX, HX2, FD, VOR, FE, CDR\} \ . \tag{3}$$

The function set F, on the other hand, contains common primitive functions used for symbolic regression problems in GP research; these are

$$F = \left\{+, -, *, /, \sqrt{\cdot}, sin, cos, log, x^y\right\} \ , \tag{4}$$

with protected functions for $/$, log, $\sqrt{\cdot}$ and x^y.

4.2   Fitness Function

The goal of the GP search is to find a solution that satisfies the optimization problem of Equation 1. Therefore, fitness is posed as a cost function, using the RMSE computed on a set of n training samples, given by

$$f(K) = \sqrt{\frac{\sum_{i=1}^{n}(K(\beta_i) - \epsilon_i)^2}{n}} \ , \tag{5}$$

where β_i is the descriptive vector of problem p_i, and ϵ_i the expected classification error.

5   Experiments

5.1   Experimental Setup and Data-Sets

The experimental setup follows the same general sequence described in Section 3. First, we randomly generate 500 two-class classification problems using Gaussian mixture models (GMMs), see Figure 2. By using GMMs, we can generate unimodal or multimodal classes, thus producing a set of problems with a varied degree of difficulty. All problems were set in the $\mathbb{R}^2$ plane, with $x, y \in [-10, 10]$, and 200 sample points were randomly generated for each class. The parameters for the GMM of each class were also randomly chosen within the following ranges of values: (a) for the number of Gaussian components $\{1, 2, 3\}$; (b) for the median of each Gaussian component $[-3, 3]$; (c) for each element of the co-variant matrix of each Gaussian component $(0, 2]$; (d) for the rotation angle of the covariance matrix $[0, 2\pi]$; and (e) the proportion of sample points from each class generated with each Gaussian component $[0, 1]$.

Second, for each problem we compute its corresponding set of descriptive features β as defined in Section 4. Third, for every problem we find the best performance achieved by an ANN using an exhaustive search over various network topologies. In this step, we use a fully connected feed-forward MLP with one hidden layer, hyperbolic tangent sigmoid transfer functions, and trained using backpropagation with adaptive learning rate and momentum, with an initial learning rate of 0.5, a momentum coefficient of 0.9 and 50 training epochs. We test different network topologies in search of the one that achieves the best performance by varying the number of nodes in the hidden layer within the range of $[1, 30]$. In each case, at different topologies (number of hidden nodes), we determine the performance of the ANN classifier using 3-fold cross-validation, and perform 30 independent runs that gives a total of 90 (30×3) performance estimates. For each topology we take the median performance as the representative value and then select the best performance among all topologies. In this way, we establish the ground truth for the GP search regarding the expected performance of the ANN classier on each problem.

In Figure 2, we show six of the randomly generated classification problems, each with a different expected performance expressed by the classification error ϵ. Notice, by visual inspection, that the expected performance of the ANN classifier appears to be related to the amount of overlap between both classes and the compactness of each, as should be expected [2].

In order to explore the above qualitative observation, in Figure 3 we present scatter plots between all of the eleven descriptive measures and the expected classification error on each problem, in order to illustrate the relationship between them. Additionally, the figure also presents the value of Pearson's correlation coefficient ρ in each case. From these plots we can see that some measures are more strongly correlated with classification error than others. For instance, FE, VOR, CDR and SD show the best correlation, while others like SKEW and CORR have a correlation coefficient close to zero. However, even the best correlation values are in fact quite low, except for FE (see Section 4.1 for a description of each measure).

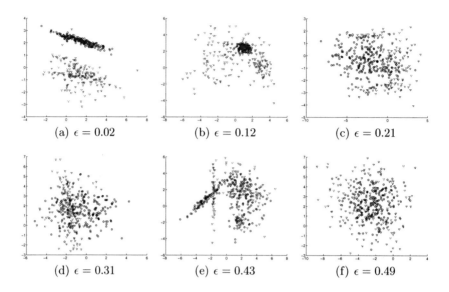

(a) $\epsilon = 0.02$ (b) $\epsilon = 0.12$ (c) $\epsilon = 0.21$

(d) $\epsilon = 0.31$ (e) $\epsilon = 0.43$ (f) $\epsilon = 0.49$

Fig. 2. Scatter plots of six different classification problems. In each case we indicate the expected best performance of the ANN classifier.

Therefore, it appears that the relationship between our descriptors and classifier performance, if any, will be non-linear, and an accurate estimator will probably need to consider several measures simultaneously.

The parameters of the GP system are presented in Table 1. We employ a standard Koza-style GP with bloat control using the dynamic depths method proposed in [13] and lexicographic parsimony pressure during tournament selection [4]. The GP was executed 30 times, thus the results presented in the following subsection are statistics computed over all of the runs. In each GP run, seventy percent of the 500 classification problems were used for fitness evaluation (training) and the rest was used as an independent test set. The training and test sets were randomly and independently chosen at the beginning of each run. Finally, all of our code was developed using Matlab 2009a, and the GP algorithm uses the GPLAB toolbox [12][1].

5.2   Results and Comparisons

Figure 4 presents the results of the evolutionary search. In Figure 4(a), we present the evolution of the best individual fitness, as well as the fitness of the best individual computed with the test set, showing median values over all the runs. From this plot we can see that the training process is successful, with training and testing fitness being practically equivalent across all of the generations. Moreover, both reach low values, which corresponds with a low predictive error. Figure 4(b) shows the amount of bloat incurred by the GP. We compute bloat

---

[1] GPLAB downloaded from http://gplab.sourceforge.net/

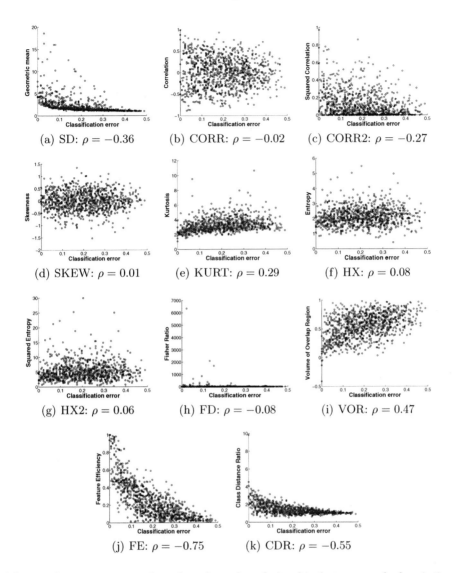

Fig. 3. These are scatter plots that show the relationship between each descriptive measure of problem data (y-axis) and classification error (x-axis). The legend of the plots also show the value of Pearson's correlation coefficient ρ.

using the measure proposed in [15], the only measure in current literature that quantifies bloat based on the increase in program size and fitness relative to that of the initial population. These results show that the bloat control strategies (dynamic limits and lexicographic parsimony pressure) included in the GP system fail to eliminate bloat from our runs.

One of the possible advantages of using GP to build symbolic expressions is that it implicitly performs a feature selection step by deciding which terminal

Table 1. GP parameters used in our runs

Parameter	Description
Population size	500 individuals.
Generations	200 generations.
Initialization	Ramped Half-and-Half, with 10 levels of maximum depth.
Genetic operator probabilities.	Crossover $p_c = 0.8$; Mutation $p_\mu = 0.2$.
Bloat control	Dynamic depth control.
Maximum dynamic depth	11 levels.
Hard maximum depth	30 levels.
Selection	Lexicographic parsimony pressure.
Survival	Keep best elitism.
Runs	30

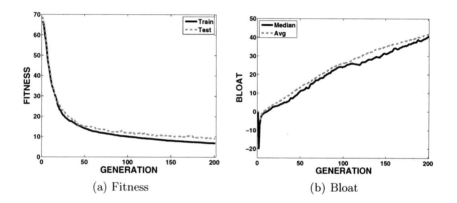

(a) Fitness   (b) Bloat

Fig. 4. The evolution statistics computed over all thirty runs. (a) Shows the median of the best training fitness and median of the test fitness for the best individual at each generation. (b) Shows the median and average of bloat over the thirty runs.

elements to include. In our case, it is informative to see which descriptive measures are chosen by the GP, and evaluate if they correspond with those suggested in Figure 3. Therefore, we take the best individual from each run and inspect which terminals are included in the program tree. Figure 5(a) summarizes these results in a bar plot that shows the proportion of individuals that contain each terminal, where a proportion of 1 means that a given terminal appears in the best individual from every run. For instance, CDR appears in all of the best individuals, while SKEW only appears in 70% of them. Indeed, when we compare these results with the correlation plots of Figure 3, we have some agreement. For example, FE and CDR, obtain the highest correlation values and are also widely used as terminal elements. Moreover, SKEW and CORR have the lowest correlation and appear less frequently in the best solutions from each run. However, some terminals, such as HX, HX2 and FD, appear very frequently despite their low correlation scores. In fact, most individuals use a majority, if not all, of the available terminal elements to estimate classifier performance. Therefore, it appears that linear correlation gives only a limited indication regarding which terminal elements are the most useful to predict classifier performance. Yet the above observation might be misleading, given that our GP runs are bloated, as seen in Figure 4(b). Since the best individuals are bloated program trees, then it is reasonable to assume that a large number of the terminals they contain are intron nodes. Hence, it is still not easy to assess which terminal elements provide the most relevant information to predict classifier performance.

Table 2 summarizes the performance of the GP estimators showing the average, median and standard deviation of the RMSE (prediction error) achieved on the test sets from each run. In summary, the estimators produced by GP can predict the performance of the ANN classifier with a median prediction error of 0.09, and an average of 0.10. While these estimates are obviously not perfect, they do provide useful bounds regarding the appropriateness of the classifier to a particular problem. These results are difficult to compare with [14], for instance, because in [14] only 19 problems are used to build the meta-model, and because most of those problems can be solved with a low classification error; i.e., the problems are easy. In our tests, on the other hand, we use a large variety of 500 multi-modal problems, which range from the extremely easy to almost random datasets, as seen in Figure 2.

Therefore, in order to compare the performance of the GP estimators, we perform the same thirty training sequences with another estimation method, using the same training and testing sets as those used in each GP run. For these tests we use another MLP, with one hidden layer of 20 neurons, hyperbolic tangent sigmoid transfer functions, trained using backpropagation with adaptive learning rate and momentum with 2000 training epochs. The results of the ANN

Table 2. Statistical comparison of the RMSE on test cases for the GP estimation of classifier performance and an ANN estimator, obtained over thirty independent runs

	Median	Average	Std.
GP	0.0914	0.1027	0.0502
ANN	0.1508	0.2007	0.1105

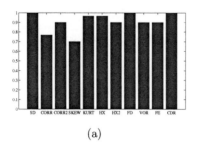

(a)

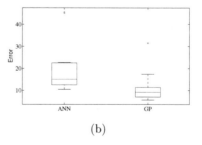

(b)

Fig. 5. (a) Bar plot that shows the proportion of GP runs that produced a final solution which contained each terminal. (b) Boxplot comparison between the GP estimators and an ANN estimator of classifier performance.

estimator are also summarized in Table 2, and a boxplot comparison is shown in Figure 5(b). In both cases we can appreciate that the GP approach achieves the best performance, considering each of the statistics shown. Moreover, in order to validate these results, a Wilcoxon non-parametric rank-sum statistical test was performed with $\alpha = 0.01$, for which the null-hypothesis was rejected with a p value of 0.00.

6   Summary and Concluding Remarks

This paper studies the problem of accurately estimating the performance of a classifier given the data of a particular classification problem. The approach we follow is to extract a set of descriptive measures from the problem data, and use this information as input for an operator produces an estimate of the best performance that the classifier can achieve. The measures we extract are based on statistical properties [14] and on geometric relationships between the problem data [2]. The classifier we study is a feed-forward fully connected neural network, a common tool for data classification [1,8].

In order to generate estimators of classifier performance, we pose an optimization problem and solve it using GP. The experimental results show that GP is capable of evolving accurate estimators for ANN classifiers. Moreover, we believe that the approach could be extend to predict the performance of various classification methods, and in this way simplify the process by which a classifier is chosen for a particular problem instance. However, it not yet clear if our proposed GP approach represents the best possible method to solve this task, a comprehensive comparison with other models, such as basic linear regression or SVM, is required to contextualize the experiments presented here.

Nonetheless, given the encouraging results, we suggest further empirical research based on our work. For instance, while GP did evolve accurate estimators it was not able to produce simple, or compact, symbolic expressions. Such expressions are desirable, since they could show which descriptive measures are most relevant, and illustrate how they ought to be combined. However, the GP was hampered by substantial bloat, and only a limited inference could be drawn from the terminal elements chosen in each GP run. Additionally, even though we evaluated our approach using a large set of 500 classification problems, future work will have to consider real-world data in order to assess the limitations of the proposed methodology. Also, the approach should be used to build performance estimators for other classification methods.

Acknowledgements. Second author was supported by scholarship 298654 from Consejo Nacional de Ciencia y Tecnología (CONACyT) of México.

References

1. Cantú-Paz, E., Kamath, C.: An empirical comparison of combinations of evolutionary algorithms and neural networks for classification problems. IEEE Trans. on Syst., Man, and Cyber., Part B 35(5), 915–927 (2005)

2. Ho, T.K., Basu, M.: Complexity measures of supervised classification problems. IEEE Trans. Pattern Anal. Mach. Intell. 24, 289–300 (2002)
3. Hordijk, W.: A measure of landscapes. Evol. Comput. 4, 335–360 (1996)
4. Luke, S., Panait, L.: Lexicographic parsimony pressure. In: Proceedings of GECCO 2002, pp. 829–836. Morgan Kaufmann, San Francisco (2002)
5. Mansilla, E.B., Ho, T.K.: On classifier domains of competence. In: Proceedings of ICPR 2004, vol. 1, pp. 136–139. IEEE Computer Society, Washington, DC, USA (2004)
6. McDermott, J., Galvan-Lopez, E., O'Neill, M.: A fine-grained view of GP locality with binary decision diagrams as ant phenotypes. In: Schaefer, R., Cotta, C., Kołodziej, J., Rudolph, G. (eds.) PPSN XI. LNCS, vol. 6238, pp. 164–173. Springer, Heidelberg (2010)
7. Michie, D., Spiegelhalter, D.J., Taylor, C.C., Campbell, J. (eds.): Machine learning, neural and statistical classification, NJ, USA (1994)
8. Ou, G., Murphey, Y.L.: Multi-class pattern classification using neural networks. Pattern Recogn. 40, 4–18 (2007)
9. Poli, R., Graff, M.: There is a free lunch for hyper-heuristics, genetic programming and computer scientists. In: Vanneschi, L., Gustafson, S., Moraglio, A., De Falco, I., Ebner, M. (eds.) EuroGP 2009. LNCS, vol. 5481, pp. 195–207. Springer, Heidelberg (2009)
10. Poli, R., Graff, M., McPhee, N.F.: Free lunches for function and program induction. In: Proceedings of FOGA 2009, pp. 183–194. ACM, New York (2009)
11. Poli, R., Vanneschi, L.: Fitness-proportional negative slope coefficient as a hardness measure for genetic algorithms. In: Proceedings of GECCO 2007, pp. 1335–1342. ACM, New York (2007)
12. Silva, S., Almeida, J.: Gplab–a genetic programming toolbox for matlab. In: Proceedings of the Nordic MATLAB Conference, pp. 273–278 (2003)
13. Silva, S., Costa, E.: Dynamic limits for bloat control in genetic programming and a review of past and current bloat theories. Genetic Programming and Evolvable Machines 10(2), 141–179 (2009)
14. Sohn, S.Y.: Meta analysis of classification algorithms for pattern recognition. IEEE Trans. Pattern Anal. Mach. Intell. 21, 1137–1144 (1999)
15. Vanneschi, L., Castelli, M., Silva, S.: Measuring bloat, overfitting and functional complexity in genetic programming. In: Proceedings of GECCO 2010, pp. 877–884. ACM, New York (2010)
16. Vanneschi, L., Tomassini, M., Collard, P., Vérel, S., Pirola, Y., Mauri, G.: A comprehensive view of fitness landscapes with neutrality and fitness clouds. In: Ebner, M., O'Neill, M., Ekárt, A., Vanneschi, L., Esparcia-Alcázar, A.I. (eds.) EuroGP 2007. LNCS, vol. 4445, pp. 241–250. Springer, Heidelberg (2007)
17. Whitley, D., Watson, J.: Complexity theory and the no free lunch theorem, ch. 11, pp. 317–339 (2005)
18. Wolpert, D., Macready, W.: No free lunch theorems for optimization. IEEE Trans. Evol. Comput. 1(1), 67–82 (1997)

Investigation of the Performance of Different Mapping Orders for GE on the Max Problem

David Fagan, Miguel Nicolau, Erik Hemberg, Michael O'Neill,
Anthony Brabazon, and Sean McGarraghy

Natural Computing Research & Applications Group
University College Dublin, Ireland
{david.fagan,miguel.nicolau,erik.hemberg}@ucd.ie,
{m.oneill,anthony.brabazon,sean.mcgarraghy}@ucd.ie

Abstract. We present an analysis of how the genotype-phenotype map in Grammatical Evolution (GE) can effect performance on the Max Problem. Earlier studies have demonstrated a performance decrease for Position Independent Grammatical Evolution (πGE) in this problem domain. In πGE the genotype-phenotype map is changed so that the evolutionary algorithm controls not only what the next expansion will be but also the choice of what position in the derivation tree is expanded next. In this study we extend previous work and investigate whether the ability to change the order of expansion is responsible for the performance decrease or if the problem is simply that a certain order of expansion in the genotype-phenotype map is responsible. We conclude that the reduction of performance in the Max problem domain by πGE is rooted in the way the genotype-phenotype map and the genetic operators used with this mapping interact.

1   Introduction

The use of a genotype-phenotype map in Genetic Programming (GP) [13,22] has been used by many within the field [1,5,7,10,11,12,15,23] and a number of variants to the standard tree-based form of GP exist, amongst which some of the most popular are Linear GP [2], Cartesian GP [16] and Grammatical Evolution (GE) [3,20]. GE is a grammar-based form of GP which takes inspiration from DNA Transcription in adopting a mapping from a linear genotype to phenotypic GP trees. The reason for adopting a genotype-phenotype map for GP where presented by O'Neill [17] as a series of arguments while also noting it can provide a number of advantages. These include a generalised encoding that can represent a variety of structures allowing GP to generate structures in an arbitrary language, efficiency gains for evolutionary search (e.g. through neutral evolution), maintenance of genetic diversity through many-to-one maps, preservation of functionality while allowing continuation of search at a genotypic level, reuse of genetic material potentially allowing information compression, and positional independence of gene functionality.

Previous work [4] showed a significant reduction in the performance of GE on the Max problem once the πGE genotype-phenotype map was applied. This was

S. Silva et al. (Eds.): EuroGP 2011, LNCS 6621, pp. 286–297, 2011.
© Springer-Verlag Berlin Heidelberg 2011

a very surprising result as previous studies by O'Neill et al. [18] had not shown a similar type of performance reduction in any of the benchmark problems in that study. This poor outcome on the Max problem sparked a lot of questions as to what was causing this result, which went against previous findings that πGE would match if not out perform standard GE on the majority of problem domains tested. The result of this examination is hoped to lead to a deeper understanding of what is going on during a πGE run and try to use this knowledge to further guide the development of different genotype-phenotype maps in GE, leading to the ultimate goal of designing the most effective, efficient version of GE.

The remainder of the paper is structured as follows. A brief overview of the essentials of GE are provided in Section 2 before an explanation of the different genotype-phenotype maps used in the study in Section 3. The next part of the paper describes the experimental setup (Section 4), the results found (Section 5) and a discussion (Section 6) before drawing conclusions and pointing to future work.

2   Grammatical Evolution Essentials

GE marries principles from molecular biology to the representational power of formal grammars. GE's rich modularity gives it a unique flexibility, making it possible to use alternative search strategies, whether evolutionary, or some other heuristic (be it stochastic or deterministic) and to radically change its behaviour by merely changing the grammar supplied. As a grammar is used to describe the structures that are generated by GE, it is trivial to modify the output structures by simply editing this grammar. The explicit grammar allows GE to easily generate solutions in any language (or a useful subset of a language). For example, GE has been used to generate solutions in multiple languages including Lisp, Scheme, C/C++, Java, Prolog, Postscript, and English. The ease with which a user can manipulate the output structures by simply writing or modifying a grammar in a text file provides an attractive flexibility and ease of application not as readily enjoyed with the standard approach to GP. The grammar also implicitly provides a mechanism by which type information can be encoded thus overcoming the property of closure, which limits the traditional representation adopted by GP to a single type. The genotype-phenotype mapping also means that instead of operating exclusively on solution trees, as in standard GP, GE allows search operators to be performed on the genotype (e.g., integer or binary chromosomes), in addition to partially derived phenotypes, and the fully formed phenotypic derivation trees themselves. As such, standard GP tree-based operators of subtree-crossover and subtree-mutation can be easily adopted with GE. By adopting the GE approach one can therefore have the expressive power and convenience of grammars, while operating search in a standard GP or Strongly-Typed GP manner. For the latest description of GE please refer to Dempsey et al. [3].

Previous work by Hemberg et al. [8] where three grammar variants were examined in the context of Symbolic Regression is also directly relevant to this study. The language represented by each of the grammars were all semantically equivalent in terms of the phenotypic behaviour of the solutions that could be generated. The only difference was syntactical. That is, postfix, prefix and infix notations were adopted for the same set of terminal symbols of the language. Performance advantages were observed on the problems examined for the postfix notation over both alternatives. If one examines the behaviour of postfix notation it amounts to a postorder expansion of the tree. In terms of a generative grammar this means that the contents of subtrees are determined before the operator at the root of the subtree.

3   Genotype-Phenotype Maps - GE, πGE

In order for a full examination of the performance on the Max Problem to be carried out four separate mappings for GE were required, details of which are explained in this section. In GE we begin the mapping process by finding the start symbol in the grammar. This non terminal (NT) in the case of the example grammar shown in Fig. 1, <e> is then evaluated using Eq. 1. By taking the first codon value of the GE chromosome (12) and the number of expansions possible for the state <e> (2), we get the first expansion of the tree, where <e> expands to <e><o><e> (12%2) . From this point on the leftmost NT is always expanded first in the derivation process. This action will continue to be performed until no NTs remain to be expanded in the derivation tree. An example of this mapping

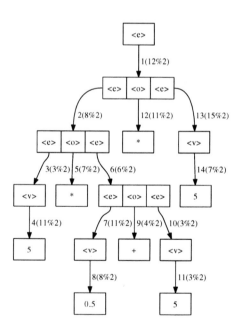

<e> ::= <e> <o> <e> | <v>

<o> ::= + | *

<v> ::= 0.5 | 5

Chromosome ::= 12,8,3,11,7,6,11,8,4,3,
               3,11,15,7,9,8,10,3,7,4

Fig. 1. Example Grammar and Chromosome

Fig. 2. Standard GE Genotype to Phenotype Mapping

is shown in Fig. 2 based on the example grammar shown in Fig. 1 where the order of expansion is indicated by a set of numbers on the arrows between the blocks on the diagram, in the form of 1(12%2) where 1 is the expansion order and 12%2 is the application of Eq. 1.

$$New\ Node\ =\ Codon\ value\ \%\ Number\ of\ rules\ for\ NT \tag{1}$$

The only difference between standard GE and πGE in its purest form is in the mapping process from genotype to phenotype. πGE's mapping process differs from that of GE in that each expansion of a NT requires two codons. The standard GE chromosome is essentially split into pairs of values where the first codon of the pair is used to choose which NT to expand and the second is used to choose what to expand the NT to, based on the rules available for a NT of that type. The chromosome shown in Fig. 1 can be viewed as a list of paired values such as ((12,8),(3,11).......), where the first value of the pair (The Order Codon) is used to determine the next NT to expand by using Eq. 2 and this will return which NT to choose from a list of unexpanded NTs. Once the NT to be expanded has been chosen, the second codon (Content Codon) is used in conjunction with Eq. 1 (the standard GE expansion rule) to determine what the NT expands to; and if this node happens to be an NT, it is added to the list of unexpanded NTs. Figs. 3 and 4 show the expansion of the example grammar in Fig. 1 using the πGE mapping process. The number associated with each branch of the tree is a reference to the numbered steps shown in Fig. 3 which show how each choice of NT to expand comes about. It is interesting to note the different shape and size of the examples based on just a change in mapping.

$$NT\ to\ expand\ =\ Codon\ value\ \%\ Number\ of\ NT's \tag{2}$$

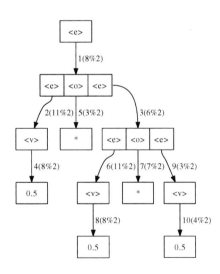

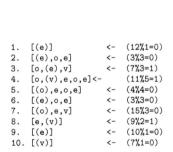

```
1.  [(e)]             <-  (12%1=0)
2.  [(e),o,e]         <-  (3%3=0)
3.  [o,(e),v]         <-  (7%3=1)
4.  [o,(v),e,o,e]<-       (11%5=1)
5.  [(o),e,o,e]       <-  (4%4=0)
6.  [(e),o,e]         <-  (3%3=0)
7.  [(o),e,v]         <-  (15%3=0)
8.  [e,(v)]           <-  (9%2=1)
9.  [(e)]             <-  (10%1=0)
10. [(v)]             <-  (7%1=0)
```

Fig. 3. NT selection process in πGE

Fig. 4. Standard πGE Genotype to Phenotype Mapping

In addition to the πGE mapper above a slightly tweaked version of this mapping was also required. The experiments that were planned required a version of πGE in which the order codons of the mapping were fixed across the whole population and also to be not affected by crossover or mutation during evolution. This requires the addition of an order chromosome to standard GE and then an edit to the πGE mapper so it will work with the new setup. This setup is referred to as Fixed-Order mapping and is necessary to see if the constantly changing order of expansion in the πGE mapping process is the cause of the performance decrease. The mapping also serves as an interesting experiment by itself as it will show if any randomised order might do as well as πGE . This way of mapping can be seen as fixed order πGE mapping or as a standard GE mapping but with a randomised order of mapping.

Finally Right-Most mapping is needed, this mapping is just a variant of the standard GE mapping process. Rather than taking the left-most non terminal and expanding as in standard GE, in Right-Most we simple always select the right-most non terminal for expansion. This way of mapping is required for this study as it should provide a similar performance to the standard GE mapping, whilst also allowing further insight into what is happening within the evolution of a solution to the Max problem in GE. An overview of the different Mappings can be seen in Table 1.

Table 1. Overview of the different Mappings

	GE	**Right-Most**	**πGE**	**Fixed-Order**
Expansion Order	Left-Most NT	Right-Most NT	Equation 2	Equation 2
Expansion Choice	Equation 1	Equation 1	Equation 1	Equation 1
Fixed Order Choice	Yes	Yes	No	Yes

4 Experimental Setup

We wish to test the hypothesis that the evolvability of the order codons in πGE is responsible for the performance decrease observed, as evolution has to try and cope with a changing tree structure across each member of the population. By removing this degree of freedom from the evolutionary process it is hoped that the performance decrease will be accounted for. We will measure performance in terms of examining the average best fitness.

We adopted GEVA v1.1 [19] for the experiments conducted in this study. The evolutionary parameters adopted on all problems are presented in Table 2. Note that we deliberately use a relatively small population size of 200 compared to the standard 500 that would typically be adopted for these problem instances. This was to make it harder for the mappers to find a perfect solution, and therefore allow us to discriminate more clearly performance differences on the various setups. For each setup we performed 100 runs using the same range of random seeds, in the case of the Fixed-Order setup we examined 100 different random expansion orders and for each order we then did 100 runs.

Table 2. Parameter settings adopted on all problems examined

Parameter	Value
Generations	100
Population size	200
Replacement strategy	Generational with elitism (10%)
Selection	Tournament size=5
Mutation probability	0.01 (integer mutation)
Crossover probability	0.9 (variable single point)
Initial chromosome length	200 codons (random init)
Initial πGE chromosome length	400 codons (random init)

```
<prog> ::= <expr>

<expr> ::= <op> <expr> <expr>
         | <var>

<op> ::= +
       | *

<var> ::= 0.5
```

Fig. 5. The grammar adopted for the Max problem instance

4.1   The Max Problem

The problem used in this paper is the Max problem first implemented for GE in [9] which is based upon the problem as described in [14] and [6]. The grammar adopted used for the problem appears in Fig. 5.

The aim of the problem is to evolve a tree that returns the largest value within a set depth limit (8 in this study). A minimal function set of addition and multiplication is provided alongside a single constant (0.5). The optimal solution to this problem will have addition operators towards the leaves of the tree to create as large a variable as possible greater than 1.0 in order to exploit multiplication operators towards the root of the tree. This problem is considered difficult for GP as solutions tend to converge on suboptimal solutions which can be difficult to escape from [14].

4.2   Mappers

For the purposes of this study we examined four different mapping strategies. The standard mapper adopted in GE we refer to as **Depth-first**. The name reflects the path this mapper takes through the non-terminal symbols in the derivation tree always selecting the left-most NT. The next mapping to consider was the opposite of the standard GE mapping, in which a depth first expansion of the derivation tree is done, but working on the right-most non terminal nodes first; we refer to this as **Right-most**. The πGE mapper as first described by O'Neill et al. [18] is the third mapper used. πGE lets the evolving genome decide which non-terminal to expand at each step in the derivation sequence. Finally a mapping was required that fixed the order codons of πGE to a random order known as **Fixed-Order**. This mapping was used to fix the πGE order codons for the whole population; unlike πGE we will have the same random order of expansion for the whole population. This mapping strategy will examine the influence of different fixed mapping orders on the Max problem, instead of only Depth-first and Right-most. Fixed-Order will also be used to examine if the evolvability of the order in πGE , or just the fact that it is a different order gives the performance decrease observed in the Max problem.

5   Results

The first set of experiments tested the hypothesis that the evolvability of the order codons in πGE were responsible for the observed performance decrease. For this the mapping order of πGE was fixed for the whole population and was not subject to crossover or mutation. One hundred different orders were generated and each order was then run one hundred times. The five best and five worst performing orders where then extracted and compared to a run of standard πGE, the results of which are shown in Table 3 in which the results shown are minimised so the optimal solution is zero. It is worth noting that all of the runs are within a standard deviation and thus show that the evolvable order is not where the performance decrease is caused. Further examination of the results show a slight performance difference between πGE and the fixed order mappers, which shows that the extra search performed by πGE for good representations is not negatively affecting performance.

After examination of these results it was hypothesised that perhaps the answer lay in how operations such as crossover and mutation would effect performance. Previous work on what is known as the ripple effect of crossover [21] and the ripple effect of mutation [9] have shown how the GE derivation is trimmed by these operators. In these papers it was shown how a single change in the genotype can lead to a change in the expansion of every codon following on from that

Table 3. Average Best Fitness Values after 100 generations over 100 independent runs. For the five best and worst performing mapping orders.

Mapping	Avg.Best (std. dev.)
πGE	29.78 (24.03)
Best 1	32.59 (24.67)
Best 2	33.21 (23.92)
Best 3	33.42 (28.97)
Best 4	35.26 (29.09)
Best 5	35.56 (25.84)
Worst 1	52.20 (39.93)
Worst 2	49.82 (35.01)
Worst 3	49.59 (40.03)
Worst 4	49.35 (34.08)
Worst 5	49.15 (36.45)

Table 4. Average Best Fitness Values after 100 generations over 100 independent runs

Mapping	Avg.Best (std. dev.)
GE	2.50 (5.42)
Right-Most	9.14 (15.54)
πGE	29.78 (24.03)

position in the genotype. This led to the introduction of the right-most mapping which is essentially a transpose of the traditional GE mapper. The results for this new comparison of mappings is shown below in Table 4 These results will be explained more fully in the discussion section, Section 6.

6   Discussion

The main focus of this study started out as an investigation to see if the evolvability of the order codons in πGE is responsible for the performance decrease observed in the Max problem. By examining the results in Table. 3 it can be seen that the evolvability of πGE's mapping order is not responsible for the performance decrease. This is clearly visible from comparing the fixed orders and πGE , that the fixed orders more closely resemble the results achieved by πGE and as the only difference between the two is that πGE 's order is subject to change that this change cannot be blamed for the observed performance decrease.

The real key to the performance by examining the results in Table 4 is in the operators and how they interact with the mapper, by interact with the mapper we mean how changes in the genotype (upon which all GE operations such as crossover and mutation act) lead to the mapping of a different phenotype. The Right-Most mapping was chosen so that we could examine if the performance on the max problem was down to burrowing down in a depth first manner. From the results in Table 4 we can see that it is the standard GE that achieves the best performance, this is due to the fact the the Max problem used is in prefix notation. Thus in the standard GE mapping we would be developing the arithmetic component of the equation first as we always expand the left-most NT; unlike the other mappings presented. Now consider when using GE how both mutation and crossover get this ripple effect as outlined in detail in [9], [21], which means that from the chosen point on the chromosome for the genetic operation all codons following that point are subject to a context change and in the majority of cases this causes a trimming of the tree as can be seen in Fig. 6. In Fig. 7 we can see that the same ripple effect is in the right-most setup of GE. Note how in the right-most tree the only remaining part of the derivation tree contains structure and terminals that describe the end of a possible max solution. As we are using a prefix notation in the grammar we can say this will not contain the important additions or multiplications, yet it is imperative that we start our expression with a multiplication to maximize the possible value. GE can thus preserve this most important part of the possible solution throughout

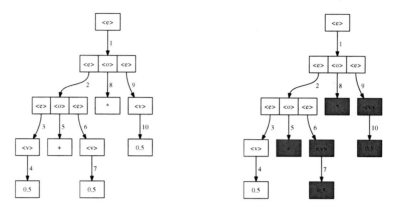

Fig. 6. An illustration of the ripple effect of crossover and mutation on a GE derivation tree, Note the tree before crossover on the left and then on the right once crossover is applied after the fourth expansion how we are left with structure and solution on the left only as the shaded nodes will be replaced

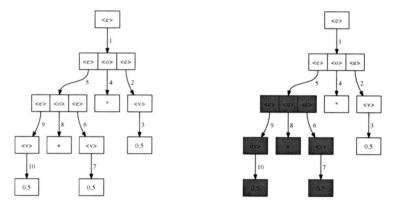

Fig. 7. An illustration of the ripple effect of crossover and mutation on a Right-Most derivation tree, Note the tree before crossover on the left and then on the right once crossover is applied after the fourth expansion how we are left with structure and solution on the right only as the shaded nodes will be replaced

the evolutionary process, where as Right-Most mapping will have to discover this after nearly every mutation or crossover.

The big clue as to where the performance loss in πGE and Max comes from lies in Fig. 8. It can be seen that the ripple effect of πGE crossover will in general leave a saw tooth cut in the tree that will leave few if any of the leaf nodes of the trees. This means that in general a πGE mapping will have to try and evolve the whole solution at once rather than having a starting point like the other mappings have. If we compare the results in the previous section to these figures, we can say that there exists a preferential bias towards evolving a solution in a left to right manner as is performed by the standard GE mapping.

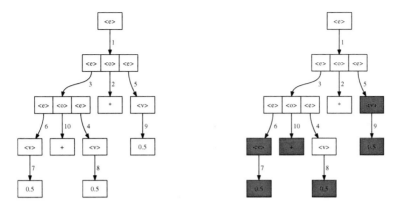

Fig. 8. An illustration of the ripple effect of crossover and mutation on a πGE or Fixed-Order derivation tree, Note the tree before crossover on the left and then on the right one crossover is applied after the fourth expansion how we are left with structure and a very tiny amount of a solution as the shaded nodes will be replaced

There is a performance drop seen in the Right-Most mapping too, this is due to the reason stated in the previous paragraph where the ability to retain knowledge of the starting part of the expression is deemed most important as the difference between adding 20+20 and multiplying 20*20 is huge and so can be deemed to be the most important symbol in the equation.

Looking again at [8] and the results above we would like to further examine if a certain order of notation is better suited to a certain mapper. From what we have seen above the prefix notation used in this grammar for Max it is the reason for the performance decrease and should also be further examined to see if changing to a infix notation might level the playing field and remove all bias towards a certain type of mapping.

7   Conclusions and Future Work

We presented an investigation into the different performance levels of genotype-phenotype maps in GE on the Max problem in order to determine the respective levels of performance of each mapping. This was done to investigate whether the evolvability of πGE was responsible for the discrepancy. By fixing the order to be a single non evolvable order across the entire population the results indicated that the difference in performance was in some other aspect of the mapping.

Upon further examination of how the mappers worked we decided to examine how the genetic operators effect the performance specifically the ripple effect of crossover and mutation in GE. The idea of a new rightmost first expansion order was tested and from this it became apparent that to achieve the best performance in the Max problem domain, with a prefix grammar, such as the one used here, a left-most genotype-phenotype map is the best mapping order for GE. This study has helped us to gain a better understanding of how πGE works and also surfaced the idea that there is a mapping order suited for each type of problem. This is of great interest and is worth further investigation to identify what mapper is best for a specific type of problems based on criteria such as grammar notation, as well as which is the best general mapper for a range of problems. Once this has been established we intend to see if this trend holds true as we progress our research into the dynamic problem domain, or if what we now consider drawbacks to certain mappings become advantages in the dynamic problem domain.

Acknowledgments

This research is based upon works supported by the Science Foundation Ireland under Grant No. 08/IN.1/I1868.

References

1. Banzhaf, W.: Genotype-phenotype-mapping and neutral variation – A case study in genetic programming. In: Davidor, Y., Männer, R., Schwefel, H.-P. (eds.) PPSN 1994. LNCS, vol. 866. Springer, Heidelberg (1994)
2. Brameier, M.F., Banzhaf, W.: Linear Genetic Programming. Springer, Heidelberg (2007)
3. Dempsey, I., O'Neill, M., Brabazon, A.: Foundations in Grammatical Evolution for Dynamic Environments. SCI. Springer, Heidelberg (2009)
4. Fagan, D., O'Neill, M., Galván-López, E., Brabazon, A., McGarraghy, S.: An Analysis of Genotype-Phenotype Maps in Grammatical Evolution. In: Esparcia-Alcázar, A.I., Ekárt, A., Silva, S., Dignum, S., Uyar, A.Ş. (eds.) EuroGP 2010. LNCS, vol. 6021, pp. 62–73. Springer, Heidelberg (2010)
5. Fernandez-Villacanas Martin, J.-L., Shackleton, M.: Investigation of the importance of the genotype-phenotype mapping in information retrieval. Future Generation Computer Systems 19(1) (2003)
6. Gathercole, C., Ross, P.: An adverse interaction between crossover and restricted tree depth in genetic programming. In: Proceedings of the First Annual Conference on Genetic Programming, pp. 291–296. MIT Press, Cambridge (1996)
7. Harding, S., Miller, J.F., Banzhaf, W.: Evolution, development and learning using self-modifying cartesian genetic programming. In: GECCO 2009: Proc. of the 11th Annual Conference on Genetic and Evolutionary Computation. ACM, New York (2009)
8. Hemberg, E., McPhee, N., O'Neill, M., Brabazon, A.: Pre-, in- and postfix grammars for symbolic regression in grammatical evolution. In: IEEE Workshop and Summer School on Evolutionary Computing (2008)
9. Byrne, J., O'Neill, M., McDermott, J., Brabazon, A.: An analysis of the behaviour of mutation in grammatical evolution. In: Esparcia-Alcázar, A.I., Ekárt, A., Silva, S., Dignum, S., Uyar, A.Ş. (eds.) EuroGP 2010. LNCS, vol. 6021, pp. 14–25. Springer, Heidelberg (2010)
10. Kell, D.B.: Genotype-phenotype mapping: genes as computer programs. Trends in Genetics 18(11) (2002)

11. Keller, R.E., Banzhaf, W.: Genetic programming using genotype-phenotype mapping from linear genomes into linear phenotypes. In: Genetic Programming 1996: Proc. of the First Annual Conference. MIT Press, Cambridge (1996)
12. Keller, R.E., Banzhaf, W.: Evolution of genetic code on a hard problem. In: Proc. of the Genetic and Evolutionary Computation Conference (GECCO 2001). Morgan Kaufmann, San Francisco (2001)
13. Koza, J.R., Keane, M.A., Streeter, M.J., Mydlowec, W., Yu, J., Lanza, G.: Genetic Programming IV: Routine Human-Competitive Machine Intelligence. Kluwer Academic Publishers, Dordrecht (2003)
14. Langdon, W., Poli, R.: An analysis of the MAX problem in genetic programming. In: Genetic Programming (1997)
15. Margetts, S., Jones, A.J.: An adaptive mapping for developmental genetic programming. In: Miller, J., Tomassini, M., Lanzi, P.L., Ryan, C., Tetamanzi, A.G.B., Langdon, W.B. (eds.) EuroGP 2001. LNCS, vol. 2038, p. 97. Springer, Heidelberg (2001)
16. Miller, J.F., Thomson, P.: Cartesian genetic programming. In: Poli, R., Banzhaf, W., Langdon, W.B., Miller, J., Nordin, P., Fogarty, T.C. (eds.) EuroGP 2000. LNCS, vol. 1802, pp. 121–132. Springer, Heidelberg (2000)
17. O'Neill, M.: Automatic Programming in an Arbitrary Language: Evolving Programs with Grammatical Evolution. PhD thesis, University Of Limerick (2001)
18. O'Neill, M., Brabazon, A., Nicolau, M., Garraghy, S.M., Keenan, P.: πgrammatical evolution. In: Deb, K., et al. (eds.) GECCO 2004. LNCS, vol. 3103, pp. 617–629. Springer, Heidelberg (2004)
19. O'Neill, M., Hemberg, E., Gilligan, C., Bartley, E., McDermott, J., Brabazon, A.: GEVA: Grammatical evolution in java. SIGEVOlution 3(2) (2008)
20. O'Neill, M., Ryan, C.: Grammatical Evolution: Evolutionary Automatic Programming in a Arbitrary Language. In: Genetic Programming. Kluwer Academic Publishers, Dordrecht (2003)
21. O'neill, M., Ryan, C., Keijzer, M., Cattolico, M.: Crossover in grammatical evolution. Genetic Programming and Evolvable Machines 4(1), 67–93 (2003)
22. Poli, R., Langdon, W.B., McPhee, N.F.: A field guide to genetic programming (2008), Published via http://lulu.com and freely available at http://www.gp-field-guide.org.uk (With contributions by J. R. Koza)
23. Stephens, C.R.: Effect of mutation and recombination on the genotype-phenotype map. In: Proc. of the Genetic and Evolutionary Computation Conference, vol. 2. Morgan Kaufmann, San Francisco (1999)

A Self-scaling Instruction Generator Using Cartesian Genetic Programming

Yang Liu[1], Gianluca Tempesti[1], James A. Walker[1], Jon Timmis[1,2], Andrew M. Tyrrell[1], and Paul Bremner[3]

[1] Department of Electronics, University of York, UK
{yl520,gt512,jaw500,jt517,amt}@ohm.york.ac.uk
[2] Department of Computer Science, University of York, UK
[3] Bristol Robotics Laboratory, University of the West of England, UK
paul.bremner@brl.ac.uk

Abstract. In the past decades, a number of genetic programming techniques have been developed to evolve machine instructions. However, these approaches typically suffer from a lack of scalability that seriously impairs their applicability to real-world scenarios. In this paper, a novel self-scaling instruction generation method is introduced, which tries to overcome the scalability issue by using Cartesian Genetic Programming. In the proposed method, a dual-layer network architecture is created: one layer is used to evolve a series of instructions while the other is dedicated to the generation of loop control parameters.

1 Introduction

Automatic instruction generation using genetic programming (GP) based techniques has been studied for a decade or more. Several approaches have been developed. Examples include GP with ADFs [1], LinearGP [2] and PushGP [3], based on various processor architectures. However, these approaches typically suffer from a lack of *self-scalability*. In the context of computer programs, self-scalability refers to the ability to operate correctly across a wide range of parameters (which we will call the *scale factors*), for example with arrays of different sizes.

It is well known, one of the major problems with evolutionary computation (EC) techniques is that they can only really evolve solutions to small problems due to the sheer computational power and time required. Therefore, by evolving a self-scaling program for a feasible small problem, we can easily scale the program to produce a solution that would have been an infeasible EC problem without any further evolution or fitness evaluations. This solution is called a *general solution*, which can be applied to all instances of a problem [4].

In order to have general solutions using GP, previous work, such as [5], [6] and [7], presented various methods. However, none of them is dedicatedly designed for instruction generation. In this paper, a novel self-scaling instruction generation method is introduced, which tries to overcome the scalability issue by using a dual-layer extension of the Cartesian Genetic Programming technique (CGP) [8],

S. Silva et al. (Eds.): EuroGP 2011, LNCS 6621, pp. 298–309, 2011.
© Springer-Verlag Berlin Heidelberg 2011

[9]. The evolved programs (genotypes) are capable of, when executed, expanding into a sequence of operations (phenotype) that depends on the value of the scale factor. The experiments present a simple demonstration of the operation and capabilities of the approach, which could potentially be extended and applied to more general robot control applications.

2   CGP Decode for Instruction Generation

CGP [8], [9] was originally developed for the purpose of evolving digital circuits. It represents a program as a directed graph (acyclic for feed-forward functions), as shown in Fig. 1. This type of representation allows the implicit re-use of nodes, as a node can be connected to the output of any previous node in the graph, thereby allowing the repeated re-use of sub-graphs. Originally, CGP used a topology defined by a rectangular grid of nodes with a user-defined number of rows and columns. However, later work on CGP showed that it is more effective when the number of rows is chosen to be one [10]. CGP does not require all of the nodes to be part of an output pathway, and this results in candidate circuits (phenotypes) whose size is bounded but variable.

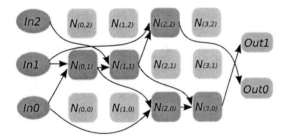

Fig. 1. An example of the CGP representation

The approach described in [11] modifies the original CGP algorithm and applies it to the generation of sequential instructions, which is also exploited by the approach described in this paper. As depicted in Fig. 2, each genotype node contains three genes, representing an instruction and two subroutine calls, the latter are represented as arrows pointing to connected nodes. In order to decode a CGP genotype, a recursive approach is used, in which the first subroutine node is always invoked before the second, which in turn is always invoked before the instruction in the node is executed. Therefore, the CGP genotype represents by a series of nested subroutine calls, from right to left.

As illustrated in Fig. 3, node I_5 is the entrance of the program, it invokes node I_1 first. Immediately, node I_1 invokes node I_0. Node I_0 invokes a terminal (denoted as T) for each of its subroutine calls. Terminals in CGP do not represent actual entities, but simply act as a method for terminating the current subroutine. As node I_0 has no connections but the terminal, it simply decodes

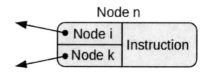

Fig. 2. A closer view of a CGP node for instruction generation

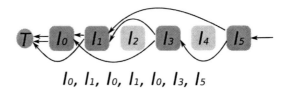

$$I_0, I_1, I_0, I_1, I_0, I_3, I_5$$

Fig. 3. An example of the decoding procedure

its own instruction, and then returns to node I_1, where, since the second call is to a terminal, the instruction is executed. The program then returns back to node I_5, and node I_5 starts to invoke its second call to node I_3, and so forth.

It is worth pointing out that the number of terminals affects the depth of the nested subroutine calls: statistically, for a genotype of a given length, the more terminals are present, the less likely it will be to evolve programs (phenotypes) with deeply nested calls. Empirical knowledge about the structures of target problems would be helpful to set up this number.

In Fig. 3, completely decoding the program reveals the occurrence of repetitive patterns ("I_0, I_1"). A clearer example is illustrated in Fig. 4. Since node I_1 is continuously invoked by nodes I_3, I_4 and I_5, the pattern of "I_0, I_1" can be seen as a subroutine repeatedly called within the instruction sequence. A subroutine can be called an arbitrary number of times by the nodes on its right, implying a capability of instruction re-use and a form of structurally modularised programming. More interestingly, when a subroutine is invoked by a series of cascaded nodes, this is functionally equivalent to being called in a 'virtual loop'. The more often the same node is addressed, the more often a subroutine is repeated in a program.

The implementation of loops is a crucial feature for the evolution of real-world programs. However, relying on cascaded nodes to implement the looping

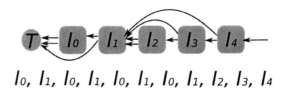

$$I_0, I_1, I_0, I_1, I_0, I_1, I_0, I_1, I_2, I_3, I_4$$

Fig. 4. An example of the structural scalability

behaviour obviously impairs the scalability of the system, since the evolution of a loop implies the evolution of as many cascaded nodes as the number of iterations within the loop. The purpose of the approach introduced in this paper is to overcome this hurdle in the evolution of complex programs by introducing the concept of parametrically defined scalability, implemented through the introduction of a special function node within the CGP algorithm and by the introduction of a second CGP network dedicated to the evolution of loop parameters. The use of variable-length loops (loops where the number of iterations depends on the scale factors) provides a simple and straightforward solution.

3 Dual-Layer CGP Architecture

3.1 Special Function Node

As previously explained, a self-scaling solution should not change the genotype structure, only the phenotype. There must be a mechanism which translates the same genotype to different phenotypes at various scales. To achieve self-scalability in loops, a special function node is added to the set of existing nodes. This *Loop* node invokes its subroutine a number of times specified by a parameter encoded in the node itself. In a two-arc *Loop* node scenario, as shown in Fig. 5, only the first arc is used for calling a subroutine, while the second is used to specify the number of iterations. The way to generate this iteration value will be detailed in Section 3.2. In a chromosome, *Loop* nodes are encoded in the same way as other nodes, and under the same evolution operators control, i.e, mutation can change a regular node to a *Loop* node, and thus the second arc becomes to control the iteration.

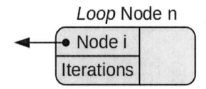

Fig. 5. A prototype of the *Loop* node

In order to create the same instruction sequence as shown in Fig. 4, we can simply use a *Loop* node instead of 3 cascaded calls, and set the number of iteration to be 4. Fig. 6 shows an example. By altering the iteration value, the subroutine can be repeatedly invoked a number of times. It can be observed that without changing the structure of the network (genotype), the execution sequence (phenotype) can be dynamically modified according to the iteration parameter. In particular, when the iteration value is set to 0, the *Loop* node acts as a terminal. This feature provides the system with additional flexibility in defining subroutines.

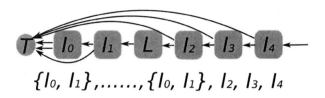

Fig. 6. An example of the parametric scalability

Generally, the parametrically-defined scalability of a program largely depends on the *loop* node(s). In many instances, real-world algorithms rely on loops that are executed a number of times, where the number of executions depends on the scale factor. Further, the relationship between the scale factor and the number of iterations can potentially be an arbitrarily complex function. In this case, the number of iterations cannot be directly encoded into the *Loop* node as a constant, which would imply the need to re-evolve the network at every single scale: the relationship between the scale factor and the number of iterations must be established parametrically through a functional mapping. This problem can be considered as symbolic regression, where the functional mappings are explicitly expressed after evolution. In the following subsection, a symbolic regression method is incorporated within our approach by re-defining the second arc of a *Loop* node.

3.2   Extended Parametric CGP

In order to efficiently determine the correct mapping from the scale factor to the number of iterations, a second CGP network, which is evolved in parallel with the first, is introduced. The purpose of this second CGP network is to implement symbolic regression [9]. The input to the second CGP is the scale factor and the outputs are the corresponding number of iterations. In order to distinguish the two CGP networks, we will rename the original CGP for instruction generation 'I-CGP', and the supplemental CGP for parameter regression 'P-CGP'.

Performing as a conventional symbolic regression method, the P-CGP network uses a set of mathematical functions that is separate from the I-CGP: where the I-CGP nodes represent instructions (which can be arbitrarily complex and

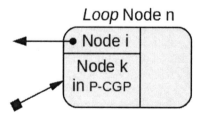

Fig. 7. A *Loop* node connecting I-CGP and P-CGP

depend on the target application) that make up the evolved program, the P-CGP relies on a set of operations that can be applied to the scale factor to determine the number of iterations of the loops contained in the program (for example, basic arithmetic operations).

The connection between the two networks is one-directional, that is, the output of a P-CGP node can feed into a I-CGP node, but not vice-versa. As shown in Fig. 7, the second arc of a *Loop* node in the I-CGP is modified to allow it to refer to a node label in the P-CGP, implying that the value of the P-CGP node determines the number of iterations of the *Loop* node. Those nodes within the P-CGP that are connected to nodes in the I-CGP network are defined as the outputs of the P-CGP network. The P-CGP does not therefore have any pre-encoded outputs within its chromosome, rather the outputs are dynamically determined by the *Loop* nodes of the I-CGP network. As a consequence, if there are no active *Loop* nodes in the I-CGP, the entire P-CGP is essentially unconnected, and since the number of the *Loop* nodes in the I-CGP network varies during the evolutionary process, the number of the P-CGP outputs varies as well. After completely decoding the I-CGP, all output nodes of the P-CGP are marked and the P-CGP decode procedure is initialised.

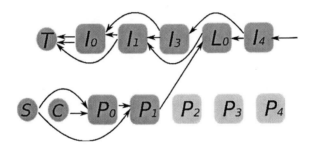

Fig. 8. The interconnection of two CGP networks

An example of the entire structure is illustrated by Fig. 8. The upper network is the I-CGP, where the *Loop* node is denoted as L_x; the lower network is the P-CGP, where S and C are the inputs, representing the scale factor and a constant[1] respectively. In this case, node P_1 is addressed by L_0, so that the value of node P_1 feeds into node L_0 as the number of iterations. Suppose nodes P_x stands for a two-input mathematical operator, defined as $P_x(*, *)$, the output value of node P_1 and also the iteration value in node L_0 can be expressed by a nested formula $P_1(P_0(S, C), S)$. Via this connection, the relationship between the number of iterations and the scale factor is established. Since no other nodes in the P-CGP are addressed by the I-CGP, nodes P_2, P_3 and P_4 are inactive. Note that the numbers of scale factors and constants can be more than one in various scenarios.

---

[1] The constant is encoded in the chromosome and under evolutionary control.

4   Experiments

The classic lawnmower problem [1] was used to demonstrate the operation and features of the approach described in this paper. As shown in Fig. 9, a piece of lawn is divided into a square two-dimensional grid. A mower is required to cut the entire lawn as efficiently as possible, which means every location in the grid must be visited once and only once. The motion of the mower is controlled by the decoded execution sequence: one single movement, one operation. In order to find a general solution, the program must be capable of scaling its execution sequence according to the size of the grid. Therefore, in this problem, the length of the side of the lawn (5 in the figure) is used as the scale factor. The scaling issue in this problem is representative of a wide variety of problems and the performance can be very easily visualised and assessed.

Fig. 9. An illustration of the lawnmower problem

This section described two separate experiments. The first experiment is the evolution of a general solution to the lawnmower problem. The system is evolved to solve the problem at N scale levels (specifically, from lawn size 2^2 to size $(N+1)^2$). The more scale levels cause more evolutionary computation, but we hypothesise that *given a number of independent runs with a fixed number of generations, evolution at more scale levels will find more general solutions than at less scale levels*. The choice of N is a trade-off between computational overhead and the chance to find a general solution.

Once a general solution was found, the second experiment aims at minimizing the length of instruction sequence that solves the problem. The second hypothesis is that *in proportion to the entire program, loops in shorter solutions contain more operations than in longer ones*. In other words, a shorter solution tends to contain more repetitive patterns. From an engineering perspective, shorter solutions are also preferable, because it implies a smaller amount of memory required to store the program.

4.1   Settings

As previously stated, an optimal solution is described as: every square in the grid must be visited it once and only once. Therefore, the fitness at each scale is measured with respect to two objectives: efficiency and completion. Efficiency is defined as the number of visited squares divided by the length of the instructions; completion is defined as the number of visited squares divided by the entire number of squares, so that the value of both objectives is in the range $[0, 1]$. Since the program is evolved for N scale levels, the total number of objectives for evolution is $2N$. In order to deal with this multi-objective optimisation problem, we customised a light-weight variant of the well-known NSGA-II algorithm [12], specifically, embedded into a (10+4) evolutionary strategy (selected empirically as a function of the problem size). This algorithm is not detailed in this paper due to space limitations. In order to accelerate the evolution, a promotion of neutral drift within the search space is adopted, as addressed in previous work [13]: when a tie occurs in the fitness score between a parent and its offspring on all objectives, the offspring is classed as dominating the parent, and is therefore ranked higher than the parent. The parameters used are shown in Table 1.

Table 1. The overall parameters

Parameter	Value
Population size	100
Mutation rate (% of genes)	2
Run length (generations)	10,000,000
I-CGP nodes	100
I-CGP terminals	1
Max length of executions	$10 \times n^2$
P-CGP nodes	25
P-CGP constant range	$[0, N]$

The regular nodes in the I-CGP can contain one of 4 predefined operations: turn *Left*, turn *Right*, go *Straight*, and a pseudo *Nop*, which represents no physical operation, and is only used to increase the evolutionary diversity. Compared to the NOP instruction in real assembly languages, it costs no execution time and thus will be simply skipped during decoding. For the sake of simplicity, we intentionally limited the size of the instruction set. Moreover, there must be a limit for the maximum length of execution sequences at each level, because the number of loop iterations could be any possible value generated by the P-CGP network. A chromosome will be treated as invalid when the length of execution sequence exceeds the predefined limit. In this experiment, the maximum length is proportional to the size of the grid, as defined in Table 1, where n is defined as the side length of the grid.

The function set of the P-CGP was chosen to be low order arithmetic operations: addition (+), absolute of subtraction (|-|), multiplication (*), and protected division (/) and modulus (%)[2]. In the lawnmower problem, the travel

---

[2] Protected division and modulus mean that the result equals to the dividend when the divisor is zero.

Table 2. A general solution

P-CGP		I-CGP				Pseudo code	
Node	Expression	Node	Operation	Arc1	Arc2	Order	Instruction
$S =$	n	$L_0:$		T	$<S>$	[0]	Straight
$C =$	1	$I_1:$	Nop	L_0	T	[1]	Left
$P_0 =$	$\lvert S - C \rvert$	$L_2:$		T	$<S>$	[2]	$Loop \rightarrow [3]$: P_1
$P_1 =$	$\lvert C - P_0 \rvert$	$I_3:$	Right	L_2	T	[3]	Straight
$P_2 =$	$\lvert P_1 - C \rvert$	$I_4:$	Straight	L_2	T	[4]	Right
$P_4 =$	$\lvert P_0 - P_2 \rvert$	$I_5:$	Left	L_2	I_4	[5]	Right
$P_6 =$	P_1/P_4	$I_7:$	Left	L_0	L_2	[6]	$Loop \rightarrow [7]$: P_1
$P_7 =$	$\lvert P_1 - P_6 \rvert$	$I_9:$	Left	I_1	I_7	[7]	Straight
$P_9 =$	$\lvert P_7 - P_6 \rvert$	$L_{11}:$		I_4	$<P_1>$	[8]	$Loop \rightarrow [16]$: P_6
		$I_{13}:$	Nop	I_9	L_{11}	[9]	Left
		$I_{15}:$	Right	I_{13}	I_3	[10]	Left
		$I_{21}:$	Nop	I_{15}	L_{11}	[11]	$Loop \rightarrow [12]$: P_1
		$I_{30}:$	Right	I_5	L_{11}	[12]	Straight
		$L_{35}:$		I_{13}	$<P_9>$	[13]	Right
		$L_{39}:$		I_{21}	$<P_6>$	[14]	Right
		$I_{43}:$	Nop	L_{11}	L_{39}	[15]	$Loop \rightarrow [16]$: P_1
		$I_{55}:$	Nop	I_3	I_{43}	[16]	Straight
		$I_{56}:$	Nop	I_{55}	L_{35}	[17]	$Loop \rightarrow [21]$: P_9
		$I_{60}:$	Nop	I_{30}	I_{56}	[18]	Left
						[19]	Left
						[20]	$Loop \rightarrow [21]$: P_1
						[21]	Straight

pattern in odd and even scales is different, and a modulus operator would help to distinguish even and odd numbers. In order to save the computational overhead from the P-CGP, the number of its nodes is quarter that of the I-CGP, a size chosen empirically. Consequently, the integer on the second arc of a *Loop* node should be divided by 4 when addressing the P-CGP node, so that the label of the corresponding P-CGP node will not exceed the limit of the available nodes. Additionally, a constant is introduced as an input as well; it is evolved in a range as defined in Table 1.

4.2   Results

In the first experiment, two cases were considered: the number N of scale levels used for evolution was set to 4 and 6 respectively. They are both evolved for 100 independent runs. In case 1, the program was evolved with side lengths n ranging from 2 to 5; in case 2, it was evolved from 2 to 7. When $N=4$, during the training session, 100% of the runs found solutions. However, only 1% of the solutions were analytically proven to be a general solutions. However, when N was increased to 6, 87% of the runs evolved a valid solution, and 74% of these were proven to be general solutions. This confirmed our hypothesis that the more scale levels are used in the evolutionary run, the higher the likelihood to find a general solution. Further statistics of the solutions, such as the confidence level, are going to be fully detailed in the future.

A general solution is shown in Table 2. We use the same notations as in Fig.8 to represent various CGP nodes. In the I-CGP table, the arc2 in *Loop* nodes, marked by brackets, points to the node label in the P-CGP table. In the pseudo code table, a *Loop* function is represented by showing the number of iterations (a reference to a P-CGP output) and the end of the loop. For example, $\{Loop \rightarrow [16]: P_6\}$ means that from the current position to instruction [16], this procedure is repeated P_6 times.

In the second experiment, once a solution was found, the program kept searching for a shorter instruction sequence until the maximum number of iterations was reached. The shortest solution is presented in Table 3 and the path followed by the lawnmower for $n=4$ and $n=5$ is shown in Fig. 10. Compared with Fig. 2, this solution contains longer repetitive patterns in proportion to the entire program length. Namely, almost all operations are within loops. This fact reveals the importance of the *Loop* nodes and confirms our second hypothesis.

Table 3. The shortest general solution

P-CGP		I-CGP				Pseudo code	
Node	Expression	Node	Operation	Arc1	Arc2	Order	Instruction
$S =$	n	I_0 :	Right	T	T	[0]	Straight
$C =$	2	I_1 :	Straight	T	T	[1]	$Loop \rightarrow [4]: P_2$
$P_0 =$	C/C	L_3 :		I_1	$< S >$	[2]	$Loop \rightarrow [3]: P_1$
$P_1 =$	$|S - P_0|$	I_4 :	Left	L_3	T	[3]	Straight
$P_2 =$	$P_1 \% C$	I_7 :	Right	I_0	T	[4]	Left
$P_3 =$	$|C - S|$	L_{20} :		I_4	$< P_1 >$	[5]	Left
$P_5 =$	P_1/C	L_{30} :		I_1	$< P_2 >$	[6]	$Loop \rightarrow [14]: P_5$
		I_{32} :	Left	I_7	L_{30}	[7]	$Loop \rightarrow [8]: P_3$
		I_{35} :	Left	L_{30}	I_{32}	[8]	Straight
		I_{37} :	Left	I_1	L_{20}	[9]	Right
		L_{42} :		I_{35}	$< P_4 >$	[10]	Right
		I_{43} :	Nop	I_{37}	L_{42}	[11]	$Loop \rightarrow [12]: P_3$
		I_{92} :	Nop	I_{43}	L_{30}	[12]	Straight
						[13]	Left
						[14]	Left
						[15]	$Loop \rightarrow [16]: P_3$
						[16]	Straight

Fig. 10. A general solution with the shortest program

5   Conclusions and Future Work

In this paper, a self-scaling instruction generator is introduced, based on a pair of coupled CGP networks: the first is used to evolve a sequence of instructions to solve a given task, while the task of second is to evolve iteration parameters

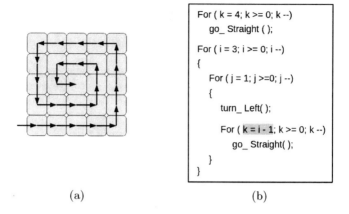

```
For ( k = 4; k >= 0; k --)
    go_ Straight ( );

For ( i = 3; i >= 0; i --)
{
    For ( j = 1; j >=0; j --)
    {
        turn_ Left( );

        For ( k = i - 1; k >= 0; k --)
            go_ Straight( );
    }
}
```

(a)   (b)

Fig. 11. A hypothetical trace in a spiral form

for loop instructions. The two CGP networks are decoded in a different manner, but evolved in parallel. Without human intervention, the approach is able to express various execution sequences (phenotypes) from a fixed generated program (genotype), according to a set of scale factors. Experiments show that the dual-layer CGP is capable of evolving self-scaling general solutions to the lawnmower problem, a testbench chosen as representative of a wide range of scale-dependent problems. Only being evolved at a small number of scale levels, general solutions were found and able to be correctly executed at any other larger scales.

However, by further analysing the results we found that none of the general solutions generated a spiral trace (Fig. 11(a)), which, if written in a C-like high level programming language, can be generated using tightly-coupled nested loops (Fig. 11(b)) and probably represents the most efficient (from the point of view of the number of instructions) general solution to the problem. The current implementation of our algorithm is not capable of evolving this solution because the limit of the inner loop counter is a variable determined not only by the scale factor, but also by the current value of the outer loop counter. It has to be clarified that a spiral trace could be possibility found without using nested loops, but it would not be a general solution.

In order to establish the relationship between the outer and inner loop counters, the P-CGP should be able to use the current value of the outer loop counter as an input. In terms of the dual-layer network graph, this can be understood as changing the connection between the two networks from one-directional to two-directional, allowing values from the I-CGP to be used in the P-CGP. This transformation could theoretically provide the system with a more dynamic scaling capability, but it is also likely to dramatically increase the size of the search space and is the subject of ongoing work.

Further work to be carried includes verifying the potential of the approach and applying it to a real-world application, for instance, when targeting machine code generation for a specific processor architecture, as trialled in [11].

Additionally, the lawnmower problem is considered as a simplified example of a robot control task, and thus the same mechanism will be extended to more general robotics scenarios.

Acknowledgement

The SABRE (Self-healing cellular Architectures for Biologically-inspired highly Reliable Electronic Systems) project is funded by the EPSRC under grant No. FP/F06219211.

References

1. Koza, J.R.: Genetic Programming II: Automatic Discovery of Reusable Programs. MIT Press, Cambridge (1994)
2. Nordin, P.: Evolutionary Program Induction of Binary Machine Code and its Applications. PhD thesis, der Universitat Dortmund am Fachereich Informatik (1997)
3. Spector, L., Robinson, A.: Genetic programming and autoconstructive evolution with the push programming language. Genetic Programming and Evolvable Machines 3, 7–40 (2002)
4. Kushchu, I.: Genetic programming and evolutionary generalization. IEEE Transactions on Evolutionary Computation 6(5), 431–442 (2002)
5. Yu, T.: Hierachical processing for evolving recursive and modular programs using higher order functions and lambda abstractions. Genetic Programming and Evolvable Machines 2, 345–380 (2001)
6. Wong, M.L., Mun, T.: Evolving recursive programs by using adaptive grammar based genetic programming. Genetic Programming and Evolvable Machines 6, 421–455 (2005)
7. Harding, S., Miller, J.F., Banzhaf, W.: Developments in cartesian genetic programming: Self-modifying cgp. Genetic Programming and Evolvable Machines 11, 397–439 (2010)
8. Miller, J.F.: An empirical study of the efficiency of learning boolean functions using a cartesian genetic programming approach. In: Proceedings of the 1st Genetic and Evolutionary Computation Conference, pp. 1135–1142 (1999)
9. Miller, J.F., Thomson, P.: Cartesian genetic programming. In: Poli, R., Banzhaf, W., Langdon, W.B., Miller, J., Nordin, P., Fogarty, T.C. (eds.) EuroGP 2000. LNCS, vol. 1802, pp. 121–132. Springer, Heidelberg (2000)
10. Yu, T., Miller, J.F.: Neutrality and the evolvability of boolean function landscape. In: Miller, J., Tomassini, M., Lanzi, P.L., Ryan, C., Tetamanzi, A.G.B., Langdon, W.B. (eds.) EuroGP 2001. LNCS, vol. 2038, pp. 204–217. Springer, Heidelberg (2001)
11. Walker, J.A., Liu, Y., Tempesti, G., Tyrrell, A.M.: Automatic code generation on a MOVE processor using cartesian genetic programming. In: Tempesti, G., Tyrrell, A.M., Miller, J.F. (eds.) ICES 2010. LNCS, vol. 6274, pp. 238–249. Springer, Heidelberg (2010)
12. Deb, K., Pratap, A., Agarwal, S., Meyarivan, T.: A fast and elitist multiobjective genetic algorithm: NSGA-II. IEEE Transactions on Evolutionary Computation 6(2), 182–197 (2002)
13. Walker, J.A., Hilder, J.A., Tyrrell, A.M.: Towards evolving industry-feasible intrinsic variability tolerant cmos designs. In: IEEE Congress on Evolutionary Computation, pp. 1591–1598 (2009)

Exploring Grammatical Modification with Modules in Grammatical Evolution

John Mark Swafford[1], Michael O'Neill[1],
Miguel Nicolau[1], and Anthony Brabazon[2]

[1] Natural Computing Research & Applications Group
Complex and Adaptive Systems Laboratory
School of Computer Science & Informatics
University College Dublin, Ireland
[2] Natural Computing Research & Applications Group
Complex and Adaptive Systems Laboratory
School of Business
University College Dublin, Ireland

Abstract. There have been many approaches to modularity in the field of evolutionary computation, each tailored to function with a particular representation. This research examines one approach to modularity and grammar modification with a grammar-based approach to genetic programming, grammatical evolution (GE). Here, GE's grammar was modified over the course of an evolutionary run with modules in order to facilitate their appearance in the population. This is the first step in what will be a series of analysis on methods of modifying GE's grammar to enhance evolutionary performance. The results show that identifying modules and using them to modify GE's grammar can have a negative effect on search performance when done improperly. But, if undertaken thoughtfully, there are possible benefits to dynamically enhancing the grammar with modules identified during evolution.

1 Introduction

Modularity in evolutionary computation has been studied in a variety of contexts. The range of research covering this topic starts with principles taken from biology [17] and extends to the empirical analysis of performance of different approaches to enabling and exploiting modularity. The work presented here is classified under the latter. As modularity has been shown to be extremely useful for the scalability of evolutionary algorithms (Koza shows this for genetic programming [12]), it is important to understand the effects of different methods of encapsulating and exploiting modularity in these stochastic search methods.

As genetic algorithms (GAs) [8] and genetic programming (GP) [10] have been studied fairly extensively in this context, this work focuses on grammatical evolution (GE) [14]. Studying modularity in the context of GE is especially interesting because of the genotype-to-phenotype mapping process employed. The context-free grammar used in this process provides one of the easiest methods of

S. Silva et al. (Eds.): EuroGP 2011, LNCS 6621, pp. 310–321, 2011.

reusing information in GE, incorporating that information into the grammar. By examining the derivation trees created by this mapping process, "good" information found by GE can be encapsulated and placed directly in the grammar to be used in the correct context. However, modifying the grammar in this way comes with possible drawbacks, which will be discussed later, in Sect. 3. As modularity in GP is one of the key open issues [15], the research presented here aims to add to the knowledge base of enabling modularity in grammar-based GP. This is accomplished by exploring initial methods of identifying, encapsulating, and reusing modules in GE by modifying the grammar during an evolutionary run.

In this paper, a module refers to a sub-derivation tree of an individual which is considered to contain beneficial information. It is also important to point out that there are a few differences between this approach and using Koza's automatically defined functions (ADFs) [12]. Individuals using ADFs each have their own, local, ADFs, which are may be modified by genetic operators. Modules in this study are stored in the universal grammar for all individuals. Once encapsulated in the grammar, they may not be modified by genetic operations.

The rest of the paper is structured as follows. Next, Section 2 outlines some previous work relating to modularity in GP and GE. Section 3 presents the experimental setup used here, and Sect. 4 details the results of this work, and their meaning. Finally, Sect. 5 gives the conclusions and avenues for future work.

2   Previous Work

Some of the earliest and most relevant work is Angeline and Pollack's [1,2] methods for picking out modules of useful information to be passed from individual to individual during an evolutionary run. They use *compress*, *expand*, and *atomization* to encapsulate sub-trees into modules, release entire or pieces of compressed sub-trees, and make compressed modules part of the original representation. Angeline and Pollack show the first step in basic module encapsulation and how advantageous this can be, and how beneficial capturing these modules is during the course of an evolutionary run.

Another relevant body of work is that conducted by Keijzer et al. [9]. They introduce the notion of run-transferable libraries (RTLs), which are lists of modules discovered over a number of independent runs and used to seed the population of a separate run. These RTLs show an increase in performance over standard GP, and greatly enhance the scalability of GP by training the RTLs on simple problems before using them for harder problems [16].

In the context of GE, modularity has been previously studied in a variety of ways. Hemberg et al. [6] use meta-grammars, or grammars which generate grammars, which are then used to solve the given problem, called GE^2. GE^2 is shown to have increased performance and scale better in comparison to the Modular Genetic Algorithm (MGA),[3] on problems known to have regularities (one example is the checkerboard problem).

The most popular approach to enabling and exploiting modularity in GP is the use of automatically defined functions (ADFs) [12]. ADFs are parameterized functions which are evolvable and reusable sub-trees in a GP individual. GP equipped with this form of modularity is shown to out-perform standard GP on problems of sufficient difficulty [12]. There have been previous approaches to enabling ADFs, in GP [11] and GE [7] (as well as the similar Dynamically Defined Functions in GE [4]). A more in-depth review of previous work modularity in GP can be found in work by Walker and Miller [18] and Hemberg [5].

3   Experimental Setup

The purpose of this work is to examine a simple method of identifying modules in GE and using these modules to enhance GE's grammar, and subsequently performance. The first step in this process is picking the parent individual from which a module will be selected. Once the parents are selected, candidate modules are picked from these parents and evaluated. After the modules have been evaluated, they may (or may not) go through a replacement operation which throws away modules based on a fitness value assigned when each module is evaluated. Finally, the collected modules are added to the grammar and the main evolutionary loop continues. Figure 1 shows an example of this process.

Module Identification: There are a number of steps involved in picking out suitable modules. The first step in identifying a module is to pick a parent individual which will provide a candidate module. All experiments presented in this work use an iterative approach where each candidate module was taken from each individual in the population. A roulette selection for the parents was tested in preliminary experiments, but the results of these tests showed no difference between the roulette and iterative approaches.

The next step in the module identification process is picking a candidate module and evaluating it. To do this, a random node on the parent individual's derivation tree is picked. The sub-derivation tree starting at this node is the candidate module, and original fitness of the individual, f_0, is recorded. Next, 50 new derivation tree starting with the same root as the candidate module are randomly created and take turns replacing the candidate module in the parent. After each replacement, the fitness of the new parent individual is calculated, $f_{1...50}$. If f_0 is a better fitness than 75% of $f_{1...50}$, the candidate module is saved for later use. When this is the case, the difference between f_0 and each of $f_{1...50}$ is taken and the average of these is the module's fitness value. No attempt was

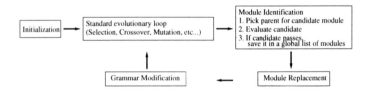

Fig. 1. Evolutionary loop with module identification and grammar modification

Table 1. Experimental setup for all evolutionary runs unless otherwise noted

Parameter	Value
Generations	100
Population	500
Selection	Tournament (Size 5)
Wrapping	None
Crossover	Context-sensitive (80%)
Mutation	Context-sensitive (20%)
Elites	5
Initialization	Ramped Half and Half
Replacement	Generational
Max. Derivation Tree Size	50

made to optimize these parameters. This approach to module identification was inspired by that taken by Majeed and Ryan [13].

If left unchecked, the list of modules could grow to unreasonable sizes. In order to remedy this, only the 20 best modules are kept after each module identification step, before using them to modify the grammar. The problem used in this work was the Santa Fe ant trail as it has been shown to benefit from other approaches to modularity, both in standard GP [12] and GE [7]. The experimental setup can be found in Table 1. The context-sensitive crossover and mutation in Table 1 are similar to GP's sub-tree mutation and crossover, except they operate on GE derivation trees. These operators ensure that when sub-trees are crossed the appropriate portions of the genotype are also swapped.

Grammar Enhancement by Modules: Once modules have been identified, they must somehow be incorporated back into the evolving population. This was done by simply adding the modules to the appropriate rules in the grammar. This was done two different ways. First, consider the simple grammar in Fig. 2(a), an individual producible by that grammar (Fig. 2(b)), and a module selected from that individual (Fig. 2(c)). This module is incorporated into the grammar by taking the module's phenotype (move move) and making it a production of the rule matching the module's root symbol. Since the module's root symbol is <acts>, the module will be added to the rule

<acts> ::= <act> | <act> <acts>.

The updated rule would then be

<acts> ::= <act> | <act> <acts> | <mod_0>,

and a new rule would be created:

<mod_0> ::= move move.

The complete grammar with this modification can be seen in Fig. 3(a). Given the nature of the phenotype to genotype mapping in GE, it is quite obvious that

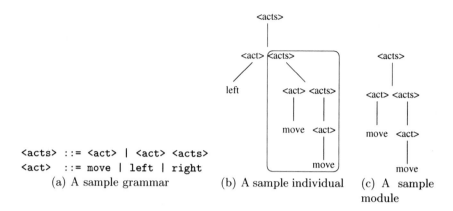

<acts> ::= <act> | <act> <acts>
<act> ::= move | left | right
  (a) A sample grammar

(b) A sample individual

(c) A sample
    module

Fig. 2. A sample grammar, an individual producible by that grammar, and a sample module from that individual. Note: the grammar in Fig. 2(a) is not the actual grammar used in this work.

<acts> ::= <act> | <act> <acts> | <mod_0>
<act> ::= move | left | right
<mod_0> ::= move move
  (a) Grammar not using module libraries

<acts>   ::= <act> | <act> <acts>
                     | <acts_mod_lib>
<act>   ::= move | left | right
<acts_mod_lib> ::= <mod_0>
<mod_0>   ::= move move
  (b) Grammar using module libraries

Fig. 3. Grammars not using and using module libraries

modifying the grammar in this manner has the potential to be extremely destructive. When a new production is added to any rule in the grammar, is it likely that individuals using that rule will no longer map to the same phenotype. An alternative, less destructive, method was also used to modify the grammars. In this approach, module library non-terminals were added to the grammar instead of adding the module directly. Again, consider the initial grammar (Fig. 2(a)), individual (Fig. 2(b)), and module (Fig. 2(c)). Using the module library method, a new non-terminal will be added to the rule matching the module's root node:

<acts> ::= <act> | <act> <acts> | <acts_mod_lib>,

and a new rule is created for the actual module phenotypes:

<acts_mod_lib> ::= <mod_0>,

and a new rule will be created:

<mod_0> ::= move move.

The grammar modified in this way can be seen in Fig. 3(b). By using the module library non-terminals, the addition and subtraction of modules is localized to those library non-terminals. This should have the effect of less disruption to the individuals when the grammar is being modified and they are remapped.

Modifying GE's grammar based on sub-trees which are considered valuable is similar to Whigham's [19] work, where he also identifies what are considered to be useful sections of parse trees and alters the grammar with additional productions containing the terminals and possibly non-terminals from these parse trees. However, his grammar-based form of GP does not employ the genotype-to-phenotype mapping used in GE.

4   Results and Discussion

This section explains how the different approaches to module identification and grammar modification impact GE's search ability. The effects of grammar modification using modules are shown and analyzed in terms of their impact on the fitness of the population over an evolutionary run.

4.1   Grammar Modification

One of the factors that can make a large impact on how destructive or beneficial grammar modification can be is the manner in which the grammar is modified. In this work, there are two techniques for doing this: 1) adding modules as productions to their respective rules in the grammar, and 2) using module library rules to incorporate modules into the grammar (See Sect. 3).

The graphs in Figs. 4 and 5 show the differences between four approaches to adding modules to the grammar. First, only the best 20 modules were added to the grammar using module libraries. Next, all modules were added using module libraries. The same setups were repeated, but without the use of module libraries. The purpose for showing these particular variations is to see how adding small and large numbers of modules to the grammar in different ways affect the population's fitness. Figures 4(a) and 4(b) suggest that when the grammar is modified frequently (every 5 generations), approaches using module libraries to modify the grammar are more resistant to the dis-improvement in fitness (both the population's average and best fitness) that can come with remapping the population, but when the grammar is modified less frequently (Figs. 4(c) – 4(d) show grammar modification every 20 generations), approaches that do not use module libraries perform better.

Another noteworthy feature of Fig. 4 is the consistent way in which the fitness of the population is affected between the different setups when the grammar is modified at different intervals. The best fitness of runs where the grammar is modified every 5 generations is not as good as the runs where the grammar is only modified every 10 generations. The same applies for runs where the grammar is modified every 10 and 20 generations (The figures showing the fitness of modifying the grammar every 10 generations was omitted for lack of space). This can be thought of in terms of evolution not having enough time to exploit the new grammar when grammar modification intervals are short.

When modules are found and added to the grammar, individuals must be re-mapped in order to use them. With GE's mapping process the addition of

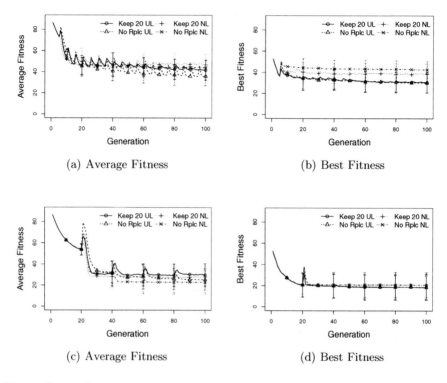

(a) Average Fitness

(b) Best Fitness

(c) Average Fitness

(d) Best Fitness

Fig. 4. Best and average population fitness values for the Santa-Fe ant trail problem. Figures 4(a) and 4(b) show the grammar changing every 5 generations. Figures 4(c) and 4(d) show the grammar changing every 20 generations.

just one production to any rule in the grammar has the possibility to cause an individual to map to a completely different phenotype, thus losing any valuable information that individual might have developed so far. One way to think of this is the population going into a state of "shock" when the grammar is modified. This shock has a negative impact on the fitness of all the individuals, and the population needs time to recover in order for the best fitness to continually improve. As can be seen in the graphs in Figs. 4(a) and 4(c) the longer the gap between shocks the longer the population has time to recover and improve its fitness, but when it is shocked more regularly, this recovery period is shortened and the population is unable to improve to its maximum potential.

To further examine these setups, Fig. 5 shows distributions of how the average fitness of the population changes when the grammar is modified. Examining this figure shows that changing the grammar is very likely to have a negative impact on the fitness of the population, especially when the grammar is changed frequently. As there is no consistent significant difference in the approaches here in terms of the degradation of fitness when the grammar is modified, this raises the question if there is some way to modify the grammar in order to minimize this loss of good information. To help explain the above results, one aspect of the

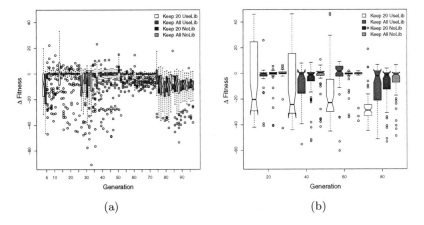

Fig. 5. Change in the population fitness when the grammar is modified. A negative change shows the fitness getting worse, while a positive change shows the fitness improving. The graph showing the change in average fitness every 10 generations was omitted for lack of space.

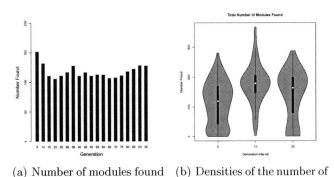

(a) Number of modules found   (b) Densities of the number of
per ID step   modules found with the dif-
                                ferent ID steps

Fig. 6. Figures 6(a) shows the average number of modules found per module identification step. Figure 6(b) shows the distribution of modules found with the different module identification intervals.

different approaches to consider is how many modules are found per identification step as this affects the number of productions being added to the grammar. Note that this is not how many modules are actually used to enhance the grammar, just how many are identified before the module replacement takes place. Figures 6(a) shows the average number of modules found at each module identification step. Figures for module identification steps 10 and 20 were omitted as they showed a similar trend, with the most modules being discovered at the first identification step and less at the subsequent identification steps. The violin plots in Fig. 6(b) show the distribution of these values.

Results so far show the largest differences between grammar modification with and without module libraries occur when the grammar is modified frequently. Since modifying the grammar with module libraries was the better of the two approaches (it is only significantly better when the grammar is modified frequently), it will be discussed in future sections, unless otherwise noted.

4.2   Grammar Enhancement and Fitness

Now, with some understanding of how modifying the grammar can change GE's search potential, the module-grammar enhancement technique is compared to standard GE. Fig. 7 shows the best and average fitness of the grammar modification runs and the standard GE runs. This shows that modifying the grammar frequently is significantly detrimental to the performance of GE. But, if given longer intervals between grammar modification, the impact on average and best fitness is less notable. To further test the intervals between grammar modification and how they might impact search performance, longer runs were used with larger intervals between modifying the grammar. Fig. 8 shows runs with 500 generations and grammar modification steps of 20 and 100 generations.

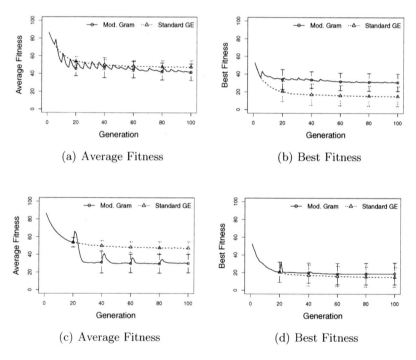

(a) Average Fitness   (b) Best Fitness

(c) Average Fitness   (d) Best Fitness

Fig. 7. Comparisons of the average and best fitness values found by modifying the grammar every 20 and 5 generations and by standard GE. Figures 7(a) and 7(b) show the average and best fitness when the grammar is modified every 5 generations. Figures 7(c) and 7(d) show the average and best fitness when the grammar is modified every 20 generations.

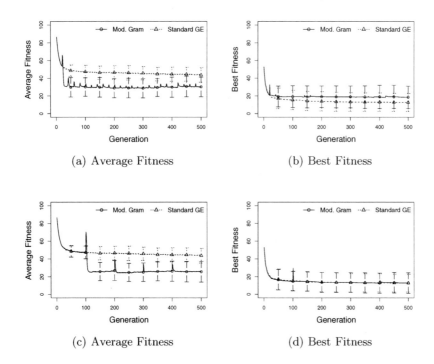

Fig. 8. Comparisons of the average and best fitness values found by modifying the grammar every 20 and 100 generations and by standard GE. Figures 8(a) and 8(b) show the average and best fitness when the grammar is modified every 100 generations. Figures 8(c) and 8(d) show the average and best fitness when the grammar is modified every 100 generations.

Figs. 8(a) and 8(b) show how maintaining the grammar modification step from earlier runs, but over a longer number of generations, does not offer much improvement in the best of average fitness of the individuals. But, when longer intervals between grammar modification are used, a performance increase in the grammar modification approach can be seen. Figures. 8(c) and 8(d) show that the best performance, both for the average and best fitness of individuals, come when the grammar is modified at the longest interval of 100 generations. The difference in fitness values between approaches at the last generation suggests that modifying the grammar under these particular settings can have a beneficial effect on the population as a whole, without losing the valuable information in the best individual.

This data demonstrates that there are some benefits to adding modules to the grammar, but when the grammar is modified too frequently, or if the method of modifying the grammar does not make new and useful information easily accessible, there is no benefit in modifying the grammar, even with information that could be useful.

5 Conclusion and Future Work

This research examined one approach to identifying modules and using those modules to enhance the grammar in different ways with hopes of improving search in GE. There were some very distinct performance changes when the grammar was modified, which suggests that there is some merit to enhancing the grammar with information identified over the course of an evolutionary run. The results show that there is definitely a balancing act between modifying the grammar more or less frequently. It is also apparent that care must be taken when modifying the grammar in order to preserve the valuable information that already exists in the population. However, to truly understand the effects of modifying the grammar additional problems must be studied.

One of the major downfalls of this approach is that modifying the grammar creates a large negative impact on the fitness of the population and valuable information that has been learned will likely be lost. The first promising venue for future work includes identifying and implementing less destructive ways of modifying the grammar. Another possibility is making the modules found parameterized and/or capable of being modified by crossover and mutation operators. Even more possibilities for future work exist in the manner in which modules are identified. Here, only one approach was examined, but there are numerous other possibilities for module identification which may lead to discovering better modules and minimizing the loss of information when these modules are used to modify GE's grammar. This paper on presents the results for the Santa Fe ant trail problem. Future work will also include using more benchmark problems.

Acknowledgements. The authors would like to thank members of the Natural Computing Research and Applications group for their support, comments, and discussions, especially Dr. Erik Hemberg. This research is based upon works supported by the Science Foundation Ireland under Grants No. 08/RFP/CMS1115 and 08/IN.1/I1868.

References

1. Angeline, P.J., Pollack, J.: Evolutionary module acquisition. In: Fogel, D., Atmar, W. (eds.) Proceedings of the Second Annual Conference on Evolutionary Programming, La Jolla, CA, USA, February 25-26, pp. 154–163 (1993)
2. Angeline, P.J., Pollack, J.B.: The evolutionary induction of subroutines. In: Proceedings of the Fourteenth Annual Conference of the Cognitive Science Society, pp. 236–241. Lawrence Erlbaum, Bloomington (1992)
3. Garibay, O., Garibay, I., Wu, A.: The modular genetic algorithm: Exploiting regularities in the problem space. In: Yazıcı, A., Şener, C. (eds.) ISCIS 2003. LNCS, vol. 2869, pp. 584–591. Springer, Heidelberg (2003)
4. Harper, R., Blair, A.: Dynamically defined functions in grammatical evolution. In: Proceedings of the 2006 IEEE Congress on Evolutionary Computation, July 6-21, pp. 9188–9195. IEEE Press, Vancouver (2006)
5. Hemberg, E.: An Exploration of Grammars in Grammatical Evolution. Ph.D. thesis, University College Dublin (2010)

6. Hemberg, E., Gilligan, C., O'Neill, M., Brabazon, A.: A grammatical genetic programming approach to modularity in genetic algorithms. In: Ebner, M., O'Neill, M., Ekárt, A., Vanneschi, L., Esparcia-Alcázar, A.I. (eds.) EuroGP 2007. LNCS, vol. 4445, pp. 1–11. Springer, Heidelberg (2007)

7. Hemberg, E., O'Neill, M., Brabazon, A.: An investigation into automatically defined function representations in grammatical evolution. In: Matousek, R., Nolle, L. (eds.) 15th International Conference on Soft Computing, Mendel 2009, Brno, Czech Republic, June 24-26 (2009)

8. Holland, J.H.: Adaptation in natural and artificial systems. The University of Michigan Press, Ann Arbor (1975)

9. Keijzer, M., Ryan, C., Cattolico, M.: Run transferable libraries—learning functional bias in problem domains. In: Deb, K. et al. (eds.) GECCO 2004. LNCS, vol. 3103, pp. 531–542. Springer, Heidelberg (2004)

10. Koza, J.R.: Genetic Programming: on the Programming of Computers by Means of Natural Selection. MIT Press, Cambridge (1992)

11. Koza, J.R.: Architecture-altering operations for evolving the architecture of a multi-part program in genetic programming. Tech. rep., Stanford, CA, USA (1994)

12. Koza, J.R.: Genetic Programming II: Automatic Discovery of Reusable Programs. MIT Press, Cambridge (1994)

13. Majeed, H., Ryan, C.: Context-aware mutation: a modular, context aware mutation operator for genetic programming. In: GECCO 2007: Proceedings of the 9th Annual Conference on Genetic and Evolutionary Computation, pp. 1651–1658. ACM, New York (2007)

14. O'Neill, M., Ryan, C.: Grammatical Evolution: Evolutionary Automatic Programming in an Arbitrary Language. Kluwer Academic Publishers, Dordrecht (2003)

15. O'Neill, M., Vanneschi, L., Gustafson, S., Banzhaf, W.: Open issues in genetic programming. Genetic Programming and Evolvable Machines 11, 339–363 (2010)

16. Ryan, C., Keijzer, M., Cattolico, M.: Favourable biasing of function sets using run transferable libraries. In: Genetic Programming Theory and Practice II, pp. 103–120 (2005)

17. Simon, H.A.: The sciences of the artificial, 3rd edn. MIT Press, Cambridge (1996)

18. Walker, J., Miller, J.: The automatic acquisition, evolution and reuse of modules in cartesian genetic programming. IEEE Transactions on Evolutionary Computation 12(4), 397–417 (2008)

19. Whigham, P.: Inductive bias and genetic programming. In: First International Conference on Genetic Algorithms in Engineering Systems: Innovations and Applications, GALESIA (Conf. Publ. No. 414), September 1995, pp. 461–466 (1995)

Multi-objective Genetic Programming
for Visual Analytics

Ilknur Icke[1] and Andrew Rosenberg[1,2]

[1] The Graduate Center, The City University of New York,
365 Fifth Avenue 10016
New York, NY
iicke@gc.cuny.edu
[2] Queens College, The City University of New York,
65-30 Kissena Blvd. Flushing
NY 11367−1575
andrew@cs.qc.cuny.edu

Abstract. Visual analytics is a human-machine collaboration to data modeling where extraction of the most informative features plays an important role. Although feature extraction is a multi-objective task, the traditional algorithms either only consider one objective or aggregate the objectives into one scalar criterion to optimize. In this paper, we propose a Pareto-based multi-objective approach to feature extraction for visual analytics applied to data classification problems. We identify classifiability, visual interpretability and semantic interpretability as the three equally important objectives for feature extraction in classification problems and define various measures to quantify these objectives. Our results on a number of benchmark datasets show consistent improvement compared to three standard dimensionality reduction techniques. We also argue that exploration of the multiple Pareto-optimal models provide more insight about the classification problem as opposed to a single optimal solution.

1 Introduction

The amount of data poses both theoretical and practical challenges for machine learning. According to the *curse of dimensionality* [4] theorem, the number of samples needed for a classification task increases exponentially as the number of dimensions (variables, features) increases. Moreover, irrelevant and redundant features might hinder classifier performance. On the other hand, it is costly to collect, store and process data. In exploratory analysis settings, high dimensionality prevents the users from exploring the data visually. Feature extraction is a two-step process that seeks suitable data representations that would help us overcome these challenges. Feature construction step creates a set of new features based on the original features and feature selection is the process of finding the best features amongst them. Feature selection techniques are divided into filter, wrapper and embedded methods depending on the assessment method [10].

S. Silva et al. (Eds.): EuroGP 2011, LNCS 6621, pp. 322–334, 2011.

In this paper, we focus on feature extraction methods for visual analytics in classification problems. Visual analytics is a human-machine collaborative approach to data modeling where visualization provides means to understand the structure of the data such as revealing the clusters formed within the observed data and the relationships between the features. Various linear (such as principal components analysis (PCA), multiple discriminants analysis (MDA), exploratory projection pursuit) and non-linear (such as multidimensional scaling (MDS), manifold learning, kernel PCA, evolutionary constructive induction) techniques have been proposed for dimensionality reduction and visualization. Traditionally, these algorithms aim to optimize a scalar objective and return a single optimal solution. However, recently it has been claimed that the analysis of multiple Pareto-optimal data models provides more insight about the classification problem as opposed to a single optimal solution [13].

Our algorithm is a novel adaptive approach to feature extraction for visual analytics that consists of Pareto-based multi-objective evolutionary constructive induction for feature construction and a hybrid filter/wrapper method for feature selection. Genetic programming approaches for feature selection/extraction in various classification tasks have been reported in the literature ([14,17,16,9,6]) where a single or scalar objective is optimized. A multi-objective method that optimizes accuracy of a simple classifier along with tree size as the complexity measure was described in [21]. Our algorithm differs from these methods in the sense that we aim to optimize human interpretability and discrimination power simultaneously and study visualization as a tool for interpretability. Our previous work studied a single objective (classifier accuracy) genetic programming method for data visualization([12]) and a preliminary multi-objective implementation has been discussed in [11]. This paper further extends our work into a multi-objective setting with various measures for interpretability and also proposes ideas for model selection by analyzing the multiple Pareto-optimal solutions.

The remaining sections are organized as follows: section 2 introduces the MOG3P algorithm. Section 3 explains the optimization criteria used in a number of standard dimensionality reduction techniques versus the proposed criteria for the MOG3P algorithm. Experiment results on a number of benchmark datasets are reported in section 4. Section 5 presents the conclusions and future work.

2 The Multi-objective Genetic Programming Projection Pursuit (MOG3P) Algorithm

We utilize the genetic programming (GP) framework in order to simultaneously evolve data transformation functions that would project the input data into a lower dimensional representation for visualization and classification (figure 1).

Each data transformation function is represented as an expression tree which is made up of a number of functions over the initial features and represents a 1D projection of the data. Each individual contains two such expression trees that evolve independently and generate a 2D view of the N-dimensional dataset. The algorithm is named Multi-Objective Genetic Programming Projection Pursuit

(MOG3P) since it searches for *interesting* low dimensional projections of the dataset where the measure of interestingness consists of three equally important objectives. These objectives are: 1) classifiability: the generated data representation should increase the performance of the learning algorithm(s), 2) visual interpretability: easily identifiable class structures on 2D scatterplots, 3) semantic interpretability: the relationships between the original and evolved features should be easy to comprehend.

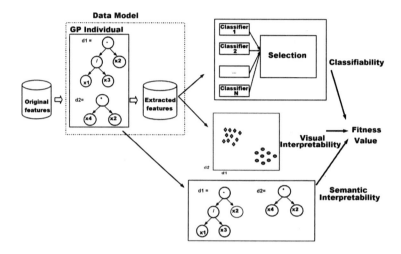

Fig. 1. Components of the MOG3P algorithm

3   Feature Extraction Criteria

In this section, we briefly cover the criteria utilized by the three standard dimensionality reduction methods and our MOG3P algorithm. Each standard method presented here optimizes a single criterion whereas the MOG3P utilizes multiple objectives simultaneously.

3.1   Standard Methods

The following three methods are amongst the most widely used dimensionality reduction techniques [5] for data visualization:

Principal Components Analysis (PCA). PCA finds a lower dimensional representation such that maximum variation in the data is still retained. More specifically, 2-dimensional PCA representation can be generated by projecting the data onto the two eigenvectors of the data covariance matrix corresponding to the largest two eigenvalues. Each new feature is a linear combination of the original features.

Multidimensional Scaling (MDS). MDS algorithm searches for a mapping of the input data onto a k-dimensional space that preserves the inter-point distances. Given a matrix of inter-point distances between the original data points $(\mathbf{x}_i, \mathbf{x}_j)$ where $d_{ij} = ||\mathbf{x}_i - \mathbf{x}_j||$, a variant of this algorithm named Sammon's mapping minimizes: $\sum_{i \neq j} \frac{(d_{ij} - ||\mathbf{z}_i - \mathbf{z}_j||)^2}{d_{ij}}$ where $\mathbf{z}_i, \mathbf{z}_j$ are the new feature vectors in k-dimensional space. This algorithm does not explicitly define functions that transform the original features into the new features.

Multiple Discriminants Analysis (MDA). For a c-category classification problem, MDA finds a $c-1$ dimensional representation of the data such that the separation between the categories is maximized. This is achieved by maximizing the ratio of between-class scatter matrix to within-class scatter matrix. Similar to PCA, each new feature is a linear combination of the original features.

3.2   MOG3P Multi-objective Criteria

The MOG3P algorithm aims to optimize the discriminative power and interpretability of the generated features for data classification problems. We examine the discriminative power under the classifiability criterion where the goal is to find a data representation that will increase the performance of the learning algorithm(s). For interpretability, we define two separate criteria.

Classifiability. We aim to optimize discriminative power of the constructed features by maximizing the classification accuracy. The classifiability criterion can be computed in a variety of ways, including measuring the accuracy of a single classifier, and aggregating the performance of an ensemble of classifiers. In this paper, we measure classifiability by the accuracy of a single classifier selected at random from an ensemble. This way, we aim to ensure that the features will not be biased towards one classification paradigm.

Visual Interpretability. For a data classification task, our goal is to find visualizations (scatterplots) of the dataset such that class structures are easily identifiable. Humans can easily judge the quality of a visualization regardless of the shape of the class boundaries (linear or nonlinear) or the shapes of clusters formed by members of the same class. Since there are a large number of candidate visualizations, it is not feasible to utilize human judgement. A number of numerical measures have been proposed to assess the quality of groupings automatically. Although the ultimate automated quality measure is not known, recent user studies in visual analytics literature have shown that measures of separation and compactness reflect human perception to some extent ([18,19]). In this paper, we utilize the following measures used in exploratory projection pursuit ([8], [15]):

- *LDA Index* (I_{LDA}) is based on Fisher's discriminant analysis. $I_{LDA} = \frac{|W|}{|W+B|}$
  $B = \sum_{i=1}^{k} n_i (\bar{V}_{i.} - \bar{V}_{..})(\bar{V}_{i.} - \bar{V}_{..})'$ and $W = \sum_{i=1}^{k} \sum_{j=1}^{n_i} (V_{ij} - \bar{V}_{i.})(V_{ij} - \bar{V}_{i.})'$
  where V_{ij} are data points, $\bar{V}_{i.}$ and $\bar{V}_{..}$ are group and dataset centroids, n_i is the number of points in group i, B is the between-group sum of squares and

W is the within-group sum of squares. Smaller values of I_{LDA} indicate more compact and better separated class structures.

- C $Index(I_C)$ is a cluster validation index: $I_C = \frac{SD - SD_{min}}{SD_{max} - SD_{min}}$ where SD is the total sum, for all classes, of pairwise distances between samples of the same class (total p distances), SD_{min} and SD_{max} are the sums of p smallest/largest pairwise distances across the whole dataset. Smaller values of I_C indicate more compact and better separated class structures.

- $Davies$-$Bouldin$ $Index(I_{DB})$ is a measure of compactness and well separation: $I_{DB} = \frac{1}{n} \sum_{i=1}^{n} min_{i \neq j} \{ \frac{\delta(X_i, X_j)}{\Delta(X_i) + \Delta(X_j)} \}$ where $\Delta(X_i)$ is intra-class distance for class i and $\delta(X_i, X_j)$ is inter-class distance for classes i and j. Smaller values of I_{DB} indicate more compact and better separated class structures.

- $Dunn's$ $Index$ (I_{Dunn}) is a measure of compactness and well separation: $I_{Dunn} = min_{1 \leq i \leq n} \{ min_{1 \leq j \leq n, j \neq i} \{ \frac{\delta(X_i, X_j)}{max_{1 \leq k \leq n} \{ \Delta(X_k) \}} \} \}$ where δ, Δ are defined as above. Smaller values of I_{Dunn} indicate more compact and better separated class structures.

Semantic Interpretability. The transformation functions (GP expressions) represent relationships between the original and the extracted features. Complex relationships are harder to comprehend and generalize. The most common measure of complexity used in previous research is the expression size without considering the nature of the data transformation functions. For example, a nonlinear relationship can be considered more semantically complex than a linear relationship between the features. The use of order of non-linearity as a complexity measure has been proposed in [20]. In this paper, we experiment with three measures for complexity: 1) expression size (I_{TS}), 2) expressional complexity(I_{EC}) which is the total number of nodes in all subtrees (from [20]), 3) a weighted version of the expressional complexity (I_{WEC}) where each arithmetic operator is assigned a weight related to its linearity/non-linearity. An expression is considered more complex if the nonlinear operators are closer to the root.

4   Experiments

In this paper, we report results on a number of well-known datasets (table 1) from the data mining literature. We first create 2D representations of each dataset using principal components analysis (PCA), multidimensional scaling (MDS) and multiple discriminants analysis (MDA).

Table 1. Datasets

Name	#features	# samples	# classes
Bupa Liver Disorders (BUPA) [2]	6	345	positive:145, negative:200
Wisconsin Diagnostic Breast Cancer (WDBC) [2]	30	569	malignant:212, benign:357
Crabs [3]	5	200	4 (50 each)
Pima Indians Diabetes (PIMA) [2]	7	786	diabetic:286, non-diabetic:500

Table 2. MOG3P Settings

Population size/Generations	400 / 100
Crossover/Mutation operators	One-point subtree crossover/subtree mutation
Crossover/Reproduction/Mutation Probability	0.9 / 0.05 / 0.05
Multi objective selection algorithm	Strength Pareto Evolutionary Algorithm (SPEA2 [22])
Archive Size	100
Function set symbols	$\{+, -, *, protected/, min, max, power, log\}$
Ephemeral Random Constants (ERC)	$[-1, 1]$
Terminal symbols	Variables in the dataset, ERC
Classifiability Objective (C)	maximize hit rate (minimize misclassification rate) of a random classifier per individual $\{I_{Random}\}$
Visualization objective (V)	minimize $\{ I_{LDA}, I_C, I_{DB}, I_{Dunn} \}$
Semantic objective (S)	minimize $\{I_{TS}, I_{EC}, I_{WEC} \}$
Cross validation	10 times 10-fold cross validation (total 100 runs)

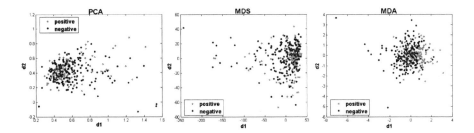

Fig. 2. PCA (I_{TS}:46), MDS and MDA (I_{TS}:46) visualizations of BUPA dataset

We report the 10-fold cross-validation performance of each classifier on these 2D representations as well as the original dataset. PCA and MDA construct new features based on linear transformations on the set of all original features and they can not uncover non-linear relationships. MDS does not construct an explicit mapping between the constructed and original features. In MOG3P, we utilize the ECJ toolkit([7]) and experiment with WEKA([1]) implementations of the following classifiers: Naive Bayes, Logistic Regression, SMO (support vector machine), RBF Network, IBk (k-Nearest Neighbors), Simple Cart and J48 (decision tree). Classifiability and visualization objectives can be used together or one at a time. Semantic interpretability objective is mandatory (table 2).

We implement a nested 10-fold cross validation scheme in order to assess generalization of the extracted features to unseen data. Each dataset is divided into 10-folds and 10 training and test set pairs are created. Each run uses one training set and only the optimal individuals are tested using the corresponding test set after each run. This process is repeated 10 times resulting in a total of 100 runs. In this paper, the quality of the two new features generated by each individual are compared to PCA, MDS and MDA methods in terms of the mean test set classification accuracy of the above mentioned classifiers.

Figures 2, 4, 6, 8 show visualizations of each dataset using the standard dimensionality reduction techniques. The visualizations reveal that the criteria optimized by the PCA and MDS algorithms (section 3) do not always generate

Table 3. Results on BUPA dataset

Classifier	PCA (2D)	MDS (2D)	MDA (2D)	All features	MOG3P (2D) I_{Random} I_{LDA},I_{TS}	MOG3P (2D) I_{Random} I_{C},I_{TS}	MOG3P (2D) I_{Random} I_{DB},I_{TS}	MOG3P (2D) I_{Random} I_{Dunn},I_{TS}
N. Bayes	54.20	59.13	64.35	55.36	77.06	76.72	76.98	77.41
Logistic R.	57.97	58.84	68.12	68.12	77.3	77.50	76.98	77.82
SMO	57.97	57.97	59.42	58.26	74.04	73.30	75.3	75.85
RBF	61.45	59.13	64.06	64.35	77.85	78.05	77.44	77.41
kNN	55.65	57.68	60.29	62.9	78.16	78.8	77.87	77.52
CART	58.55	59.71	68.7	67.54	77.39	77.25	77.12	76.4
J48	57.97	56.82	69.28	68.7	76.95	76.97	76.66	76.08
Avg	57.68	58.47	64.89	63.60	**76.96**	**76.94**	**76.91**	**76.93**
(std)	(2.29)	(1.01)	(4.0)	(5.15)	**(5.63)**	**(5.71)**	**(5.45)**	**(5.53)**

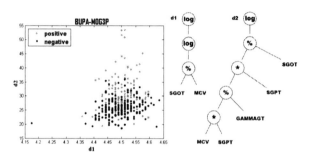

Fig. 3. One selected MOG3P model for the BUPA dataset (I_{TS}:15). Variables: 6 numerical attributes related to liver disorders: mean corpuscular volume (MCV), alkaline phosphotase (ALKPHOS), alamine aminotransferase (SGPT), aspartate aminotransferase (SGOT), gamma-glutamyl transpeptidase (GAMMAGT) and the number of half-pint equivalents of alcoholic beverages drunk per day (DRINKS). This model uses 4 of the 6 original variables.

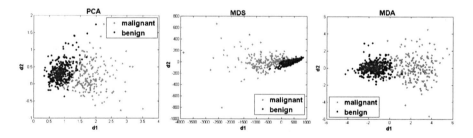

Fig. 4. PCA (I_{TS}:238), MDS and MDA (I_{TS}:238) visualizations of WDBC dataset

Table 4. Classification results on WDBC dataset

Classifier	PCA (2D)	MDS (2D)	MDA (2D)	All features	MOG3P (2D) I_{Random} I_{LDA},I_{TS}	MOG3P (2D) I_{Random} I_C,I_{TS}	MOG3P (2D) I_{Random} I_{DB},I_{TS}	MOG3P (2D) I_{Random} I_{Dunn},I_{TS}
N. Bayes	92.09	90.69	97.19	92.79	97.98	98.21	97.89	97.96
Logistic R.	94.38	93.15	97.72	94.73	97.89	98.28	98.3	98.37
SMO	93.32	87.52	96.84	98.07	97.61	97.87	97.54	97.66
RBF	94.73	92.79	98.24	93.5	98.21	98.28	98.08	98.21
IBk	91.21	90.86	96.66	94.73	98.17	98.40	98.3	98.3
CART	92.79	91.74	96.66	92.97	97.95	98.03	97.86	98.03
J48	92.97	92.44	97.19	93.67	97.98	98.14	97.89	98.01
Avg	93.07	91.31	*97.21*	94.35	**97.97**	**98.17**	**97.98**	**98.08**
(std)	(1.22)	(1.9)	*(0.6)*	(1.8)	**(2.14)**	**(1.81)**	**(1.83)**	**(1.75)**

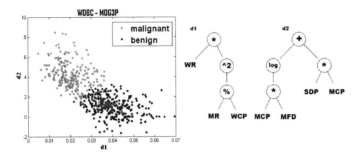

Fig. 5. One selected MOG3P model for the WDBC dataset (I_{TS}:14). Variables (total of 30): the mean (M), standard deviation (SD) and worse (W) values of 10 measurements for breast cancer diagnosis: radius (R), texture (T), perimeter (P), area (A), smoothness (SM), compactness (CPT), concavity (C), concave points (CP), symmetry (SYM) and fractal dimension (FD). This model uses only 6 of the 30 original variables.

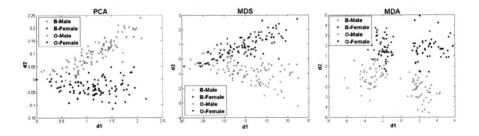

Fig. 6. PCA (I_{TS}:38), MDS and MDA (I_{TS}:38) visualizations of Crabs dataset

data representations that facilitate classification. This effect is clearly evident on the Crabs dataset.

Classifier performance on these new feature representations are compared to the performance on the original dataset (tables 3, 4, 5, 6). For each dataset, the result for a dimensionality reduction method is italicized if it shows significant

Table 5. Classification results on Crabs dataset

Classifier	PCA (2D)	MDS (2D)	MDA (2D)	All features	MOG3P (2D) I_{Random} I_{LDA},I_{TS}	MOG3P (2D) I_{Random} I_C,I_{TS}	MOG3P (2D) I_{Random} I_{DB},I_{TS}	MOG3P (2D) I_{Random} I_{Dunn},I_{TS}
N. Bayes	57.5	67	93.5	38	97.4	96.7	96.25	96.25
Logistic R.	59.5	63	94.5	96.5	98	96.95	96.6	96.25
SMO	54.5	59	94.5	63.5	97.15	96.45	96.15	95.55
RBF	67	69	96	49	97.65	96.75	96.7	96.25
IBk	57	67.5	93	89.5	97.5	97.35	96.8	96.65
CART	57.5	61	94	75.5	97.05	96.6	96.2	96.1
J48	56.5	59	92.5	73.5	97.35	97	96.25	96.45
Avg	58.5	63.64	*94*	69.36	**97.44**	**96.83**	**96.42**	**96.24**
(std)	(4.03)	(4.19)	*(1.16)*	(20.93)	**(3.62)**	**(4.13)**	**(4.2)**	**(4.47)**

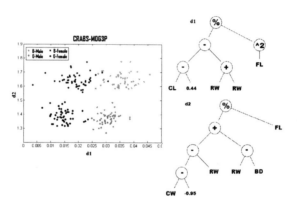

Fig. 7. One selected MOG3P model for the Crabs dataset (I_{TS}:21). Variables: frontal lobe (FL), rear width (RW), carapace width (CW), claw length (CL) and body depth (BD). Feature d1 separates males from females, d2 separates the two species of crabs.

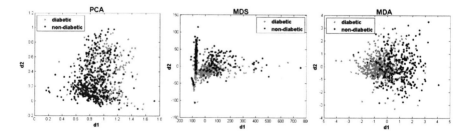

Fig. 8. PCA (I_{TS}:54), MDS and MDA (I_{TS}:54) visualizations of PIMA dataset

Table 6. Classification results on PIMA dataset

Classifier	PCA (2D)	MDS (2D)	MDA (2D)	All features	MOG3P (2D) I_{Random} I_{LDA},I_{TS}	MOG3P (2D) I_{Random} I_{C},I_{TS}	MOG3P (2D) I_{Random} I_{DB},I_{TS}	MOG3P (2D) I_{Random} I_{Dunn},I_{TS}
N. Bayes	71.22	74.22	76.43	76.30	82.03	81.90	81.68	81.86
Logistic R.	72.00	74.35	78.26	77.21	81.54	81.39	81.47	81.75
SMO	72.14	74.74	77.73	77.34	81.51	81.25	81.30	81.24
RBF	72.4	73.44	76.82	75.39	82.17	82.41	81.55	81.79
IBk	61.07	64.45	68.36	70.18	82.01	82.44	80.79	81.32
CART	68.49	73.7	75.39	75.13	81.79	81.96	81.15	81.41
J48	69.40	71.88	75.65	73.83	81.58	81.81	80.60	80.93
Avg	69.53	72.4	75.52	75.06	**81.81**	**81.87**	**81.22**	**81.47**
(std)	(4.01)	(3.62)	(3.32)	(2.48)	**(4.32)**	**(4.19)**	**(4.26)**	**(4.32)**

improvement over the original features. In this paper, all significance tests are reported based on a two-tailed t-test at α =0.05. As it is evident from the results, dimensionality reduction using the standard methods does not guarantee significantly better classifier performance compared to using the original features. On the other hand, for all four datasets, MOG3P consistently finds better data representations using any of the four visualization criteria in conjunction with the classifiability and semantic interpretability objectives.

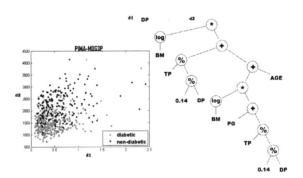

Fig. 9. One selected MOG3P model for PIMA dataset (I_{TS}:22). Variables: times pregrant (TP), plasma glucose concentration (PG), diastolic blood pressure (BP), triceps skin fold thickness (FT), body mass index (BM), diabetes pedigree function (DP), age (AGE). This model uses 5 out of the 7 original variables.

We also compare the effects of different combinations of classifiability and visualization objectives for each dataset. Figure 10 shows the MOG3P results using the tree size (I_{TS}) as the semantic interpretability measure. The most significant finding here is that, using any visualization objective along with the classifiability objective consistently outperforms the cases where either of these objectives were used alone. The other two complexity measures I_{EC} and I_{WEC} demonstrate similar results to I_{TS}, therefore they are not included here.

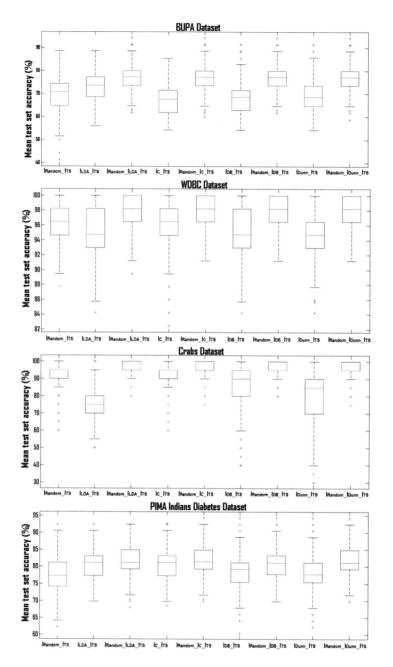

Fig. 10. Comparison of MOG3P objectives (semantic interpretability measure: I_{TS}). Using classifiability and visual interpretability objectives together outperforms either objective alone.

5   Conclusions

We present a visual analytics approach to data modeling that seeks to simultaneously optimize the human interpretability and the discriminative power using genetic programming. We introduce visual interpretability as an explicit objective for exploratory data analysis in classification problems. Different measures of interpretability and discriminative power can easily be incorporated into the algorithm in a multi-objective manner without forcing the user to make a-priori decisions on relative importance of these measures. Our experiments show that using visualization objectives along with the discrimination objectives, MOG3P finds better data representations for classification problems compared to the three standard dimensionality reduction techniques or wrapper techniques used alone. Moreover, the MOG3P algorithm is a data model mining tool providing the users with multiple optimal models aiming to help them discover the set of most informative features or select a classification algorithm by examining classifier performance across multiple models. Model selection can be performed either by choosing one best model or an ensemble of good models. Our ongoing work is concentrating on interactive and automated methods to analyze the pareto optimal solutions for model selection on higher dimensional problems.

Acknowledgements

This research was supported in part by a grant of computer time from the City University of New York's High Performance Computing Research Center under NSF Grants CNS-0855217 and CNS - 0958379.

References

1. http://www.cs.waikato.ac.nz/~ml/weka/
2. http://www.ics.uci.edu/~mlearn
3. http://www.stats.ox.ac.uk/pub/PRNN
4. Bellman, R.: Adaptive Control Processes. Princeton University Press, Princeton (1961)
5. Duda, R.O., Hart, P.E., Stork, D.G.: Pattern Classification. Wiley, Chichester (2000)
6. Estèbanez, C., Aler, R., Valls, J.M.: A method based on genetic programming for improving the quality of datasets in classification problems. International Journal of Computer Science and Applications 4, 69–80 (2007)
7. Luke, S., et al.: Ecj: A java-based evolutionary computation research system, http://cs.gmu.edu/~eclab/projects/ecj/
8. Garcia-Osorio, C., Fyfe, C.: Comparing exploratory projection pursuit artificial neural networks. In: 3rd International Workshop on Practical Applications of Agents and Multiagent Systems, IWPAAMS 2004, pp. 201–210 (2004)
9. Guo, H., Nandi, A.K.: Breast cancer diagnosis using genetic programming generated feature. Pattern Recognition 39(5), 980–987 (2006)
10. Guyon, I., Gunn, S., Nikravesh, M., Zadeh, L. (eds.): Feature Extraction, Foundations and Applications. STUD FUZZ. Physica-Verlag, Springer, Heidelberg (2006)

11. Icke, I., Rosenberg, A.: Multi-objective genetic programming projection pursuit for exploratory data modeling. In: 5th Annual Machine Learning Symposium, New York Academy of Sciences (2010)
12. Icke, I., Rosenberg, A.: Symbolic regression for dimensionality reduction. In: Late Breaking Abstracts, GECCO 2010 (2010)
13. Jin, Y., Sendhoff, B.: Pareto-based multiobjective machine learning: An overview and case studies. IEEE Transactions on Systems, Man, and Cybernetics, Part C 38(3), 397–415 (2008)
14. Krawiec, K.: Genetic programming-based construction of features for machine learning and knowledge discovery tasks. Genetic Programming and Evolvable Machines 3(4), 329–343 (2002)
15. Lee, E.-K., Cook, D., Klinke, S., Lumley, T.: Projection pursuit for exploratory supervised classification. Journal of Computational and Graphical Statistics 14(4), 831–846 (2005)
16. Muharram, M., Smith, G.D.: Evolutionary constructive induction. IEEE Trans. on Knowl. and Data Eng. 17(11), 1518–1528 (2005)
17. Otero, F.E.B., Silva, M.M.S., Freitas, A.A., Nievola, J.C.: Genetic programming for attribute construction in data mining. In: Ryan, C., Soule, T., Keijzer, M., Tsang, E.P.K., Poli, R., Costa, E. (eds.) EuroGP 2003. LNCS, vol. 2610, pp. 384–393. Springer, Heidelberg (2003)
18. Sips, M., Neubert, B., Lewis, J.P., Hanrahan, P.: Selecting good views of high-dimensional data using class consistency. In: Hege, H.-C., Hotz, I., Munzner, T. (eds.) EuroVis 2009: Eurographics/ IEEE-VGTC Symposium on Visualization 2009. Computer Graphics Forum, vol. 28, pp. 831–838. Blackwell, Berlin (2009)
19. Tatu, A., Bak, P., Bertini, E., Keim, D.A., Schneidewind, J.: Visual quality metrics and human perception: an initial study on 2d projections of large multidimensional data. In: AVI-Proceedings of AVI Working Conference on Advanced Visual Interfaces, pp. 49–56 (2010)
20. Vladislavleva, E., Smits, G., den Hertog, D.: Order of nonlinearity as a complexity measure for models generated by symbolic regression via pareto genetic programming. IEEE Trans. Evolutionary Computation 13(2), 333–349 (2009)
21. Zhang, Y., Rockett, P.: A generic multi-dimensional feature extraction method using multiobjective genetic programming. Evolutionary Computation 17(1), 89–115 (2009)
22. Zitzler, E., Laumanns, M., Thiele, L.: Spea2: Improving the strength pareto evolutionary algorithm. Technical report, ETH Zentrum, Gloriastrasse 35, CH-8092 Zurich, Switzerland (2001)

A Continuous Approach to Genetic Programming

Cyril Fonlupt and Denis Robilliard

LISIC — ULCO
Univ Lille Nord de France
BP 719, F-62228 Calais Cedex, France
{fonlupt,robilliard}@lisic.univ-littoral.fr

Abstract. Differential Evolution (DE) is an evolutionary heuristic for continuous optimization problems. In DE, solutions are coded as vectors of floats that evolve by crossover with a combination of best and random individuals from the current generation. Experiments to apply DE to automatic programming were made recently by Veenhuis, coding full program trees as vectors of floats (Tree Based Differential Evolution or TreeDE). In this paper, we use DE to evolve linear sequences of imperative instructions, which we call Linear Differential Evolutionary Programming (LDEP). Unlike TreeDE, our heuristic provides constant management for regression problems and lessens the tree-depth constraint on the architecture of solutions. Comparisons with TreeDE and GP show that LDEP is appropriate to automatic programming.

1 Introduction

In 1997, Storn and Price proposed a new evolutionary algorithm called Differential Evolution (DE) [7] for continuous optimization. DE is a stochastic search method akin to Evolution Strategies, that uses information from within the current vector population to determine the perturbation brought to solutions (this can be seen as determining the direction of the search).

To our knowledge O'Neill and Brabazon were the first to use DE to evolve programs with the use of the well known grammatical evolution engine [5]. A diverse selection of benchmarks from the literature on genetic programming were tackled with four different flavors of DE. Even if the experimental results indicated that the grammatical differential evolution approach was outperformed by standard GP on three of the four problems, the results were somewhat encouraging. Recently, Veenhuis [9] also introduced a successful application of DE for automatic programming, mapping a continuous genotype to trees: Tree based Differential Evolution (TreeDE).

TreeDE improves somewhat on the performance of grammatical differential evolution, but it requires an additional low-level parameter, the number of tree levels, that has to be set beforehand, and it does not provide constants.

In this paper, we propose to use a Differential Evolution engine in order to evolve not program trees but directly linear sequences of imperative instructions,

S. Silva et al. (Eds.): EuroGP 2011, LNCS 6621, pp. 335–346, 2011.
© Springer-Verlag Berlin Heidelberg 2011

as in Linear Genetic Programming (LGP) [1]. We can thus implement real-valued constants management inspired from the LGP literature. The tree-depth parameter from TreeDE is now replaced by the maximum length of the programs to be evolved: this is a lesser constraint on the architecture of solutions and it also has the benefit of avoiding the well known bloat problem (uncontrolled increase in solution size) that plagues standard GP.

This paper is organized in the following way. Section 2 presents the main DE concepts. In section 3, our scheme Linear Differential Evolutionary Programming (LDEP) is introduced while section 4 and 5 give experimental results and comparisons with TreeDE and standard GP. Future works and conclusions are discussed in section 6.

2 Differential Evolution

This section only introduces the main *DE* concepts. The interested reader might refer to [7] for a full presentation. DE is a search method working on a set of N $d-$dimensional vector solutions X_i, f being the fitness function.

$$X_i = (x_{i1}, x_{i2}, \ldots, x_{id}) \quad i = 1, 2, \ldots, N \tag{1}$$

DE can be roughly decomposed into an initialization phase, where the genes of the initial population are randomly initialized with a Gaussian law, and four vary steps that are iterated on: mutation, crossover, evaluation and selection.

Mutation. In the initial DE [7], a so-called variant vector $V_j = (v_{j1}, v_{j2}, \ldots, v_{jd})$ is generated for each vector $X_j(t)$ of the population, according to Equation 2:

$$V_j(t + 1) = X_{r_1}(t) + F \times (X_{r_2}(t) - X_{r_3}(t)) \tag{2}$$

where r_1, r_2 and r_3 are three mutually *different* randomly selected indices that are also different from the current index j; F is a real constant which controls the amplification of the differential evolution and avoids the stagnation in the search process — typical values for F are in the range $[0, 2]$; t indicates the number of the current iteration (or generation). The expression $(X_{r_2}(t) - X_{r_3}(t))$ is often referred to as a difference vector.

Many variants were proposed for Equation 2, including the use of three or more individuals. According to [9,6], the mutation method that leads to the best results on most problems is the method DE/best/2/bin, with:

$$V_j(t + 1) = X_{\text{best}}(t) + F \times (X_{r_1}(t) + X_{r_2}(t) - X_{r_3}(t) - X_{r_4}(t)) \tag{3}$$

$X_{\text{best}}(t)$ being the best individual in the population at the current generation. The DE/best/2/bin is the method used throughout the paper.

Crossover. As explained in [7], the crossover step ensures to increase or at least maintain the diversity. Each trial vector is partly crossed with the variant vector. The crossover scheme ensures that at least one vector component will be crossovered.

The trial vector $U_j = (u_{j1}, u_{j2}, \ldots, u_{jd})$ is generated using Equation 4:

$$u_{ji}(t+1) = \begin{cases} v_{ji}(t+1) & \text{if} \quad (rand \leq CR) \quad \text{or } j = rnbr(i) \\ x_{ji}(t) & \text{if} \quad (rand > CR) \quad \text{and } j \neq rnbr(i) \end{cases} \tag{4}$$

where x_{ji} is the jth component of vector x_i; v_{ji} is the jth component of the variant vector $V_j(t+1)$; $rand$ is drawn from a uniform random number generator in the range $[0, 1]$; CR is the crossover rate in the range $[0, 1]$ which has to be determined by the user; $rnbr(i)$ is a randomly chosen index in the range $[1, d]$ which ensures that $U_j(t+1)$ gets at least one parameter from the variant vector $V_j(t + 1)$, and t is the number of the current iteration (or generation).

Selection. The selection step decides whether the trial solution $U_i(t+1)$ replaces the vector $X_i(t)$ or not. The trial solution is compared to the target vector $X_i(t)$ using the greedy criterion. If $f(U_i(t+1)) < f(X_i(t))$, then $X_i(t+1) = U_i(t+1)$ otherwise the old value $X_i(t)$ is kept.

These four steps are looped over until the maximum number of evaluations/iterations is reached or until a fit enough solution is found. DE is quite simple as it only needs three parameters, the population size (N), the crossover rate (CR), and the scaling factor (F).

3   Linear Differential Evolutionary Programming

We propose to generate linear programs from DE, which we call Linear Differential Evolutionary Programming (LDEP). The real-valued vectors that are evolved by the DE engine are mapped to sequences of imperative instructions. Contrary to the usual continuous optimization case, we can not deduce the vector length from the problem, so we have to set this parameter quite arbitrarily. This length will determine the maximum number of instructions allowed in the evolved programs.

3.1   Representation

For the representation of programs, our work is based on LGP [1], with slight modifications.

For regression, we use mainly 3-register imperative instructions that includes an operation on two operand registers, one of them could be holding a constant value, and then assigns the result to a third register: $r_i = (r_j|c_j)\text{op} (r_k|c_k)$; where r_i is the destination register, r_j and r_k are calculation registers and c_j, c_k are constants. Only one constant is allowed per instruction as in LGP. Of course, even if the programming language is basically a 3-register instruction language, we can drop the last register/constant to include 2-register instructions like $r_i = \sin(r_j|c_i)$ if needed.

Instructions are executed by a virtual machine with floating-point value registers. A subset of the registers contains the inputs to our problem. Besides this required minimal number of registers, we use an additional set of registers for calculation purpose and for storing constants. In the standard case, one of the calculation registers (usually named r_0) is used for holding the output of the program after execution. Evolved programs can read or write into these registers, with the exception of the read-only constant registers. The use of calculation registers allows a number of different program paths, as explained in [1].

All the constants available for computation are initialized at the beginning of the run with values in a range defined by the user, then stored once for all in read-only registers that will be accessed in mode from now on. The number of constants has to be set by the user, and we will also use a constant probability parameter to control the probability of occurrences of constants p_c, as explained below.

3.2   Implementation

LDEP splits the population vectors into subsets of consecutive 4 floating point-values (v_1, v_2, v_3, v_4). v_1 will encode the operator, v_2, v_3 and v_4 will encode the operands. The real values will be cast into integers to index the set of operators or registers as needed, using the following equation :

- Let $n_{\text{operators}}$ be the number of operators, the operator index is:

$$\#\text{operator} = (\lfloor (v_j - \lfloor v_j \rfloor) \times n_{\text{operators}} \rfloor) \tag{5}$$

- If the decimal part of an operand value, returned by the expression $(v_j - \lfloor v_j \rfloor)$, is greater than the fixed constant probability parameter p_c, then a register must be used. Let $n_{\text{registers}}$ be the number of registers, the register index is:

$$\#\text{register} = (\lfloor (v_j - \lfloor v_j \rfloor) \times n_{\text{registers}} \rfloor) \tag{6}$$

- If $(v_j - \lfloor v_j \rfloor) \le p_c$ then a constant must be used, provided no other constant was already used as first operand for this expression. Let C be the number of constants, the constant registers index is:

$$\# \text{ constant} = \lfloor ((v_j) \times C) \rfloor \bmod C \tag{7}$$

Example: Let us suppose LDEP works with the 4 following operators $\{0 : +, 1 : -, 2 : \times, 3 : \div\}$, that 6 calculation registers (r_0 to r_5) are available, 50 constant registers and $p_c = 0.1$. Our example solution vector is made of 8 values (2 imperative instructions): $\{0.17, 2.41, 1.84, 1.07, 0.65, 1.22, 1.22, 4.28\}$.

The first vector value denotes one operator among the four to choose from. To make this choice, this first value is turned into an integer using Equation 6, operator $= \lfloor 0.17 \times 4 \rfloor = 0$, meaning that the first operator will be $+$.

The second value $v_2 = 2.41$ is turned into the destination register. According to Equation 5, operand $= \lfloor 0.41 \times 6 = 2.4 \rfloor = 2$, meaning that r_2 will be used as the destination register.

The next value $v_3 = 1.84$ is a register since the decimal part is greater than the constant probability: $\lfloor 0.84 \times 6 \rfloor = 5$, r_5 will be the leftmost operand.

The decimal part of the next value is: $(v_4 - \lfloor v_4 \rfloor) = 0.07$, as it is less than the constant probability, v_4 will be used to index a constant register. The index number will be $\lfloor 1.07 \times 50 \rfloor \bmod 50 = 3$.

So with the 4 first values of the genotype, we now have the following:

$$r_2 = r_5 + c_3$$

Let us assume c_3 holds the value 1.87. The mapping process continues with the four next values, until we are left with the following program:

$$r_2 = r_5 + 1.87$$
$$r_1 = r_1 \times r_1$$

The rest of LDEP follows the DE general scheme.

Initialization. During the initialization phase, 50 values in the range $[-1.0, +1.0]$ are randomly initialized and will be used as constants for regression experiments. In DE, an upper and a lower bounds are used to generate the components of the vector solution X_i. LDEP uses the same bounds

Iteration. We tried two variants of the iteration loop described: either generational replacement of individuals as in the original Storn and Price paper [7], or steady state replacement, which seems to be used by Veenhuis [9]. In the generational case, new individuals are stored upon creation in a temporary, and once creation is done, they replace their respective parent if their fitness is better. In the steady state scheme, each new individual is immediately compared with its parent and replaces it if its fitness is better, and thus it can be used in remaining crossovers for the current generation. Using the steady state variant seems to accelerate convergence as it is reported in the results section 4.

During the iteration loop, the DE vector solutions are decoded using Equations 5 and 6. The resulting linear programs are then evaluated on a set of fitness cases (training examples).

4   Symbolic Regression and Artifical Ant Experiments

In order to validate our scheme against TreeDE [9], we used the same problems as benchmarks (4 symbolic regression and the artificial ant problems), and we also added two regression problems with constants. We also ran all problems with standard GP, using the well-known ECJ library (http://cs.gmu.edu/~eclab/projects/ecj/). For all problems we measured the average best fitness of 40 independent runs. We also computed the ratio of so-called "hits", i.e. perfect solutions, and average number of evaluations to reach a hit. These last two figures are less reliable than the average fitness, as shown in [4], however we included them to allow a complete comparison with [9] that used these indicators.

- We used the standard values for the control parameters for DE namely : $F = 0.5, CR = 0.1$. We work with a set of $N = 20$ vectors (population size) for regression and for the artificial ant. As said before, the update method for DE is the so-called DE/best/2/bin. Results are shown for both generational and steady state LDEP variant.
- The size of vector (individual) that DE works with was set to 128 for regression, meaning that each program is equal to $128 \div 4 = 32$ imperative instructions. It is reduced to 50 for the artificial ant, since in that case we need only one float to code one instruction, as explained in Section 4.2.
- When needed in the symbolic regression problems, the constant probability was set to 0.05.
- For regression 6 read/write registers were used for calculation (from r_0 to r_5), r_0 being the output register. They were all initialized for each training case (x_k, y_k) with the input value x_k.
- 1500 iterations on a population of 20 vectors were allowed for regression in the TreeDE experiments [9]. Runs were done for every tree depth in the range $\{1, \ldots, 10\}$, thus amounting to a total of $300,000$ evaluations, among these only the runs with the best tree depth were used to provide the figures given in Veenhuis paper. We could not apply this notion of best tree depth in our heuristic, and thus decided as a trade-off to allow $50,000$ evaluations.
- For the Santa Fe Trail artificial ant problem, the same calculation gives a total of $450,000$ evaluations in [9]. We decided as a trade-off to allow $200,000$ evaluations.
- the GP parameters were set to 50 generations and respectively 1000 individuals for the regression problems, and 4000 individuals for the artificial ant, in order to have the same maximum number of evaluations than LDEP. Genetic operator rates were tuned according to the usual practice: 80% for crossover, 10% for sub-tree mutation and 10% for duplication. The maximum tree depth was set to 11, and we kept the best (elite) individual from one generation to the next. For the regression problems, we defined 4 "input" terminals (reading the input value x_k for each training case (x_k, y_k)) against only one ephemeral random constant (ERC) terminal, thus the probability to generate a constant was lower than the usual 50% and thus closer to LDEP (this improves sensibly the GP results in Table 2).

4.1   Symbolic Regression Problems

The aim of a symbol regression problem is to find some mathematical expression in symbolic form that associates input and output on a given set of training pairs. In our case, 20 evenly distributed data points x_k in the range $[-1.0, +1.0]$ are chosen as inputs, the outputs being given by the following test functions from [9]:

$$f_1 = x^3 + x^2 + x$$
$$f_2 = x^4 + x^3 + x^2 + x$$
$$f_3 = x^5 + x^4 + x^3 + x^2 + x$$
$$f_4 = x^5 - 2x^3 + x$$

As TreeDE benchmarks were run without constants in [9], we run LDEP both without and with constants. While this allows us to assess the impact of constant management, anyway we strongly believe that constants should be included in any regression problem, since in the general case one can not know in advance whether or not they are useful. For that same reason we add two benchmarks:

$$f_5 = \pi \quad \text{(constant function)}$$
$$f_6 = \frac{x}{\pi} + \frac{x^2}{\pi^2} + 2x\pi$$

The set of operators is $\{+, -, \times, \div\}$ with $\div$ being the protected division (*i.e.* $a \div b = a/b$ if $b \neq 0$ else $a \div b = 0$ if $b = 0$).

Evaluation (or fitness computation) is done in the typical way, that is computing the sum of deviations over all points, i.e $fitness = \sum_k |f(x_k) - P(x_k)|$ where P is the evolved program and k the number of input/output pairs. A hit means that the fitness function is less than 10^{-4} on each training pair.

As it can be seen in Table 1, all three heuristics LDEP, TreeDE and GP exhibit close results on the f_1, f_2, f_3, f_4 problems, with GP providing the overall most precise approximation, and LDEP needing the largest number of evaluations (however the TreeDE figures are taken from [9] where they are given only for the best tree depth). Note that the steady state variant of LDEP converges faster than the generational, as shown by the average number of evaluations for perfect solutions. It seems safe to conclude that this increased speed of convergence is the explanation for the better result of the steady state variant versus generational, in this limited number of evaluations framework.

When running the heuristics with constants (thus ruling out TreeDE) on all problems f_1 to f_6 in Table 2, we again observe that the steady state variant of LDEP is better than the generational. For its best version LDEP is comparable to GP, with a slightly higher hit ratio and better average fitness (except on f_6), with more evaluations on average.

These results confirm that DE is an interesting heuristic, even when the continuous representation hides a combinatorial type problem, and thus the heuristic is used outside its original field. The LDEP mix of linear programs and constant

Table 1. Results for symbolic regression problems without constants

For each heuristic, the column Fit. gives the average of the best fitness over 40 independent runs (taken from [9] for TreeDE); then we have the percentage of hits, then the average number of evaluations for a hit.

Problem	generational LDEP			steady state LDEP			TreeDE			standard GP		
	Fit.	% hits	Eval.	Fit.	% hits	Eval.	Fit.	% hits	Eval.	Fit.	% hits	Eval.
f_1	0.0	100%	4297	0.0	100%	2632	0.0	100%	1040	0.0	100%	1815
f_2	0.0	100%	12033	0.0	100%	7672	0.0	100%	3000	0.0	100%	2865
f_3	0.28	72.5%	21268	0.08	85%	21826	0.027	98%	8440	0.03	97%	6390
f_4	0.20	62.5%	33233	0.13	75%	26998	0.165	68%	14600	0.01	80%	10845

Table 2. Results for symbolic regression problems with constants

For each heuristic, the column Fit. gives the average of the best fitness over 40 independent runs; then next column gives the percentage of hits, then the average number of evaluations for a hit (if any).

Problem	generational LDEP			steady state LDEP			standard GP		
	Fit.	%hits	Eval.	Fit.	%hits	Eval.	Fit.	%hits	Eval.
f_1	0.0	100%	7957	0.0	100%	7355	0.002	98%	3435
f_2	0.02	95%	16282	0.0	100%	14815	0.0	100%	4005
f_3	0.4	52.5%	24767	0.0	100%	10527	0.02	93%	7695
f_4	0.36	42.5%	21941	0.278	45%	26501	0.33	23%	24465
f_5	0.13	2.5%	34820	0.06	15%	29200	0.07	0%	NA
f_6	0.59	0%	NA	0.63	0%	NA	0.21	0%	NA

management seems interesting enough, when compared to standard GP, to deserve further study.

4.2 Santa Fe Ant Trail

The Santa Fe ant trail is a quite famous problem in the GP field. The objective is to find a computer program that is able to control an artificial ant so that it can find all 89 pieces of food located on a discontinuous trail within a specified number of time (either 400 or 600 time steps)s. The trail is situated on a 32×32 toroidal grid. The problem is known to be rather hard, at least for standard GP (see [2]), with many local and global optima, which may explain why the size of the TreeDE population was increased to $N = 30$ in [9].

We do not need mathematical operators nor registers, only the following instructions are available:

- MOVE: moves the ant forward one step (grid cell) in the direction the ant is facing, retrieving an eventual food pellet in the cell of arrival;
- LEFT: turns on place 45 degrees anti-clockwise;
- RIGHT: turns on place 45 degrees clockwise;
- IF-FOOD-AHEAD: conditional statement that executes the next instruction or group of instructions if a food pellet is located on the neighboring cell in front of the ant, else the next instruction or group is skipped;
- PROGN2: groups the two instructions that follow in the program vector, notably allowing IF-FOOD-AHEAD to perform several instructions if the condition is true (the PROGN2 operator does not affect *per se* the ant position and direction);
- PROGN3: same as the previous operator, but groups the three following instructions.

Each MOVE, RIGHT and LEFT instruction requires one time step.

Table 3. Santa Fe Trail artificial ant problem

The 1st columns is the number of allowed time steps, then for each heuristics, we give the average of the best fitness value over the 40 independent runs (taken from [9] for TreeDE), then the percentage of hits (solutions that found all 89 food pellets), then the average number of evaluations for a hit if applicable.

	generational LDEP			steady state LDEP			TreeDE			standard GP		
# steps	Fit.	% hits	Eval.	Fit.	% hits	Eval.	Fit.	% hits	Eval.	Fit.	% hits	Eval.
400	11.55	12.5%	101008	14.65	7.5%	46320	17.3	3%	24450	8.87	37%	126100
600	0.3	82.5%	88483	1.275	70%	44260	1.14	66%	22530	1.175	87%	63300

Programs are again vectors of floating point values. Each instruction is represented as a single value which is decoded in the same way as operators are in the regression problems, that is using Equation 5. Instruction are decoded sequentially, and the virtual machine is refined to handle jumps over an instruction or group of instructions, so that it can deal with IF-FOOD-AHEAD. Incomplete programs may be encountered, for example if a PROGN2 is decoded for the last value of a program vector. In this case the incomplete instruction is simply dropped and we consider that the program has reached normal termination (thus it may be iterated if time steps are remaining).

The Santa Fe trail being composed of 89 pieces of food, the fitness function is the remaining food (89 minus the number of food pellets taken by the ant before it runs out of time). So, the lower the fitness, the better the program, a hit being a program with fitness 0, i.e. able to pick up all food on the grid.

Results are summed-up in Table 3. Contrary to the regression experiment, the generational variant of LDEP is now better than the steady state. We think this is explained by the hardness of the problem: more exploration is needed, and it pays no more to accelerate convergence. GP provides the best results for 400 time steps, but it is LDEP that provides the best average fitness for 600 steps, at the cost of a greater number of evaluations. LDEP is also better than TreeDE on both steps limits.

5   Evolving a Stack with LDEP

As it seems that LDEP achieved quite interesting results on the previous benchmarks for genetic programming we decided to move forward and to test whether or not LDEP was able to evolve a more complex data structure: a stack. Langdon [3] successfully showed that GP was able to evolve five operations needed to manipulate the stack (push, pop, top, empty and makenull). Some of these operations are considered to be inessential (top, empty) but we chose to follow the settings introduced by Langdon's original work. Table 4 presents the five primitives that were used in our implementation with some comments.

Table 4. The five primitives to evolve (from [3])

Operation	Comment
makenull	initialize stack
empty	is stack empty?
top	return top of the
pop	return top of stack and remove it
push(x)	place x on top of stack

This is in our opinion a more complex problem as the correctness of each trial solution is established using only the values returned by the stack primitives and only three (pop, top and empty) out of the five operations return values.

Choice of primitives. As explained in [3], the set of primitive that was chosen to solve this problem is a set that a human programmer might use. The set basically consists in functions that are able to read and write in an indexed memory, functions that can modify the stack pointer and functions that can perform simple arithmetic operations. The terminal set consists in zero-arity functions (increment or decrement the stack pointer) and some constants. The following set was available for LDEP:

- arg1, the value to be pushed on to the stack (read-only argument).
- aux, the current value of the stack pointer.
- arithmetic operators + and −.
- constants 0, 1 and MAX (maximum depth of the stack, set to 10).
- indexed memory functions read and write. The write function is a two arguments function arg1 and arg2. It evaluates the two arguments and sets the indexed memory pointed by arg1 to arg2 (stack[arg1] = arg2). It returns the original value of aux.
- functions to modify the stack pointer (inc_aux to increment the stack pointer, dec_aux to decrement it, write_Aux to set the stack pointer to its argument and returns the original value of aux).

5.1   Architecture and Fitness Function

We used a slightly modified version of LDEP as the stack problem requires the evolution of the five primitives (makenull, top, pop, push and empty) simultaneously. An individual is now composed of 5 vectors, one for each primitive. Mutation and crossover are only performed with vectors of the same type (*i.e.* vectors evolving the top primitive for example). Moreover the vector associated to a primitive is decoded as Polish Notation (or prefix notation), that means an operation like (`arg1 + MAX`) is coded as `+ arg1 max`. Each primitive vector has a maximum length of 100 values (this is several times more than sufficient to code any of the five primitives needed to manipulate the stack).

We used the same fitness function that was defined by Langdon. It consists of 4 test sequences, each one being composed of 40 operations push, pop, top, makenull and empty. As explained in the previous section, the makenull and push operations do not return any value, they can only be test indirectly by seeing if the other operations perform correctly.

Results. In his original work, Langdon chose to use a population of $1,000$ individuals with 101 generations. That means that roughly $100,000$ fitness functions evaluations were used. In the LDEP case, based on our previous experience it appeared that using large population is usually inadequate and that LDEP performs better with small size population. In this case, a population of 10 individuals with $10,000$ generations were used.

Langdon wrote that with his own parameters 4 runs (out of 60), produced successful individuals. It was interesting to see that similar rates of results were obtained: 6 runs out of 100 yielded perfect solutions. An example of successful run is given in table 5 with the evolved code and with the simplified code (redundant code removed).

Table 5. Example of an evolved push-down stack

Operation	Evolved stack	Simplified stack
push	write(1 ,write(dec_aux ,arg1))	stack[aux] = arg1
		aux = aux - 1
pop	write(aux ,((aux + (dec_aux + inc_aux)) + read(inc_aux)))	aux = aux + 1
		tmp = stack[aux];
		stack[aux] = tmp + aux;
		return tmp
top	read(aux)	return sp[aux]
empty	aux	if (aux > 0) return true
		else return false
makenull	write((MAX - (0 + write_Aux(1))),MAX)	aux = 1

6   Conclusion and Future Works

This paper is a further investigation into Differential Evolution engines applied to automatic programming. Unlike TreeDE [9], our scheme allows the use of constants in symbolic regression problems and translates the continuous DE representation to linear programs, thus avoiding the systematic search for the best tree depth that is required in TreeDE.

Comparisons with GP confirm that DE is a promising area of research for automatic programming. In the most realistic case of regression problems, when using constants, steady state LDEP slightly outperforms standard GP on 5 over 6 problems. On the artificial ant problem, the leading heuristic depends on the number of steps. For the 400 steps version GP is the clear winner, while for 600 steps generational LDEP yields the best average fitness. LDEP improves on the TreeDE results for both versions of the ant problem, without the need for fine-tuning the architecture of solutions.

These results led us to try to evolve a set of stack management primitives. The experiment proved successful, with results similar to the GP literature. Many interesting questions remain open. In the beginnings of GP, experiments showed that the probability of crossover had to be set differently for internal and terminal nodes: is it possible to improve LDEP in similar ways? Which parameters are crucial for DE-based automatic programming?

References

1. Brameier, M., Banzhaf, W.: Linear Genetic Programming. Genetic and Evolutionary Computation. Springer, Heidelberg (2007)
2. Langdon, W.B., Poli, R.: Why ants are hard. Tech. Rep. CSRP-98-4, University of Birmingham, School of Computer Science (January 1998); presented at GP 1998
3. Langdon, W.B.: Genetic Programming and Data Structures = Automatic Programming! Kluwer Academic Publishers, Dordrecht (1998)
4. Luke, S., Panait, L.: Is the perfect the enemy of the good? In: Langdon, W.B., Cantú-Paz, E., Mathias, K., Roy, R., Davis, D., Poli, R., Balakrishnan, K., Honavar, V., Rudolph, G., Wegener, J., Bull, L., Potter, M.A., Schultz, A.C., Miller, J.F., Burke, E., Jonoska, N. (eds.) GECCO 2002: Proceedings of the Genetic and Evolutionary Computation Conference, July 9-13, pp. 820–828. Morgan Kaufmann Publishers, New York (2002)
5. O'Neill, M., Brabazon, A.: Grammatical differential evolution. In: International Conference on Artificial Intelligence (ICAI 2006), Las Vegas, Nevada, USA, pp. 231–236 (2006)
6. Price, K.: Differential evolution: a fast and simple numerical optimizer. In: Biennial Conference of the North American Fuzzy Information Processing Society, pp. 524–527 (1996)
7. Storn, R., Price, K.: Differential evolution – a simple and efficient heuristic for global optimization over continuous spaces. Journal of Global Optimization 11(4), 341–359 (1997)
8. Vanneschi, L., Gustafson, S., Moraglio, A., De Falco, I., Ebner, M. (eds.): EuroGP 2009. LNCS, vol. 5481. Springer, Heidelberg (2009)
9. Veenhuis, C.B.: Tree based differential evolution. In: [8], pp. 208–219 (2009)

Author Index